Henry C. Lang

The butterflies of Europe

Henry C. Lang

The butterflies of Europe

ISBN/EAN: 9783744650540

Printed in Europe, USA, Canada, Australia, Japan

Cover: Foto ©berggeist007 / pixelio.de

More available books at **www.hansebooks.com**

𝔑𝔥𝔬𝔭𝔞𝔩𝔬𝔠𝔢𝔯𝔞 𝔈𝔲𝔯𝔬𝔭𝔞

DESCRIPTA ET DELINEATA.

THE

BUTTERFLIES OF EUROPE

DESCRIBED AND FIGURED.

BY

HENRY C. LANG, M.D., F.L.S., &c.,

MEMBER OF THE ENTOMOLOGICAL SOCIETY OF LONDON.

ILLUSTRATED WITH MORE THAN EIGHT HUNDRED COLOURED FIGURES, DRAWN, MOSTLY
FROM NATURE, UNDER THE DIRECTION OF THE AUTHOR.

VOLUME 1.--TEXT.

LONDON:

L. REEVE & CO., 5, HENRIETTA STREET, COVENT GARDEN.

—

1884.

TO MY WIFE,

AND TO THE MEMORY OF MY FATHER
HENRY LANG, M.B., LOND.,

This Work is Dedicated.

IV.—Concerning the figures, I can only say that every effort has been made to secure accuracy in representing the various species. The process of chromo-lithography was chosen as being most likely to attain this end; and Mr. Knight, the artist, whose name appears on each plate, has drawn the figures under my direction, in most cases from Nature, only two or three having been copied from authentic figures. The majority of the European butterflies have not before been figured in any English work; and one or two are, I believe, represented here altogether for the first time. Only two species—one, *Lycæna Panope*, a very rare Russian species, and the second a newly-discovered member of the family Hesperidæ—are unavoidably omitted.

V.—The figures of larvæ and pupæ are mostly taken from Hübner, some few from Millière, it being impossible to obtain living specimens from which to draw them. These plates are merely intended to include typical examples of the different genera; and in most cases those species are given which do not occur in Britain, and with which British collectors are least likely to be already familiar.

VI.—In conclusion, I have great pleasure in alluding to the kind consideration and courtesy I have everywhere received from all those who have in any way assisted me in the production of this volume. To the following, especially, I wish to record my thanks:—The President and Council of the Entomological Society of London, for the unusual facilities given me in the unrestricted use of the Society's library; the official staff of the Zoological Department of the British Museum in charge of the entomological collections, especially Messrs. A. G. Butler, F.L.S., and W. F. Kirby, for their kind assistance and interest; Messrs. F. Ducane Godman, F.R.S., and O. Salvin, F.R.S., for their kindness in so readily allowing me to make use of their magnificent collection, and for the loan of specimens whenever I required them for figures or descriptions. Lastly, I thank all those gentlemen who have helped me with their valuable advice or suggestions, and I know that they will believe me when I say that I have done my best to follow them whenever practicable. Though I may not have always been successful in carrying out the wishes of correspondents, and though there are in this book many things which I, the author, would gladly see better accomplished, yet I know that those who read it will be the first to recognise the difficulty of the undertaking; and so, I may venture to hope, will look more at the useful side of the book than at its failings.

H. C. L.

Maidenhead, Berkshire.
August, 1884.

CONTENTS.

VOLUME I. — TEXT.

	PAGE
INTRODUCTION	1
PAPILIONIDÆ	5
PIERIDÆ	26
LYCÆNIDÆ	74
ERYCINIDÆ	149
LIBYTHEIDÆ	151
APATURIDÆ	153
NYMPHALIDÆ	159
DANAIDÆ	225
SATYRIDÆ	228
HESPERIDÆ	334
ADDENDA	363
BIBLIOGRAPHICAL LIST	375
A SYSTEMATIC LIST OF THE RHOPALOCERA OF THE EUROPEAN FAUNA	381
INDEX OF GENERA AND FAMILIES	387
INDEX OF SPECIES, VARIETIES, AND SYNONYMS	389

VOLUME II. — PLATES.

CONTENTS OF PLATES	v
ALPHABETICAL INDEX TO PLATES	xi
PLATES	I—LXXXII

INTRODUCTION.

The object of the following pages is to form a treatise on those species of Rhopalocera that are found in Europe, as generally understood; that is, with the Ural Mountains and River, the Caspian Sea, and the chain of the Caucasus as its boundaries towards the East, the Black Sea and Mediterranean on the South, and on the other sides its ocean boundaries as usually received; and though the region thus presented to us may be zoologically somewhat artificial, yet I think it is more convenient for our present purpose to use the term "European" as applying to the species belonging to the region thus restricted, rather than in the more extended sense in which it is now frequently used by zoologists.

Without discussing the merits of any of the various theories respecting geographical distribution, or of the zoological regions recently proposed by writers of eminence, I will say that I consider it imperative to take notice of the entire geographical range of each species, and to endeavour to become acquainted, as far as possible, with forms allied to those under consideration, especially those inhabiting the same latitude.

To carry out this object I have determined to make the well-known 'Catalog der Lepidopteren des Europäischen Faunengebiets' of Dr. Staudinger the basis of arrangement and nomenclature, only departing from this where it seems necessary, and then chiefly taking advantage of such improvements as have been suggested by

B

a perusal of Mr. W. F. Kirby's more recent 'Synonymic Catalogue of Diurnal Lepidoptera.'

Though it is my intention only to figure and describe the Butterflies of Europe, yet in following Staudinger I shall, for the sake of completeness, add to the description of each family group a short notice of the extra-European species included by him in his work, most of these being closely allied to our European forms. The region or territory of the European Fauna as understood by him extends over a very large area of the Northern Hemisphere, comprising the whole of Europe and of North Asia as far as the River Amour, Persia, Armenia, and Asia Minor, with a considerable portion of North Africa bordering on the Mediterranean, and in the Polar Regions, Labrador, and Greenland.

This zoological area corresponds very nearly to the "Palæarctic Region" of Mr. Wallace, omitting his fourth or Oriental subregion, and with the addition of that part which belongs to Polar America.

It is difficult to say whether either of these theoretical divisions has any advantage over the other; in fact, I believe that the only satisfactory way of solving this question with regard to our present subject would be to unite the Palæarctic and Nearctic regions proposed by Wallace, in his 'Geographical Distribution of Animals,' in one great Region, and so to include the greater portion of the extra-tropical Northern Hemisphere of the world, for it cannot be denied that there is a great analogy between the species belonging to North America and those of Europe and Northern Asia.

The limits of the present work would not, however, permit me to treat of such an extensive region as the one suggested, even if I were able to do so in a satisfactory manner. I hope, however, that the plan proposed with regard to Staudinger's non-European species, and occasional references to allied North American forms, will help to increase the interest of the subject.

Whenever possible these notices will be made from the examination of actual specimens; but where this is not possible I shall, of course, be indebted to the works of other observers, and generally to the original figures or descriptions, to which due reference will be made.

With regard to synonyms, only those will be given that are likely to be useful. The references to authors will be placed after the name of every species at the head of the description, and will include the principal works in which the insect is described. A short tabular notice of these will appear at the end of the volume, but in the meantime it will be found that the abbreviations used will be such as are now commonly understood by entomologists. In giving the name of a species the reference to the author following the specific name will be understood to refer to *that* name only, and not necessarily to the generic name: example, *Zegris Eupheme*, Esper. The reference made to Esper concerns the specific name alone, the species thus named being the *Papilio Eupheme* of that author. The references to works treating of the genera will be given in their proper places.

There are more than three hundred species of Rhopalocera, or Butterflies, occurring in Europe, all of which are distributed, somewhat unevenly, among ten families.

It is usual to divide these ten families into three groups, characterized by the manner in which the caterpillar changes into the chrysalis. This arrangement seems to me both convenient and natural, and will be followed in the present volume: it is shown in the following table :—

Group I.—SUCCINCTI, Boisd.

Chrysalis attached by the tail and by a silken band round the middle which supports it in an upright position. This group contains four families.

1. PAPILIONIDÆ.   3. LYCÆNIDÆ.
2. PIERIDÆ.   4. ERYCINIDÆ.

Group II.—SUSPENSI, Boisd.

Chrysalis suspended by the tail alone. Five families are included in this group.

5. LIBYTHEIDÆ.   7. NYMPHALIDÆ.
6. APATURIDÆ.   8. DANAIDÆ.
                9. SATYRIDÆ.

Group III.—INVOLUTI, Boisd.

Chrysalis enclosed in a slight cocoon. Contains one family.

10. HESPERIDÆ.

It would have been perhaps more correct to have placed Group I. between the others, and to have inverted the order of the families, thus bringing into closer relation the natural affinities of the groups. But as this is not of primary importance I have preferred to keep to the arrangement used by Staudinger.

THE BUTTERFLIES OF EUROPE.

Fam. 1.—PAPILIONIDÆ.

Larva cylindrical, not spiny, furnished with two retractile tentacles on the second segment. Imago with the abdominal or inner margin of the wings concave. Anterior legs fully developed. Antennæ distinctly clubbed.

Four genera of European butterflies belong to this family— PAPILIO, THAIS, DORITIS, and PARNASSIUS.

Genus 1.—PAPILIO, Linn. Syst. Nat. i. 2, p. 744 (1767).

Clubs of antennæ curved upwards. Head large; eyes prominent; palpi very short, scarcely reaching beyond the eyes. Abdomen moderately thick and long. Wings tolerably thick, with prominent nervures, the inferior pair generally provided with a long tail as an elongation of the hind margin.

This is a very numerously represented genus, species of it being found almost in every part of the globe. Its head-quarters, however, are the inter-tropical regions of both hemispheres. Only four species occur in Europe, and of these one is confined to certain islands of the Mediterranean, and has not been recorded as inhabiting any other region of the world.

1. P. Podalirius, Mus. Ulr. p. 208 (1764); Hüb. Eur. Schmet. I. f. 388, 389; Boisd. Sp. Gen. i. 245 (1836).

Expands 2·50 to 3·25 in. Wings pale yellow. Fore wings with hind margins black; three or four black streaks running from the costa nearly to the inner margin, the costal end being broader. Hind wings with black streaks ; hind margin black, tail long and narrow, slightly everted at the tip ; four lunules of blue scales on the border ; an orange and blue eye at the anal angle. Head, thorax, and abdomen black. Antennæ black, with recurved clubs. Pl. I., 1.

TIMES OF APPEARANCE.—April to September.

HABITAT.—North-central and South-central Europe, France, North Italy to Spain, South Russia, North Africa, Western Asia, and the Altai. It does not now occur in Great Britain, though there is little doubt that it was a British insect up to the beginning of the present century. It frequents woods, open plains, and road-sides, and may even be seen flying in the streets of towns.

The LARVA of *Podalirius* is yellowish green covered with red dots, with yellowish lines on the back and sides, and with oblique streaks. In shape it is thick in the middle, and tapering towards the extremities. Feeds on almond, sloe, plum, apple, pear, and oak. Times of appearance, June and September. Pl. V., 1.

The PUPA is light straw-colour, the wing-cases being browner.

VARIETY.

Feisthamelii, H.-S. 414, 416. Has the ground colour of the wings lighter than the type ; the markings being darker, and the border on the costa of the fore wings and the hind margins and anal angles of the hind wings, being strongly ochreous in the male and slightly so in the female, contrast more or less with the ground colour of the wings. Pl. I., 2.

HABITAT.—Spain and North Africa.

Aberrant form of the second brood. *Zanclæus,* Z., has the abdomen white in the female, and more rarely in the male. Habitat, the South of Europe.

2. **P. Alexanor,** Esp. 110, 1 (1799); Hüb. Eur. Schmet. i. 787, 788.

Expands 2·18 to 2·75 in. Wings yellow, much more decidedly so than in *Podalirius.* Base of fore wings and hind margin black ; three or four black streaks run from the costa, but only one is continued to the inner margin. Hind wings with the hind margin black powdered with blue; a black band passes across the wing, having the appearance of being prolonged from the long band of the fore wing. A conspicuous elongated black spot closes the discoidal cell. A very small eye of red, blue, and black appears at the anal angle. Head, thorax, and abdomen yellow, on the dorsal aspect black. Antennæ black, with straight clubs which are tipped with yellow. Pl. I., 3.

TIMES OF APPEARANCE.—May till the end of June.

HABITAT.—The South of France, Switzerland, Tyrol, and North Italy, Greece, Western Asia, and Persia. This is essentially an Alpine species and very local, inhabiting mountain gorges.

LARVA bright green, each segment marked anteriorly with a black band interrupted with yellow spots. The V-shaped process on the second segment reddish. Feeds on *Seseli montanum* and various alpine umbelliferous plants. Time of appearance, July. Pl. V., 2.

3. P. Machaon, Linn. Syst. Nat. i. 2, p. 750; Esp. i. 1; Hüb. 390, 391.

Expands 2·50 to 3·43 in. Wings yellow. Fore wings black at the base, dusted with yellow, the hind margin with a broad black band, powdered with yellow, with marginal yellow lunules; nervures of wing black and distinct. Hind wings with the hind margin black, powdered with blue, and a marginal row of yellow lunules; this black margin has a well-defined edge internally; discoidal mark not so conspicuous as in the last species. The eye at the anal angle is round or ovoid and well defined, being blue and dull red from above downwards. The tail straight and shorter than in the preceding species, but well formed. The body is yellow, with a black dorsal band. Antennæ black, with the clubs curved. Pl. I., 4.

TIMES OF APPEARANCE.—From May to end of August.

HABITAT.—The whole of Europe (except the Polar Regions), Syria, Egypt, and North Africa. It occurs also throughout Northern and Eastern Asia, India, and China. It frequents woods, fields, gardens, and road-sides; in fact, it is a generally-distributed species in the lowlands, but does not ascend to any great elevation in mountainous districts. As a British insect it is extremely local, being entirely confined to the fen districts of Cambridgeshire, Huntingdon and Norfolk.

So widely distributed an insect as *P. Machaon* may naturally be expected to present many local forms. The largest specimens are those from Japan; India and China seem to furnish lighter-coloured specimens than the European districts, the dark markings being less strongly expressed in these. A smaller and darker form called *Sphyrus* by Hübner is taken in Sicily and North Africa.

LARVA bright green, with deep black rings which are spotted with red; V-shaped process reddish. Feeds on fennel (*Anethum feniculum*) and wild carrot (*Daucus Carota*), and other *Umbelliferæ*. Appears from June to September. Pl. V., 3.

PUPA bright green shaded with buff, or buff shaded with brown.

4. P. Hospiton, Géné, Mem. Acc. Tor. 1839; H.-S. 249, 250.

Expands 2·50 to 3 in. The wings shorter and proportionately broader than in *Machaon*, which species the fore wings resemble in their markings, except that the band of the hind margin gets gradually broader as it approaches the inner margin; the nervures and spots are very strongly marked and the base very dark. Hind wings with the marginal band very broad and reaching to the discoidal mark (which is not nearly so conspicuous as in *P. Machaon*), its internal edge *not* clearly defined. The blue scales concentrated so as to form six or seven bright blue spots. The tails are very short. Body, head, and antennæ of the same colour as those of *P. Machaon*. Pl. II., 1.

TIMES OF APPEARANCE.—May and June.

HABITAT.—Corsica and Sardinia, frequenting mountainous districts.

LARVA.—I take my description and figure of the larva from the original paper by M. Géné, the first describer of the insect, entitled "De quibusdam insectis Sardiniæ novis aut minus cognitis." He describes it as bright green, furnished with yellow tentacles, and covered with short prickles, each segment having short interrupted black and longitudinal lines and four red spots. It feeds on *Ferula communis* in June and the beginning of July. Perhaps an extract from his paper in the author's own words may be acceptable :—"Differt a *Pap. Machaone;* 1°, Larvâ spinosâ aliterque pictâ; 2° fasciis discoidalibus in superiori alarum pagina angustioribus; 3° anguli anali maculâ tantum semilunari picto. Marem *Cisto Monspellicnsi* insidentem reperi in viciniis *di Portoli* die 20 m. maii. Fœminam legi in *Barbagia Ollolai* prope *Gavoi* ineunte junio." Pl. V., 4.

OBS.—The Corsican specimens seem to be larger than those from Sardinia. Mine, which are from Corsica, are quite as large as *P. Machaon*, though in most figures the species appears much smaller.

Genus II.—THAIS, Fab.; Lat.; Boisd.

Medium-sized butterflies having the head and eyes of moderate size, palpi straight, passing beyond the head. Clubs of antennæ curved. Wings not semi-transparent. All the known species have black spots on the costa of fore wings; the hind wings are more or less dentated and decorated along their hind margin with red and black dots. The prevailing colour of the wings is yellow, more or less intense.

The Larvæ are cylindrical, rather short, and armed with spines, which are set at the extremity of tubercular elevations. A small y-like process posterior to the head. They feed on various species of *Aristolochia*.

The species of this beautiful genus are not numerous, and all belong to the Mediterranean division of our Fauna, being peculiar to the South of Europe, North Africa, and Asia Minor.

The design of the wings is so unlike that of any other tribe of Lepidoptera, especially as regards the dentate pattern of the hind margins, that it is not difficult to recognise a *Thais* at the first glance.

1. T. Cerisyi, B.; Hüb. Eur. Schmet. i. 890-1.

Expands 2 to 2·25 in. Wings yellowish white, black at the base. Fore wings with a row of black streaks running from the costa; hind margin black, with a row of black spots placed parallel to it. Hind wings tailed as in *Papilio;* a row of black V-shaped

c

marks along the hind margin; internal to this a row of bright red spots, inside these near the base one or two oblong black spots; a bright red spot surrounded with black on the costa.   Under side:—The anterior wings have the markings similar to those above, but the ground colour is whiter.   Hind wings white, pure and lustrous; a light green border inside the black spots of the hind margin; base with large oblong black and light green spots, a green costal spot internal to the red one.   Body black, with a lateral orange band.   The female has the ground colour of the wings deeper and more inclining to ochreous; the black bands of the hind margin of the fore wings are larger and deeper, and on the hind wings the black marks form a border with small blue spots below the red ones.   Head, antennæ, thorax, and abdomen black in both sexes, each segment of the abdomen having a reddish yellow spot on either side.   Pl. II., 2.

TIMES OF APPEARANCE.—January, February, and June.

HABITAT.—Turkey and Greece, including some of the Greek Islands.   It seems to be scarce and local in Europe.

LARVA.—Greyish black, with yellow spots on the back and sides.   Feeds on *Aristolochia* in August.

VARIETIES.

a. **Caucasica**, Led. Wien. Ent. Mon. vii. 165.   Hind margin of hind wings not tailed as in the type.   The ground colour of both sexes more inclining to ochre.   Pl. II., 3.

HABITAT.—The Caucasus.

b. **Deyrollei**, Oberth. Petites Nouvelles Entom. i. 2.   Hind margin of hind wings with three projections, instead of one as in the type.   Both sexes have the ground colour nearly white, but the female is very strongly suffused with dusky blackish grey. Pl. II., 4.

HABITAT.—Asia Minor, probably not European.

The following information on the life-history of this species and its varieties is translated from Staudinger, Hor. Ent. Soc. Ross. xiv. 216:—

   "The caterpillar (*Thais Cerisyi*) is only figured by Freyer (in

every case from a prepared specimen he received through Kinder-
man from the coasts of the Sea of Marmora), who says 'the
caterpillar varies extraordinarily,' but only describes the specimen
figured, which is very dazzling yellow, striped with black, and has
red prickles. The only prepared specimen of a real *Cerisyi* which
I possess bears a single clear black dorsal stripe, laterally only
detached black marks.

"The caterpillars of the var. *Deyrollei* vary very much. I saw
between five or six hundred specimens without any black longitudinal stripes; many were bright greenish yellow or reddish,
with sharp black short lines and dots. On the dorsal aspect of
every segment are four of these, two in front of the prickles and
two behind, the posterior being most widely separated; two short
unbroken lines are placed laterally between the segments; the
stigmata are black, with a black longitudinal streak in front. The
wart-prickles (*warzendornen*) are smaller than the red prickles of
the larva of *Cerisyi* or *Caucasica*, and always yellowish green, even
in the quite dark and almost black form. . . . . Two caterpillars of the var. *Caucasica* which I have before me are dark, with
reddish prickles, and two quite black stripes on the dorsal aspect
in front of the prickles. We found the caterpillars of *T. Deyrollei*
in crowds on the *Aristolochia hastata* from the middle of May.
They grow but slowly, and take from six to eight weeks from the
egg to the chrysalis. Before changing the otherwise sluggish
caterpillar crawls about restlessly during one or two days, often
disturbing, in the breeding-cage, the suspended caterpillars, which,
like all other *Thais* larvæ, spin themselves fast with a thread round
the head and by the tail."

2. **T. Polyxena**, Schiff. S. V. 162 (1776); Hub. 392-3.—
Hypsipyle, F. Gen. 265 (1777); H.-S. 557-8.—*Rumina
alba*, Esp. 105, 1, 2.—*Aristolochiæ*, Schu. (1787).—
Cassandra, Mann. Stett. E. Z. (1844), 348.

Expands 2 to 2·75 in. Wings ochre-yellow. Fore wings with
the hind margin black, the black border being divided throughout
its length with yellow. Five black transverse bands reaching from
the costa and not passing the median nervure; beneath this two

black marks, the external one generally joining the third costal band. The fifth costal band counting from within frequently has a red spot. Hind wings black at the base, the discoidal cell containing a trifid black mark. Internal to the marginal border is a black band on which are seven carmine spots, and beneath each of these a patch of blue scales. The hind margin of all the wings much dentated, and having highly dentate black and yellow borders; marginal fringes light yellow. Under side:—The markings of the upper surface are repeated, but the fore wings have three or four of the costal spots marked with red. The ground colour of the hind wings is white, and each marginal indentation is filled up by an orange crescent. Pl. III., 1.

Time of Appearance.—April.

Habitat.—South-Eastern Germany, the South of France and Italy (part of), South-Eastern Europe, Bithynia and Armenia (Staudinger).

It frequents marshy ground and is of short duration.

Larva yellow, with a black dorsal band; six rows of spines bordered with black and with a lateral series of black points forming a triangle. The Pupa is not unlike that of *Cassandra* in form, but is greyish brown in colour. The Larva is found about August feeding on *Aristolochia* (Timins).

VARIETIES.

a. **Cassandra**, Hub. 910, 3. Ground colour lighter than in the type; the black markings are darker and more diffused. Pl. III., 2.

It will be seen that I follow Staudinger in considering *Cassandra* to be a local variety of *Polyxena*. Boisduval, however, considered it distinct, and the Rev. Douglas C. Timins, in a monographic paper on the genus *Thais*, read before the Entomological Society of London, points out that *Cassandra* differs from *Polyxena* in the larva and pupa states; he considers them to be specifically distinct. I quote from his monograph:*—"The Larva feeds on several species of *Aristolochia*: it varies much in colour, but is generally pale reddish, spotted with black. The pupa-state lasts

---

* 'Proceedings of the Entomological Society of London,' 1867, page ci.

from November till March. The pupa is reddish brown, the wing-cases yellowish. I have found this species at Cannes and Hyéres. This species usually appears on the wing in March, about the 15th, and after a fortnight few good specimens are to be seen. The time of appearance, however, varies much; in forward seasons it appears in February, but in 1864 and 1865 it was not on the wing until April."

b. **Ochracea,** Staud., differs from the type in having the ground colour dark ochre-yellow. This form occurs in the South of Europe and Asia Minor.

3. **T. Rumina,** Linn. Syst. Nat. x. 480; Hub. 633-4.

Expands 1·50 to 2·25 in. The wings are of a deeper and brighter yellow than in the foregoing species; the general arrangement of the black markings is much the same. The hind margin of the fore wings is straighter, and near the apex is a transparent spot. There are four or five red spots on the costal bands. Hind wings with a marginal row of red spots, and a red spot at the base. Pl. III., 3.

TIME OF APPEARANCE.—May and June.

HABITAT.—Waste places in Spain, Portugal, and in North Africa. This species occurs on the Rock of Gibraltar.

LARVA grey, with a reddish tint, the ventral surface paler, with parallel black streaks placed longitudinally on the anterior part of each segment. In addition to this its body is provided with six rows of short red fleshy points, armed at their extremity with small black bristles. The Chrysalis is of a dark ash-colour.* Feeds on *Aristolochia.* Pl. V., 5.

a. **Canteneri,** Staud. Cat. p. 2. This form differs from the type in having the ground colour very dark ochre; it is also somewhat larger, and seems to be common in Spain and North Africa.

* Boisd., ' Species général des Lepidoptères,' 1836, p. 388.

b. **Medesicaste**, Illig. Mag. ii. 181; Hüb. i. 632. Has the wings of a much paler colour than the typical form; in some it is nearly white. This is rather smaller, and is the French form, being found at Cannes, Hyéres, and other places in the South of France in May. Mr. Timins, who is acquainted with the larva of *Medesicaste*, describes it as specifically distinct from *Rumina*. But it seems to me that the difference is chiefly one of coloration. Pl. III., 4.

c. **Honoratii**, Boisd. Icon. p. 18, fig. 3, 4; Linn. Syst. 251, 2. This beautiful variety has the red spots enlarged and expanded so as to cover the greater part of the wings; the ground colour is pale straw-yellow, and the spots light rose-colour. This variety is very rare and local, being only found in Digne, in South-Eastern France. It is very scarce in collections. Pl. III., 5.

Genus 3.—DORITIS, Boisd.
THAIS, Lat.

Clubs of antennæ curved. Palpi shorter than in *Thais*. Body hairy. Wings rounded, with black costal spots, semitransparent. Hind wings rounded, with an even unindentated margin. Abdomen of female not furnished with a horny pouch.

This genus is intermediate between *Thais* and *Parnassius*, but seems to be nearer the latter, and perhaps nearer still to the Australian genus *Eurycus*.

1. **D. Apollinus**, Hbst. T. 250, 5-8 (1798); H.-S. 253-6.— *Thia*, Hüb. 635-6, 686-7.

Expands 2 to 2·25 in. Fore wings semitransparent, of grey tint, with small transverse black striæ; two large black spots in the discoidal cell reaching to the costa close to the hind margin, and parallel to it is a row of elongated black spots. Hind wings yellow, with the base black; discoidal cell prominent; hind margin bordered throughout its whole length with neutral grey, and inside this a row of about six eye-like spots orange-blue and black from within out. Antennæ grey, with black clubs. Under side smooth and shining, apparently scaleless, excepting as regards the black spots, and with the orange marks of the hind wings appearing as red ones, the whole design of the upper sides showing through. The female, which seems to be much rarer in collections than the male, has a transverse red mark between the discoidal cell and the marginal border on the fore wings; the hind wings being finely marked with red points, and having the marginal row of eyes more strongly marked than in the male. Pl. III., 7.

TIMES OF APPEARANCE.—February and March.

HABITAT.—Asia Minor and Syria. As a European insect very rare and local, being found in some of the Greek Islands. " Montagnes de la Calabre," *Boisd.*

LARVA.—I copy Mr. W. F. Kirby's description (from Kindermann), in his valuable little 'Manual of European Butterflies':— " Cylindrical, clothed with short hairs, black with two rows of red spots on each side, between which, on the middle segment, are a row of six red spots." The following particulars are gathered from a communication by Staudinger in the Hor. Ent. Soc. Ross. xiv. 216 :—The larva is found at the beginning of May on *Aristolochia Hastata;* it changes to a pupa in June, under moss and stones. The imago is found in sunny weather in January, but February and March are its usual times.

OBS.—Several entomologists have noticed the striking affinity that exists between the genera *Doritis* and *Parnassius* and some of the Heterocera. The present species, *Apollinus*, is perhaps the most moth-like of all the European Butterflies, reminding one, even in its imago state, of some of the dark-coloured *Bombycina*.

Genus 4.—PARNASSIUS, Lat.; Boisd.
DORITIS, Fab. ; Ochs.

Large or middle-sized Butterflies with semi-transparent whitish wings, generally with black spots in the discoidal cell reaching nearly to the costa, rounded and never dentated. Clubs of antennæ straight. Female provided with a horny chitinous abdominal pouch. Palpi reaching beyond the eyes.

The Larva is smooth and cylindrical, furnished with small tubercles, and with a y-shaped process posterior to the head.

The Chrysalis is not only supported by a silken belt, but is

spun up by numerous filaments of silk forming a rudimentary kind
of cocoon; it is rounded and covered with a purple effloresence.
Boisduval compared it to the pupa of the Heterocerous genus
Catocala.

There is no doubt that in its pupation the genus *Parnassius*
resembles the *Hesperidæ*, though the larval and imaginal states
show that it is properly placed in the present family.

The genus *Parnassius* contains about thirty known species,
and is confined to the temperate and boreal portion of the Northern
Hemisphere, all the species being found in mountainous or elevated
regions. Central Asia may be considered the head quarters of the
group, but three species occur in the mountainous parts of Europe,
some inhabit the Caucasus, others the Himalayas, whilst several
species are obtained in the mountainous regions towards the west
of North America. There are but four species that can be considered European.

1. P. Apollo, Linn. Syst. Nat. x. 465; Hüb. 396-7, 730-1; Esp. 2, f. 1.

Expands 2·50 to 3·18 in. Wings white. Fore wings semitransparent at the tip and along the hind margin. There are two
large black spots in the discoidal cell and two or three outside
these, and a well-defined black spot is always present near the
inner margin in both sexes. Parallel to the hind margin, but
separated from the semitransparent border by a white band, is a
wavy blackish band reaching from the costa to the inner margin.
Hind wings with two dusky semitransparent wavy bands running
parallel to the hind margin. Near the anal angle a chain of several
black spots more or less coalescent, and occasionally marked with
red. Two large red spots, one on the costa and one in the centre
of the wing generally white in their centres and surrounded by a
black ring. The bases of all the wings are dusky. Under side,
shining and thinly scaled, marked as on the upper surface, but the
hind wings have at the base and hind margins four large red spots
surrounded with black, and one red spot, or several smaller ones,
at the anal angle. The antennæ are black, ringed with white,
and have black clubs. The head, thorax and abdomen are black

above, the thorax being covered with white silky hair, and also the abdomen in the male; in the female, however, this is black, every segment being marked by a yellowish white ring. The ventral surface and legs are yellowish. The female has a brownish horny abdominal pouch. Pl. IV., 1.

TIMES OF APPEARANCE.—June, July, and August.

HABITAT.—Hills and mountains throughout Europe (except Great Britain and the Polar Regions), also the greater part of Eastern and Central Asia. In Europe it is common in Scandinavia and Switzerland, and is a familiar object to the summer tourist in the latter country. It occurs in France, Spain, and Russia, also in Germany, and less commonly in other parts of Central Europe; the Siberian specimens of the var. *Sibiricus*, Nordm., are very magnificent insects, expanding 3·50 to nearly 4 in., with red spots on the fore wings.

LARVA.—Velvety black, covered with reddish orange points and small blue elevations. There is a retractile y-shaped process placed posteriorly to the head. It feeds on Saxifrages and *Crassulaceœ* in May. Pl. V., 6.

PUPA smooth and covered with a purple bloom, spun up between leaves by threads of silk, as well as being attached by silken bands as in the other genera of *Papilionidæ*. This arrangement forms a sort of rudimentary cocoon.

This species has been described as British, but no authentic record of its capture can be found, and recent entomological observations in the North of Scotland have not resulted in its addition to the British list.

2. **P. Delius**, Esp. 115, 5; Hüb. 649, 684; H.-S. 317.— *Phœbus*, Hüb. 567, 8.

Expands 2·25 to 2·50 in., the average size being more uniform than that of *P. Apollo*. The wings are generally whiter and more thickly covered with scales than in that species; the fore wings are larger and not so rounded, and frequently have red spots. The male has the fore wings somewhat similar to those of *Apollo*, but the spots are smaller, and those near the costa are often marked with red, whilst the spot near the inner margin, which is always present in the

D

last species, is generally absent in *P. Delius*, male. The hind wings
are white, without any trace of a wavy band. The red spots are
smaller than in *Apollo*, and more often without than with white
centres. The spots near the anal angle on the under side are very
small and reduced to one or two, and at the most but slightly
apparent on the upper surface. The dusky hind marginal patch of
the hind wings is darker and more concentrated than in *Apollo*.
Both sexes have four red basal spots on the under side of the hind
wings. The female is much more dusky and more strongly marked
than the male. It has the hind marginal spot on the fore wings
well defined, and sometimes has the wavy dark band on the hind
wings that is found in both sexes of *Apollo*. The abdominal pouch
is small, and the abdomen is hairy. Pl. IV., 2.

TIMES OF APPEARANCE.—July to September.

HABITAT.—The Alps of Europe and Northern and Central
Asia. This species is much more local and scarcer than *P. Apollo;*
it also inhabits more elevated regions. It may be found, for
instance, in mountain gorges in the higher Alps of Switzerland,
but not in the lower valleys, where *P. Apollo* is common. I have
always taken it in very moist situations.

LARVA.—I am very pleased to be able to take advantage of a
notice by Zeller in the ' Stettiner Entomologische Zeitung' of the
earlier stages of this insect, which have hitherto been undescribed
by other observers. In the above-mentioned journal for 1875 he
says that the Butterflies suck the flowers of *Saxifraga aizoides;*
they are to be seen near springs and damp places, where this plant
is common. He considers that there is some foundation for the
suggestion that the larva lives in water during some portion of its
existence, as he found Butterflies that had but newly emerged and
never flown settling on plants of the Saxifrage growing in the
middle of water. On the 19th of July he found the larva on very
wet soil feeding on *Saxifraga aizoides*, and resting exposed upon it.
The pupa he found lying on the surface of the ground, whence he
thought it had been disturbed by the feet of cattle. In captivity
the larva underwent their transformation beneath a fresh sod in a
corner of the box. The pupa resembles that of *P. Apollo* in shape
and colour. The larva when full grown resembles that of *P. Apollo*

in size and shape; in pattern and coloration it differs, as will be seen by the description. The ground colour is satiny black; the spots are lemon-yellow; the head and prolegs are black and without lustre; the legs are shining black. The spots are large and transversely elliptical, on every segment from the third backwards there are one large and two small ones; above these lemon-yellow spots every segment has two small pale, bluish ones; close behind the second one is a shining dot.

OBS.—In dealing with European specimens of this and the preceding species, the difficulty in diagnosis is not very great, especially when one has a good series of each for comparison; those from Siberia, however, run very closely into one another. I have Siberian specimens of *P. Apollo* which are spotted with red on the fore wings, a condition which seldom or never occurs in European specimens of that species; the size, too, varies greatly, and is no guide whatever, excepting that *P. Delius* never reaches the dimensions of the larger specimens of *Apollo*. A knowledge of the habits of the insects, especially those of their earlier stages, will be of great assistance in separating the two species.

3. P. Nordmanni, Nord. Bull. Mosc. (1851), p. 453.—*Clarius*, H.-S. 257, 8 (1848).

Expands 2·50 to 2·75 in. Wings white, semitransparent, but not so much so as in other species. The fore wings have two deep black spots in the discoidal cell. The hind margin is blackish, more transparent than the rest of the wing; the dark band thus situated is broadest at the costa, and becomes narrower as it approaches the hind margin (it is not divided throughout its length by a wavy white band as in the allied Siberian *P. Clarius*, with which this species was at first confounded by Herrich-Schäffer). The hind wings are white, slightly dusky along the hind margin, and have two red spots surrounded by black rings. There are *no* red spots at the base or inner margin on the under side of the hind wings, nor any red spots near the anal angle.

TIME OF APPEARANCE.—July.

HABITAT.—The Caucasus. A rare species in collections; the

only specimens I have seen are in the British Museum (Hewitson Collection). Pl. IV., 3.

Larva unknown.

4. **P. Mnemosyne**, Linn. Faun. Suec. p. 269; Hüb. 398.

Expands 1·93 to 2·50 in. Wings white and semitransparent, the nervures black; bases of all the wings blackish. The fore wings have two black spots in the discoidal cell, and the apices are darker and more transparent than the rest of the wing. The hind wings have a dusky lozenge-shaped spot in the centre immediately below the discoidal cell. The wings are entirely without red spots either above or beneath. The female has a very large and horny abdominal pouch, convex in shape and of a light yellow colour. The head, thorax, and abdomen are black, the latter strongly marked laterally with bright yellow. Pl. IV., 4.

Times of Appearance.—June and July.

Habitat. — High mountain gorges in Switzerland, the Pyrenees, Sweden, Hungary, Sicily, Russia, and Western Asia. A local insect, occurring at considerable elevations, even as high as 8000 feet.

Larva.—Smoky black, the spaces between the segments being darker; on every segment are two reddish yellow spots. The legs are black. Feeds on *Corydalis Halleri*. Appears in April and May. Pl. V., 7.

The Pupa much resembles that of *P. Apollo*, spinning up between leaves.

OTHER SPECIES OF *PAPILIONIDÆ* INCLUDED IN STAUDINGER'S CATALOGUE, &c.

Genus *PAPILIO*.

P. Virgatus, Butler, Proc. Zool. Soc., 1865, p. 430. — It is close to *P. Podalirius*, but whiter, and the tails of the hind wings are much longer. Inhabits Syria.

P. Xuthus, Linn. Syst. Nat. xii. 751; Cr. 73 A; H.-S. 411, 413.—Expands 3 to 3·60 in. A large black and yellow butterfly, not unlike *P. Machaon* in its markings, but with a very small eye at the anal angle of the hind wings. It occurs in Eastern Siberia, in the district known as Amoorland, or that portion which borders on the River Amoor on the confines of China; it also occurs in the latter country, as well as Japan and Thibet.

P. Xuthulus, Brem. Lep. Ost. Sib. p. 4 (1864).—Rather smaller than the last, and of a yellower colour; it also occurs in Eastern Siberia, and is considered by some to be a second brood of *P. Xuthus*.

P. Maakii, Mén. Bull. Acad. Pet. xvii. p. 212, n. 1 (1859); Brem. Lep. Ost. Sib. p. 3 (1864).—Expands 4·50 in. A large and splendid insect of Oriental aspect allied to *P. Paris*. The wings are black, the fore wings powdered with shining bluish green scales at the base, costa, and hind margins, the patch of green scales being thickest along the latter. Hind wings powdered with blue and green scales; hind margins black, with a row of blue and violet crescents, one of these forming an indistinct eye at the anal angle. Habitat, Amoorland.

P. Raddei, Brem. Bull. Acad. Pet. iii. p. 462 (1861); Lep. Ost. Sib. p. 3 (1864).—Smaller than *P. Maakii*, expanding 3·50 in. Nearly similar to that species, but the green scales of more of a golden colour and less mixed with violet. There are two or three distinct brickdust-red spots beneath the hind marginal crescents of the hind wings, and an eye at the anal angle of the same colour, with a black centre. Considered by some to be the second brood of *P. Maakii*. Habitat, Amoorland.

Genus *LUEDORFIA*, Crüger, Virch. Var. Hamb. iii. p. 128.

L. Putziloi,* Ersch, Hor. Ent. Ross. viii. p. 315 (1872).— Expands 1·80 in. This is a highly interesting little species, constituting a genus, in which the characters of *Papilio* and *Thais* are curiously blended. In outline it very much resembles *Thais Cerisyi*, but the hind wings are not so dentated, so that the tail is distinct, though short, as in *Papilio Hospiton*. The wings are sulphur-yellow; the fore wings have black bands arranged almost exactly as in *Papilio Podalirius;* the hind wings have the black

---

* This species does not occur in Staudinger's Catalogue, having been described since the publication of that work. It, however, belongs to his "European" Region.

basal patches and marginal rows of red and blue spots seen in *Thais*. The under side of the hind wings is white, with black stripes at the base and anterior margin; parallel to the hind margin is a row of red spots, externally to these some black ones powdered with blue in the centres, and outside these a marginal row of ochre crescents. The head, palpi, and antennæ resemble those of *Papilio*. Habitat, Siberia.

Genus *HYPERMNESTRA*, Mén. Cat. Mus. Petr. Lep. i. p. 7 (1855).—*Ismene*, Nick. Stett. Ent. Zeit. (1846); Staudinger, Cat. p. 2. *Sed nomen prœocc.*

H. Helios, Nick.— Expands 1·50 to 1·75 in. The genus of which this species is the only representative combines the characters of *Thais* with those of *Parnassius*. The wings a yellowish white, much the same colour as *Thais Cerisyi*, and not semitransparent. The fore wings have two black spots in the discoidal cell, as in *Parnassius*, and externally to this two more with large red centres; along the hind margin from the costa to the anal angle runs a black band enclosing a row of spots of the same colour as the wings. Hind wings with a very narrow black border and black spots on the hind margin; there are two small black spots which occupy exactly the same relative position as those on the hind wings of the *Parnassii*. Under side :—Fore wings same as above, but lighter; the apices are greyish green, and the red spots are larger than on the upper surface. Hind wings white, chequered with greyish green, the mottling showing through, so as to be slightly visible above : the two black spots have orange centres, but are very minute. The female has a black spot on the upper surface of the fore wings placed near the centre of the inner margin : this sex is unprovided with an abdominal pouch as in *Parnassius*. Habitat, the region of the South-East of the Caspian Sea, Turkestan and Persia, in April and May. The Larva feeds on *Zygophyllum Turcomanum*.

Genus PARNASSIUS, Latr.

P. Apollo, var. *Sibiricus*, Nord., has been already alluded to.
P. Nomion, Fisch. Ent. Ross. ii. p. 242 (1823); H.-S. Schmet.

Eur. i. 316.—Expands 3 in. Very close to *P. Apollo*, but the hind wings have a red spot at the base, and the inner margins are more strongly black; all the wings have a hind marginal border of square black and white spots. Time of appearance, July (Boisd.). Habitat, Siberia.

P. Bremeri, Brem. Lep. Ost. Sib. p. 5, T. 1, 3, 4.—Expands 2·50 in. Fore wings with only two black spots. Hind wings white, without any marginal spots; inner margins very black; there is a red spot at the base, and two light red spots placed as in *P. Delius*, small and square in the male, but larger in the female. Habitat, Siberia, chiefly in the Amoor district.

P. Apollonius, Evers. Bull. Mosc. (1847), iii. 71; H.-S. 636, 7. —Expands 2·50 in. All the wings have a row of black spots running parallel to the hind margin. The fore wings have two black spots and three or four red ones; the hind wings have a red spot at the base, besides the usual ones found in the allied species. This is undoubtedly the handsomest species of the genus: it is rare in collections on account of its exceedingly circumscribed habitat, which is limited to the district immediately south of the Altai.

P. Delius var. *Intermedius*, Mén. En. i. p. 72. A small form of *P. Delius* from the Altai.—Expands 2·40 in. The male has a black spot in the centre of the hind margin of the fore wings and two conspicuous red ones near the costa.

P. Corybas, Fisch. Ent. Ross. ii. p. 242. —Expands 2·40 to 2·50 in. Wings more thickly clothed with scales than in others of the genus, creamy white, rather thickly speckled with black scales. The fore wings have the hind marginal and costal spots dotted with crimson. There is no red basal spot on the upper surface of the hind wings, but there is a row of black spots running parallel to the hind margins. Habitat, South and East of Siberia.

P. Actius, Evers. Bull. Mosc. (1843), p. 540; H.-S. i. 634.— Expands 2 in. Very much resembles *P. Nomion* above, but is much smaller. The hind wings have no red spot at the base, but there is a chain of black spots running parallel to the hind margin, and the two red eye-like spots have very broad black rings. Beneath, the hind wings have the red basal spot very indefinite, and are thickly clothed with hairs along the inner margin.

P. Delphius, Evers. Bull. Mosc. (1843), p. 451; H.-S. i. 638. —Expands 2·25 in.   Wings dusky white; fore wings with two black spots and three semitransparent black bands starting from the costa, the two outer ones reaching the hind margin.   Hind wings black at the base; hind margins semitransparent, and running parallel to it a wavy dark line; the two spots are present, but are yellowish red.   Beneath, the hind wings have the base blackish, without any red spots.

P. Tenedius, Evers. Bull. Mosc. (1857); H.-S. 628-631.— Expands 2·12 in.   Whiter than *Actius*.   The fore wings have three black streaks in the discoidal cell, two more external to this reaching from the costa; below the exterior one is a row of four black spots; the apex is dusky.   Hind wings with two very small red spots; a row of black spots runs parallel to the hind margin.   On the under side there is a very faint red spot at the inner margin of the hind wings. (Described from the figure in H.-S.).   Habitat, Eastern Siberia.

P. Eversmanni, Mén. En. i. p. 73.—Expands 2·56 in.   The wings are light chrome-yellow, lighter in the female.   The fore wings have three black spots, and besides these two blackish bands running from the costa to the hind margin.   The hind wings are black at the inner margin, and have the usual red spots. The female has the bands of the fore wings somewhat broader. Both sexes have the red basal and hind marginal spots on the hind wings beneath.   The abdomen is strongly marked with lines of bright yellow on the ventral surface, and the female has a large abdominal pouch very much like that seen in *P. Mnemosyne*. Habitat, Eastern Siberia and the Amur.

P. Clarius, Evers. Bull. Mosc. (1843), iii. p. 539; H.-S. 628-631.—Very like *P. Nordmanni*, and originally figured by Herrich-Schäffer as the same species; it is, however, of a purer white, and the dark hind marginal apex of the fore wing is divided by a wavy white band throughout its entire length.   The red spots of the hind wings are more square in shape, and near the anal angle there is a red spot. There are no hind marginal or basal red spots beneath. Habitat, the Altai.

P. Stubbendorfii, Mén. feu. Lehm. p. 57, pl. 6, 2; H.-S. 640. —Greatly resembles *P. Mnemosyne*, but all the black discoidal spots are wanting, and the ventral surface of the abdomen is not marked with yellow. It bears a strong superficial resemblance to *Aporia Cratœgi*. Habitat, the Altai and Eastern Siberia.

NORTH AMERICAN SPECIES ALLIED TO THOSE OF EUROPE AND NORTHERN ASIA.

Papilio Turnus, Linn., is allied to *P. Alexanor*, but very much larger. It is found in Canada and the Western States.

Papilio Ajax, Linn., is very close to *P. Podalirius*. It is found in the Valley of the Mississippi from Pennsylvania to Texas. It is figured, with its varieties, in Edwards's ' Butterflies of North America.'

There are several varieties of *P. Machaon* found in North America which have received distinct names; of these *P. Zolicaon*, Boisd., occurs in California, and *P. Aliaska*, Scudd., in Aliaska.

Parnassius Smintheus, Doubl., occurs in the Rocky Mountains, and does not seem to differ from the Siberian variety of *P. Delius* called *Intermedius*.

Parnassius Clodius, Mén., is a Californian species, and is closely allied to *P. Nordmanni* and *P. Clarius*. It was long confounded with the latter, but is quite distinct, differing in the shape of the wings, and in having red basal spots beneath.

Parnassius Eversmanni, Mén., described above, inhabits Aliaska, and does not differ from the North Asiatic form.

E

Fam. 2.—PIERIDÆ.

Larva.—Smooth or downy, more or less tapering towards the extremities, not furnished anteriorly with a retractile fork.

Pupa.—Angular, slightly compressed laterally, and tapering to a point, attached by the tail, and by a transverse median belt.

Imago.—Head moderately small; palpi distinct, longer than in the last family; antennæ long and distinctly clubbed. Six perfect legs in both sexes. Wings more or less rounded, the hind wings neither concave nor dentated. The abdomen received into a groove formed by the hind wings. Among the European species the ground colour of the wings is white or yellow, but amongst the Exotic forms there is great variety of colour. Some species of the South-American genus *Leptalis* exhibit well-marked examples of mimicry, their resemblance to certain *Heliconidæ* being very striking.

There seems to be a tendency in the family towards gregarious habits. We meet with it sometimes in the larval state, as in the case of *Aporia Cratægi*, where the caterpillar lives in company beneath a web spread over the hawthorn bushes. The Tropical genus *Callidryas* affords examples of this in the perfect state, many of the species occurring in swarms or flights of considerable numbers, especially on the sea-coasts; hence probably for purposes of migration. Our British species of *Pieris*, *P. Rapæ*, *Napi* and *Brassicæ*, have been observed both on land and at sea in similar migratory swarms.

Seven European genera are contained in this family—APORIA, PIERIS, EUCHLÖE, ZEGRIS, LEUCOPHASIA, COLIAS, and GONEPTERYX.

Genus 1.—APORIA, Hüb.
PIERIS, Schrank.
PONTIA, Fab.

Antennæ long and tolerably thick. Wings semitransparent, uniform white. Fore wings without dusky tips, the nervures very distinct and black. Hind wings rounded, and white beneath.

Larva gregarious beneath a common web.
Pupa smooth, and without angular projections.
There is but one European species.

1. A. Cratægi, Linn. Syst. Nat. x. 467; F. S. ii. 269; Hüb. 399, 400.

Expands 1·75 to 2·50 in. All the wings are white, more or less diaphanous, more so in the female than the male, without marginal fringe. The nervures are very distinct, and generally have at their marginal ends triangular patches of blackish scales. Antennæ black. Head, thorax and abdomen of the same colour, and slightly downy.

TIMES OF APPEARANCE.—May and June.

HABITAT.—The whole of Europe (except the Polar Regions) and Siberia. As a British Insect it seems to be confined to the Midland and Southern Counties of England, occurring locally, and absent from Scotland and Ireland. On the Continent—"Commune dans toutes les prairies de l'Europe" (Boisd.). Pl. VI., 1.

LARVA covered with a white down, with the sides and ventral surface lead-coloured. The dorsal surface marked with two longitudinal yellowish bands. Feeds in company on the hawthorn, sloe, wild cherry, and other fruit trees. On the Continent it is sometimes very destructive to orchards.

PUPA greenish white, with two lateral yellow lines, and numerous black points. The larva appears in the spring. Pl. XV.

Genus 2.—PIERIS, Schrank. Faun. Boic. (1801); Lat. Hist. Nat. Crust. et Insect. (1805).
PONTIA, Fab. (1807).
DANAÏ CANDIDI, Linn.

Middle-sized or large Butterflies, with the antennæ long, distinctly articulated, and terminated by a very distinct club. Abdomen moderately slender, and not passing beyond the inferior wings. All the European species have the ground colour of the wings white. The fore wings are pointed and tipped with black, and there is generally a black spot near the inner margin in the female. The black markings are always more developed in the female than in the male. The hind wings are decorated on the under side with green or yellow scales, variously disposed, but having a tendency to follow the course of the nervures, or to form quadrate or triangular blotches between them.

The Larvæ are long and cylindrical, covered with down, slightly tapering at the extremities, marked with longitudinal stripes, and more or less granulated on the surface. They feed only on low herbaceous plants, and principally on those belonging to the order CRUCIFERÆ.

The Chrysalides are angular, and terminated anteriorly by a single point.

The genus *Pieris* is very widely distributed, being found over the greater part of the globe, but chiefly in the Tropical regions of the Eastern Hemisphere. The most remarkable species are found in Africa, the Islands of the Indian Archipelago, and Australia.

Upwards of 160 species of *Pieris* are named in Mr. W. F. Kirby's ' Synonymic Catalogue of Diurnal Lepidoptera.'

Though the prevailing colour of the European species is white, many of the Tropical species are richly and brilliantly coloured.

There are seven European species, and four of these occur in Britain.

1. P. Brassicæ, Linn. (*Papilio B.*) Faun. Succ. 269; Syst. Nat. 467; Esp. 3, 1; Hüb. 401, 3; Man. i. 18.

Expands 1·75 to 2·50 in. The wings are white, with the bases dusky. The fore wings are tipped with black in both sexes; the female has, besides this, three black spots, two of them nearly circular, placed one above the other, nearly midway between the centre and the hind margin; the third is triangular, with its apex towards the base of the wing, beneath the other two spots, and generally touching the lower one. Hind margin simple. The hind wings are rounded, with the margins simple; white, with a black spot in the middle of the costal border in both sexes. On the under side the fore wings are white, with the tip ochre-yellow, and two black spots. The hind wings ochre-yellow, dusted with black.

TIMES OF APPEARANCE.—April till September.

HABITAT.—The whole of Europe, except the Arctic Region; the greater part of Asia and North Africa. As a British insect it is generally distributed, and is one of the very commonest species; it is, however, much commoner during some seasons than in others; it sometimes occurs in vast numbers, forming migratory swarms, often accompanied by individuals of *P. Rapæ* and *Napi.* Pl. VI., 2.

LARVA yellowish green, with three longitudinal yellow stripes, separated by little black points or tubercles. The head is bluish, dusted with black. It feeds, generally in groups, on various kinds of *Cruciferæ*, in many places being very destructive to the cabbage and other garden vegetables; it also feeds on Nasturtiums and Tropæolums. Larva appears in May or July.

CHRYSALIS ashy white, speckled with black and yellow. Pl. XV.

2. P. Krueperi, Staud. Wien. Ento. Mon. iv. 19.

Expands 2 to 2·25 in. Wings white, very slightly shaded at the base. Fore wings black at the tips and for some distance along the hind margins, a triangular black spot on the costa near the tip; below this, about midway between the costa and the hind margin, a distinct black spot, reniform in shape. Hind wings marked on the costa with a black spot, having the shape of a triangle with its base downwards. Under side:—Fore wings primrose, yellow at the base, also at the tip, and two-thirds of the distance along the hind margin; a triangular dusky patch extending from the costa, and below this a reniform black spot, the same as on the upper surface. Hind wings yellow, more distinctly so at the base; from the outer edge of the costa, extending downwards, is a dusky patch. Body covered with white down. Antennæ and eyes black.

TIMES OF APPEARANCE.—In the spring months of March and April, and again at the end of June.

HABITAT.— Greece, Asia Minor, and Persia, frequenting mountains. It is a very local species, but does not seem so difficult to obtain now as formerly. Pl. VI., 3.

Neither the LARVA nor PUPA have been described, so far as I am aware.

Vernalis, Staud. Staudinger describes under this name (Hor. Soc. Ent. Ross. 1870, p. 34) a variety of the spring brood, in which the hind wings are green instead of yellow on the under side.

3. **P. Rapæ,** Linn. (*Papilio R.*) Syst. Nat. 468; F. S. 270; Esp. 3, 2; Hüb. 404, 5; St. Man. i. 19.

Expands 1·50 to 1·75 in. Wings white. Fore wings dusky at the tip, but not so dark as in *Brassicæ;* two round black spots in the centre of the wing in the female, which is generally darker than the male. Hind wings rounded, white, with a small blue costal spot. Under side:—Anterior wings tipped with ochre, and with two black spots in both sexes. Hind wings pale ochre, more tinged with yellow than in *Brassicæ.*

TIMES OF APPEARANCE.—April to October.

HABITAT.—The whole of Europe and North Asia (excepting the Polar Regions), North Africa and Asia Minor. As a British insect it is universally distributed, being our commonest Butterfly. Occasionally it is very abundant, occurring in swarms. It appears to inhabit any kind of country, excepting the more elevated mountain regions. Gardens and fields are, however, its principal haunts.

Recently this insect has been imported into North America, and it seems that in the species thus acclimatised the wings are often entirely suffused with bright yellow, so that it appears more like a yellow than a white butterfly. Pl. VI., 4.

LARVA.—Green, covered with down, with one dorsal and two lateral yellow lines. Lives on *Cruciferæ,* and, like the last, often destructive in gardens.

PUPA.—Ashy, speckled with black, often tinged with reddish. Pl. XV.

4. **P. Ergane,** Hüb. (*Papilio E.*) Eur. Sch. i. 904; Dup. i. 47, 1, 2.—*Narcæa,* Fr. Beitr. Schmett. 43, 2.—*Rapæ,* Variété A. Boisduval, 1, 5, 20.

Very much resembles *P. Rapæ,* of which species it used often to be reckoned a variety. It is now, however, considered to be distinct by most authors. It is about the same size as *P. Rapæ*, or perhaps on an average a little smaller. The fore wings are rather more pointed. The dusky tip and black spots are more diffused but less black than in *P. Rapæ*. The costal spot of the hind wings is often absent, or represented by a fine crescentic line. Under side as in *P. Rapæ*, but *entirely without black markings*.

TIME OF APPEARANCE ———?

HABITAT.—Dalmatia and the Balkan Mountains. Bythinia (Standinger). Pl. VI., 5.

LARVA ———?

5. **P. Napi**, Linn.; Syst. Nat. x. 468; F. S. 271; Esp. 3, 3; Hüb. 406, 7.

Expands 1·50 to 1·87 in. Wings white, with the bases dusky. The nervures are distinct and black. The fore wings have the tips, and sometimes the ends of the nervures, dusky. Male sometimes with, but often without, a small black spot midway between the centre and the hind margins. Female with two black spots, as in *P. Rapæ*. Hind wings with a black spot on the costa. The female is usually larger than the male, and is always darker, having blackish scales running along the course of the nervures. Under side:—Fore wings white, tipped with greenish yellow, with nervures conspicuous, and with two black spots, as in the allied species. Hind wings pale yellow, with dark scales placed thickly along the course of the nervures, giving the appearance of green veins. Pl. VII., 1.

TIMES OF APPEARANCE.—April to September.

HABITAT.—Europe, and Northern and Eastern Asia; frequenting gardens, &c. As a British insect it appears to be generally distributed throughout England, Scotland and Ireland, and to be in most places a common insect.

LARVA.—Green, brighter on the sides than on the back; the spiracles are marked with red and yellow. Feeds on various kinds of *Cruciferæ* and *Resedaceæ* from June to September. It hybernates as a pupa, which is greyish or greenish yellow, speckled with black. Pl. XV.

Specimens of the second or autumnal brood often occur, in which the under side of the hind wings is paler, and has only short blackish veins, reaching about as far as the discoidal cell. This is the var. *Napææ*, Esp. 116, 5 ; Hüb. 664, 5 ; but is rather an aberrant form of one brood than a variety.

Bryoniæ, O. 1, 2, 151. This is a form of the female in which the ground colour of all the wings is greyish yellow, and the nervures and spots dilated, producing a very dark appearance. It occurs in the Alps as a local variety ; and is not uncommon in Switzerland in mountain meadows, &c. Pl. VII., 2.

6. **P. Callidice**, Esp. 115, 2, 3 ; Hüb. 408, 9, 551, 2. Expands 1·50 to 1·75 in. Wings white, dusky at the base. Male with small triangular black marks at the tip and along the hind margin of the fore wings ; in the centre, at the outer edge of the discoidal cell, a narrow oblong black spot. Hind wings white, without spots, but showing the pattern of the under side through. The female has the hind margin of the fore wings black, with large oval or lozenge-shaped white spots, and a large more or less quadrate black spot at the outer boundary of the discoidal cell. Hind wings white, with a slight yellowish tinge, and powdered with black scales ; the hind margin black, with spots similar to those on the upper wing. Under side :—Fore wings much the same as on the upper surface, but the black spots have greenish scales upon them. Hind wings dark green, with pale yellow spaces between the nervures in the shape of arrow-heads. Pl. VII., 3.

TIMES OF APPEARANCE.—July and August.

HABITAT.—This is truly a mountain species, inhabiting the Alps of Switzerland, Savoy, the Tyrol, and the Pyrenees. Menétriès, in his Catalogue of the Imperial Academy of St. Petersburg, gives Kamtschatka. It also occurs in the other mountainous districts of Northern Asia, extending as far as Cashmere and the Himalayas. It is not at all uncommon in Switzerland on the higher passes ; I have taken it at an elevation of 8000 feet, in places where the snow is perpetual.

Larva.—Dark greyish blue spotted with black; on each segment are four longitudinal stripes, marked with a light yellow spot. The spiracles are bluish white. The head is the same colour as the body, marked on each side with a yellow spot. The chrysalis, which passes the winter fastened to the rocks, is grey, finely powdered with black, and with a yellow line on the back (Boisduval, Species Gen. de Lepidopt. i. 542). The larva feeds on alpine *Cruciferæ*. There are no European varieties of this species, but Herrich-Schäffer figures a variety called *Chrysidice*, which occurs in Asia Minor and the North of Persia. It differs from the type in having the under side of the hind wing of a deeper green, and with the marginal patches square, instead of in the form of arrow-heads.

7. P. Daplidice, Linn. Syst. Nat. x. 468; Hüb. 414-5, 777-8; Frr. 533.

Expands 1·50 to 1·80 in. Wings white, marked with black and grey above; the hind wings having a chequered pattern of green beneath. The tip of the fore wings is black in both sexes, divided by four small white spots, each one sending an elongation into the marginal fringe; at the extremity of the discoidal cell is a black spot, in which the discoidal nervure appears as a fine white line; this black spot is narrow in the male, large and square in the female; the latter has also a black spot of a round or lunar shape near the hind margin. The hind wings are white and unspotted in the male, showing the pattern of the under side through. In the female they have a black border formed of crescentic patches, with the convex edge inwards, and divided by black dashes. Under side :—Pattern of the fore wings the same as above, except that the border of the tip and hind margin is powdered with green scales, also the discoidal spot; the base of the wing is tinged with greenish yellow, and there is a black spot near the inner margin in both sexes. Hind wings green, with a slight tinge of yellow, and finely powdered with black scales; on this ground-work there is an arrangement of white spots, disposed as follows :— Two or three irregularly placed near the base of the wing; outside those a row forming a band; and again, external to these, a marginal row of five spots, oval or nearly quadrate.

F

TIMES OF APPEARANCE.—From April to August, being double-brooded.

HABITAT.—Common in the greater part of Europe, North Africa, Asia Minor, North and Central Asia as far as the Himalayas; inhabiting open plains, fields, &c. It is one of the rarest of the British Butterflies, individuals of the second brood occasionally occurring on the south coast in August; sometimes two or three captures are recorded in one season. Pl. VII., 4.

LARVA.—Greyish blue, covered with small black granulations, with four longitudinal white stripes, and with a yellow spot on each segment. The legs and ventral surface are white. Chrysalis grey, speckled with black, and with reddish stripes. The larva feeds, like other species of the genus, on *Cruciferæ* and *Resedaceæ*.

VARIETY.

Bellidice, O. 1, 2, 154. Smaller than the type. The marginal band of the fore wings less extensive, and powdered with white scales. Under side:—The fore wings without the inner-marginal spot in the male. Hind wings deeper green, not so much tinged with yellow; the marginal spots are larger and narrower. This variety inhabits the same localities as the type, appearing in the spring. Pl. VIII., 1.

8. P. Chloridice, Hüb. 712-5; O. iv. 154.—*Daplidice* var. *Russiæ*, Esp. 90, 1, p. 177.

Expands 1·25 to 1·75 in. Wings white. Fore wings with the hind margin slightly concave; the marginal band consists of blackish dashes, to the number of about five, running from the margin inwards: inside these there is, in the female, a black border running downwards from the costa. The spot at the extremity of the discoidal nervure is white in the centre. The female has the usual black spot near the inner margin; it is, however, fainter than in the preceding species. The hind wings are white and unspotted in the male, but with small marginal black spots in the female. Under side:—Fore wings similar in their arrangement to those of *Callidice* and *Daplidice*, but the green colour of the hind margin is of a most beautiful and delicate tint. The hind wings have the ground colour of the same tint of green, with white spots arranged much as in *P. Daplidice*, but longer and narrower; this latter character applies especially to the marginal spots, which, instead of being roundish or trapezoid, are decidedly oblong. Pl. VII., 2.

TIME OF APPEARANCE.—July.

HABITAT.—South-Eastern Europe and West Central Asia; that is to say, the Steppes of South Russia, Turkey (Staudinger), Siberia and Persia.

This insect, which clearly belongs to the Siberian sub-region, seems to be rather rare in collections.

I have not met with any description of the larva or of its food-plant.

Genus 3.—EUCHLOË, Hüb. (1816).

ANTHOCHARIS, Boisd. Sp. Gen. i. 556.

PIERIS, Lat.

PONTIA, Fab. (1836).

Rather small butterflies. Wings rounded, and either white in both sexes, or yellow tipped with orange in the male and with white in the female. The hind wings ornamented on the under side with a mottled pattern of green or yellow. The head and eyes are of moderate size. The antennæ shorter than in the genus *Pieris*, with flattened ovoid clubs. The abdomen is slender, and of the same length as, or a little shorter than, the hind wings.

The Larvæ are small and downy, tapering towards the extremity, and, like those of *Pieris*, feed on *Cruciferæ*.

The Pupæ are arched and boat-shaped, differing much in shape from those of *Pieris*, but attached in the same way.

The species of this genus have very much the same habits as those of *Pieris*, to which genus *Euchloë* is very close. This genus is tolerably well represented in Europe, and is confined to the North Temperate regions of both Hemispheres. All the inter-tropical species that used to be included in this genus are now placed in the genus *Callosune*, of which one species, *C. Nouna*, Luc., occurs in Algeria.

The European species, like our English *E. Cardamines*, appear during the earlier months of the year, but many of these are double-brooded. *Anthocharis* is the generic name of this group in Staudinger's Catalogue, but *Euchloë* seems to have a prior claim.

1. E. Belemia, Esp. Schmet. i. 21; Hüb. Eur. Schmet. i. 412-3.

Expands from 1·50 to 1·75 in. Wings white, black spot at the base; fore wings rather pointed at the tip, which is blackish, spotted with white. Costa with small black spots. At the extremity of the discoidal cell is a large black spot with a white lunule. Hind wings somewhat angular at the hind margin, white, without spots or markings, but generally showing the pattern of the under side more or less distinctly. Under side:—Fore wings with the tips green, marked with silvery or pearly white streaks, the discoidal spot black and the costa spotted as above. Hind wings deep bright green, with transverse bands of a pearly or silvery white. Pl. VIII., 3.

TIMES OF APPEARANCE.—At the end of winter or early spring.

HABITAT.—The South of Spain, and Portugal, Algeria, and Egypt.

VARIETY.

a. **Glauce**, Staud. Cat. 3; Kirby, Cat. 506.—*Anthocharis Glauce*, Boisd. Sp. Gen. 558. Black markings rather paler than in the type, and costa of fore wings not spotted. Under side:—Tips of fore wings yellowish brown instead of green, and markings white, without any silvery or pearly lustre. Hind wings brownish olive, streaked with white, the streaks being less distinct than in *Belemia*. This variety is found in the same localities and at the same time as the type. Boisduval considers it a distinct species, and describes the LARVA as pubescent, yellow, and very finely speckled with black, with three rosy red longitudinal bands. I am not inclined to separate *Glauce* from *Belemia*, not being able to obtain any description of the larval state of the latter, but follow Staudinger and others in considering it a variety. It is quite possible, however, that it is distinct. Pl. VIII., 4.

2. **E. Belia**, Cram. Pap. Ex. pl. 397 (1782); Esp. 92, 1.— *Crameri*, Butler, Ent. Month. Mag. (1869), p. 271.

Expands from 1·25 to 1·75 in. Wings white. Fore wings with a black tip spotted with white, a large quadrate black discoidal spot, and having the costa spotted with black. Hind wings white, and without spots, but showing the pattern of the under side through the texture of the wings. Under side:—Fore wings with a white crescentic mark on the discoidal spot, the tip greenish yellow, marked with pearly or silvery touches. Hind wings dark green, mixed with yellow, and marked with a number of pearly or silvery spots, for the most part round in form. The female is rather larger than the male. Pl. VIII., 5.

Time of Appearance.—From March to June.

Habitat.—The South of Europe, North Africa, and Asia Minor. Occasionally near Paris (Boisd.).

Larva.—Yellow, with black spots and rose-coloured lateral and dorsal stripes. Feeds on *Biscutella*.

Obs.—Mr. Butler has proposed the name *Crameri* for this species, as it is probable that *Belia* was the name given by Linnæus to *E. Euphenoides* of Staudinger or to the Algerian *E. Eupheno*. *Papilio Belia* of Esper is the *Ausonia* of Hübner (var. *Ausonia* of this work). To avoid confusion, and as it is very doubtful what really was the *Belia* of Linnæus, I have, as usual, preferred to follow Staudinger's Catalogue.

VARIETIES.

a. **Ausonia**, Hüb. 582-3. Slightly larger than the type; the black markings are much paler. The costal black spots are either wanting entirely or else very inconspicuous. Under side:—The tips of the fore wings and the hind wings have the dark green colouring replaced by yellow; the streaks and spots are larger and more irregularly arranged than in *Belia*, and are white, without any silvery lustre. It is found in the same localities as *Belia*, of the second brood of which I believe it to be a variety or dimorphic condition. It appears in June. Pl. VIII., 6.

b. **Simplonia**, Frr. B. 73, 2 (1829). Very much resembles *Ausonia*, but is somewhat larger, and has the fore wings more rounded, and the costa is shaded or spotted with black. The under side is lighter than in *Ausonia*. This is a constant Alpine variety, occurring in Switzerland and the Pyrenees. Pl. VIII., 7.

3. E. Tagis, Hüb. 565-6; Esp. 117, 5, 6.

Expands from 1·25 to 1·34 in.; that is, about the size of the smallest specimens of *E. Belia*. Wings white. Fore wings with black tips, spotted with white; discoidal spot small and narrow. Hind wings white and without markings. Under side:—Tips of fore wings yellow. The discoidal spot very small and with a white lunar mark upon it. Hind wings greenish yellow, with spots arranged as in *E. Belia* and of a white colour, very rarely silvery. Pl. VIII., 8.

TIME OF APPEARANCE.—March to May.

HABITAT.—Portugal and Andalusia, also in the South-East of France.

LARVA, according to Boisduval, pubescent, green, very finely speckled with black, and with a lateral white band, and above this a bright red stripe. It feeds on *Iberis pinnata* in June and July.

PUPA.—Pale flesh-coloured, shading off posteriorly into reddish violet.

VARIETIES.

a. **Bellezina**, B. Ind. p. 2.—*Bellidice*, H. G. 929-30. This is the French form of the species, and differs from the type in having more white spots on the under side of hind wings.

b. **Insularis**, Staud. Cat. p. 4.—*Tagis* variété a., Boisd. Sp. Gen. 560. Differs from type in having the tips of the fore wings lighter. On the under side the hind wings are greener, with smaller and more numerous spots.

HABITAT.—Corsica and Sardinia.

OBS.—These islands present many points of interest in their lepidopterous fauna that are peculiar to themselves, either in the form of local varieties, as in the present case, or of distinct species, as in that of *Papilio Hospiton, Satyrus Neomyris, Argynnis Eliza*, &c.

4. **E. Cardamines,** Linn. (Pap. C.), F. S. 271, u. 1039, S. Nat.
1, 2, t. 761, n. 85; Hüb. Eur. Schmet. i. f. 419-25;
Boisd. Sp. Gen. i. p. 564.

Expands from 1·50 to 2·00 in. Wings white, rounded. The
male has the fore wings white, with a black tip and a black
discoidal spot; a large patch of brilliant orange reaches from the
tip of the wing to considerably within the discoidal spot. Hind
wings white and unspotted, but exhibiting traces of the pattern of
the under side. Under side :—Fore wings white, yellowish at the
base, and having the orange patch tipped with greenish grey
and white. Hind wings white, marbled with irregular patches
of yellow and black scales so mixed as to produce the effect of
a beautiful green; these patches follow more or less the course of
the nervures. Female similar to the male, but without the orange
patch, and the tips of the wings are marked more strongly with
black. Pl. IX., 1.

TIMES OF APPEARANCE.—April, May, and June.

HABITAT.—The whole of Europe (excepting the Polar Regions)
and Northern and Western Asia. As a British insect it is generally
common, sometimes abundant, being one of the first butterflies to
appear in spring in lanes, meadows, woods, on railway-banks, &c.,
occurring throughout England, Scotland, and Ireland.

LARVA green, finely speckled with black, with a white lateral
stripe less clearly defined at its dorsal than at its ventral edge. It
feeds on various field *Cruciferæ,* including *Cardamine pratensis,* the
pods generally forming its principal food.

PUPA boat-shaped, at first green, but changing to greyish
yellow, with clearer stripes. The eggs are laid in the summer, and
the larva changes to a chrysalis in July, hybernating in this state.
Pl. XV.

VARIETY.

a. **Turritis,** Ochs. Schmet. Eur. iv. 156. This variety, which
is found in Italy, differs from the type in having the orange
patch in the male narrower and not reaching as far as the
discoidal spot.

OBS.— Mr. Kirby ('European Butterflies and Moths,' p. 6)
gives this form the rank of a distinct species; he informs me that,

as he states in the above work, he has reason to believe that the
wing-scales differ from those of the type. I regret that I am not
able, at present, to give the result of my own observation on this
subject, but hope shortly to be able to do so.

5. E. Gruneri, H.-S. Schmet. Eur. i. 551-4.

Smaller than *E. Cardamines*, expanding from 1·25 to 1·37 in.
Male.—Fore wings white, blackish at the base and slightly tinged
with yellow near the costa; tips of wings blackish; a large orange
patch covers the outer portion as in *E. Cardamines*, but at its inner
edge there is a double black spot and just a faint trace of a black
line. Hind wings white and without pattern, excepting so far as
that of the under side shows through. Under side:—Fore wings
as in *Cardamines*, but more strongly tinged with yellow. Hind
wings almost exactly as in that species. *Female.*—Fore wings
white, with a dark tip and a large blackish discoidal spot. Hind
wings white. Under side as in the male, but without the orange
patch on the fore wings. Pl. IX., 2.

HABITAT.—Turkey and Greece, also certain parts of Asia Minor
and the Caucasus. A rare species, belonging exclusively, as will be
seen, to the Mediterranean sub-region. The female seems to be
much rarer in collections than the males, and no doubt is actually
scarcer, as in the allied forms, including our own British species.

LARVA and PUPA unknown.

6. E. Damone, Boisd. Sp. Gen. i. 564; H. G. 1010-1; H.-S. Eur. Schmet. i. 196-9.

Expands from 1·50 to 1·70 in. *Male.*—Wings bright yellow.
Fore wings with the extreme tips dusky, and having the outer
portion brilliant orange as far as the discoidal spot, which touches
the inner edge of the orange blotch. There are in some specimens
faint traces of a black line bounding the latter at its inner edge,
but in most cases it is absent. Hind wings yellow and unspotted,
but the pattern of the under side showing through gives a faint
green mottling. Under side:—Fore wings yellow, with the orange
blotch graduating into the ground colour at its inner edge; discoidal

spot visible. Hind wings almost exactly the same in pattern as those of *E. Cardamines*, but the ground colour is yellow, and the marbling more olive than green. *Female.*—Wings white, slightly tinged with yellow. Very much resembles the female of *E. Carda-mines*, but the ground colour of the hind wings on their under side is dull yellow, and the marblings are olive-green. Pl. IX. 3.[*]

TIMES OF APPEARANCE.—March and April.

HABITAT. — Turkey and Asia Minor, inhabiting mountain districts.

LARVA and PUPA undescribed.

This is another rare species, never very easy to obtain.

7. **E. Euphenoides**, Staud. Stett. Ent. Zeit. (1869), p. 92.— *Eupheno*, Esp. i. 28; Hüb. 421, 460; Boisd. Sp. Gén. 562. —*Calleuphenia*, Butler, Ent. Mo. Mag. v. p. 271 (1869).

Expands from 1·25 to 1·62 in. Wings yellow in the male, white in the female. The fore wings in the male have an orange blotch covering the outer part, broader at the costa than at the inner margin, bounded internally by a narrow but distinct black line ; the discoidal spot, being placed in the course of the black line, is scarcely visible ; the extreme tip of the wing is black and the base slightly so. Hind wings yellow and unspotted, but showing slightly the mottling of the under side. Under side :—Fore wings as above, but the discoidal spot is visible, and the orange blotch is not bounded internally by a black line. The hind wings are yellow, slightly mixed with white, and with three transverse irregular bands which coalesce and are of a dull green colour. *Female.*—Fore wings white, with a black and well-defined discoidal spot. The extreme tips orange, mixed with grey. Hind wings white and unspotted. Under side :—Fore wings white, with a black discoidal spot, yellow at the base and tips. Hind wings as in the male, but somewhat whiter in the ground colour. Pl. IX., fig. 4.

TIMES OF APPEARANCE.—April and May.

HABITAT.—The South of France, Italy, Spain, and Portugal, in open places.

[*] The right hand fig. of this species represents the female, but is erroneously marked ♂ instead of ♀ .

LARVA.—Yellowish, spotted with black, with a lateral violet stripe, above which is a row of black spots placed close together, one on the anterior part of each segment being larger than the rest; the spiracles black. Feeds on *Biscutella* and other species of *Cruciferæ.* The Chrysalis is boat-shaped, but less arched than that of *E. Cardamines*, which otherwise it appears to resemble. It is of a grey colour. The insects hybernate in the pupa state. Pl. XV.

OBS.—This species and the closely allied *E. Eupheno* from N. Africa were long confounded ; the original Linnean name, *Belia*, probably included both forms, but as there is much doubt on this point, and as the name *Belia* has been in use to designate another species (that described under this name in the present work) ever since Cramer so used it in 1782, Dr. Staudinger, in 1869, named the present species *Euphenoides*, reserving the name *Eupheno* for the North African form, which commonly had the name of *Douei*. About the same time, but shortly afterwards, Mr. Butler proposed the name *Calleuphenia ;* but Staudinger's name, having been given first, must be adopted, as it is in all the Continental lists.

8. **E. Pyrothoë**, Eversm. Nouv. Mém. Mosc. ii. p. 352 (1830).—
    Zegris Pyrothoë, Boisd. Sp. Gén. i. p. 555.—*Anthocharis Pyrothoë*, H.-S. Schmet. Eur. i. 622-3.

Expands from 1·00 to 1·25 in. Wings white. Tips of fore wings rather pointed, and having a more or less oval orange patch, bordered with black ; at the costal edge of the patch is a white spot. There is a black discoidal spot, crescentic in shape. Hind wings white, showing the markings of the under side. Under side :— Fore wings as above, but the tips are greenish and the orange scarcely shows. Hind wings dark green, with well-defined white spots, few in number, some arranged along the margin and of an oblong form, others circular or ovoid and placed towards the centre of the wing. Pl. XI., 1.

TIME OF APPEARANCE.—April (Eversmann).

HABITAT.—Mountains in the South-East of Russia (Government of Orenbourg).

LARVA unknown.

A very rare species, in fact about the most difficult to obtain of all the European butterflies. I describe it from the single specimen in my collection, and have seen only four others.

Genus 4.—ZEGRIS, Ramb. Ann. Soc. Ent. Fr. (1836), 585; Cat. Lep. And. i. p. 49; Boisd. Sp. Gén. i. p. 552.

Rather small white-winged butterflies, with orange patches at the tips of the fore wings. Antennæ proportionately shorter than in the last genus, and with larger clubs. Thorax and abdomen stouter than in *Euchloë*, and markedly covered with down.

Larva stout, pubescent, and slightly tapering towards the extremities.

Pupa short, terminating anteriorly in a blunt point, and posteriorly in an arched tail. Attached by the tail and by a transverse band, and enveloped in a slight silken web, somewhat as in *Parnassius*.

Such are the principal characters by which Rambur separated *Papilio Eupheme* of Esper from the preceding generic group.

There are only two European forms, and these are closely allied: they were long regarded as variations of the same species; even now a more extended knowledge of the earlier stages may compel us to reunite them. They were first separated by Ménétriès in 1832. These insects are very local, being confined in Europe to those districts which border upon the Mediterranean and Black Seas.

1. **Z. Eupheme**, Esper, 113, 2, 3; H.-S. 194-5. — *Erothoë*, Evers. Nouv. Mém. Mosc. (1832), p. 312. — *Tschudica*, H.-S. 451-2.

Expands from 1·50 to 1·80 in. Wings generally pure white, but sometimes with a yellowish tinge. The fore wings have a black crescentic spot at the extremity of the discoidal cell, and at the tip a black patch dusted on its outer side with yellow, and having within it a bright orange blotch of an oval shape, and above this a white spot. Hind wings white, blackish at the base, and showing the pattern of the under side. Under side :—Fore wings white, with a black discoidal spot; tips greenish yellow, rather inclining to orange about the centre, and having two small dark spots on a whitish ground on the costa. Hind wings green, mixed with yellow, with five or six white spots rather large and arranged along the margins of the wings, the result being a pattern somewhat similar to the under-side hind wings of *E. Euphenoides*. Thorax and abdomen black above and covered with white down; beneath they are light yellow, as well as the legs. The antennæ are white above and black beneath, with white clubs. The female has the tips of the wings lighter, the orange blotch smaller, and the white spot above it larger. Pl. X., 2.

TIMES OF APPEARANCE.—April and May.

HABITAT.—Plains and waste places in the South-East of Russia, including the Crimea and other places in the district bordering on the Black Sea.

LARVA yellow, with a lateral white band and large black points arranged in threes on the lateral portions of each segment (Rambur). It feeds in fields on *Sinapis incana*.

The CHRYSALIS is whitish, and not boat-shaped as in the last genus. The insect hybernates in this state.

OBS.—*Zegris Eupheme* is usually considered a rare species, and is no doubt very local; but though the continental dealers put rather a high price upon it, they seem generally to have a good stock of specimens, which would rather lead one to suppose that it is common where it occurs.

2. **Z. Menestho**, Mén. Cat. Rais. p. 245; Boisd. Sp. Gén. 555.
—*Z. Eupheme* var. *Meridionalis*, Ld. V. Zool. Bot. p. 30, 1852; Staudinger, Cat. p. 4.

Differs from *Z. Eupheme* in not having any white spot above the orange at the apex of the fore wings. The ground colour of the hind wings on their under side is white instead of yellow. I believe I am right in considering the specimens that answer to this description as belonging to *Z. Menestho* of Ménétriés. I must express my conviction that they are possibly merely local varieties of the foregoing species.

I have not seen any description of the larva or pupa.

This species appears at the same time as the last. It occurs in Spain and also in Asia Minor. Pl. X., 3.

Genus 5.—**LEUCOPHASIA**, Stev. Ill. Brit. Ent. Haust. i. 24.
PIERIS, Lat.
PONTIA, Ochs.

Head moderately large, eyes large and prominent, palpi longer than the head and covered with strong hairs. Antennæ of moderate length, and furnished with a flattened oval club. Abdomen very slender, reaching beyond the posterior wings. Wings small, white, and rounded, with their discoidal cells very small.

Larva slightly downy, and tapering at the extremities. Feeds on *Leguminosæ*.

Chrysalis angular, and not boat-shaped.

This genus is very distinct from any of the other genera of European *Pieridæ*, being particularly so in the length of the abdomen and in the extremely contracted discoidal cells. Its position, zoologically, is between the neotropical genus *Leptalis* and the genus *Pontia*, which is distributed over the intertropical regions of the Old World.

The genus *Leucophasia* is confined exclusively to Europe and North Asia, as far as the Amur.

There are but two species, and one of these, *Leucophasia Sinapis*, is not an uncommon British insect.

1. **L. Sinapis**, Linn. Syst. Nat. x. 648; F. S. p. 271, n. 1038 (1761); Hüb. Eur. Schmet. i. 410-1; Boisd. Sp. Gén. i. p. 429.

Expands from 1·37 to 1·68 in. Wings white, fore wings with dusky coloured spots at the tips of a roundish shape, and darker in the male than in the female. Hind wings white. Under side :— Fore wings white, faintly marked with greenish yellow at the tips. Hind wings with dusky shading, and one white and several greenish spots. Head, thorax and abdomen black ; antennæ black, with white flattened clubs.

TIMES OF APPEARANCE.—April and August.

HABITAT.—Woods throughout Europe (excepting the Polar Region), Western Asia and Siberia. In Britain it is generally distributed, but local, and is known by the trivial name of the "Wood White." Where it occurs it frequents open places in woods, and has a slow and languid flight, being very easy to catch. Pl. X., 4.

LARVA green, with a darker stripe on the back where the dorsal vessel shows through ; beneath this on each side is a yellow stripe.

CHRYSALIS yellowish green or grey, rusty red on the sides and on the wing-cases (Boisduval). Pl. XVI.

The larva is found in June and September, the insect being double-brooded. It feeds on *Vicia Cracca*, *Lotus*, *Lathyrus*, and several other plants of the order *Leguminosæ*.

VARIETIES.

a. **Erysimi**, Bork. Eur. Schmet. (1798). Has no dusky tips to the fore wings.

b. **Lathyri**, Hüb. 797—8. Has the hind wings greenish instead of white.

Both these varieties are met with in this country. They are rather aberrations than varieties.

c. **Diniensis**, Boisd. Gen. 6. This is a well-marked variety of the second brood, and is somewhat larger than the average size of the type. The dark markings are wholly wanting, the wings, both on the upper and under sides, being nearly pure white. It is found in the South of Europe, especially in the southern provinces of France. Pl. X., 5.

2. **L. Duponcheli**, Staud. Cat. p. 5, 1871.—*Lathyri*, Dup. i. 43, 3, 4; H.-S. 407-8.

Expands 1·50 in. It is of the same shape as *Sinapis*, but the bases of the wings are strongly tinged with greenish yellow; the tips of the fore wings are marked with an oblong spot, which is darker than that which appears in the last species. The hind wings are greenish on the under side, with lighter spots, one or two of them being white. Pl. X., 5.

Times of Appearance.—May to August.

Habitat.—The South of France, Italy, and Asia Minor.

Larva not described.

Obs.—The name *Lathyri* ought not to be given to this species, as that name was bestowed by Hübner in 1827 on one of the varieties of *L. Sinapis*, mentioned above. Dr. Staudinger has given the present species the name of *Duponcheli*, after Duponchel, who first described it in 1832.

Genus 6.—COLIAS, Fabr. Illus. Mag. vi. p. 284 (1807); Lat. Enc. Méth. ix. 10; Boisd. Sp. Gén. i. p. 634.

Butterflies of moderate size, with the margins of all the wings without angles or projections, the hind wings being always rounded. The prevailing colour of the wings is yellow of different shades, varying from greenish to the most brilliant orange. The hind margins are always more or less bordered with black, and have their fringes red. The hind wings have always on their under side a more or less conspicuous discoidal spot of a pearly colour, generally surrounded by a reddish circle. The antennæ are short and rather thick, swelling into a club at the extremity, and always of a red colour. The head is of moderate size; the eyes are naked and tolerably prominent; the palpi close together and compressed. The thorax is rather short. The abdomen moderately stout, and not reaching to the anal angle of the hind wings. The sexes are always very distinct in appearance.

The females are smaller than the males, and generally of a lighter colour; they have the black borders of the wings wider, though less definite in outline, and decorated with a row of light yellow spots. The females of those species whose wings are normally of an orange colour are liable to a dimorphic condition, in which the wings have the ground tint nearly white; the variety *Helice* of our British *C. Edusa* offers a well-known example of this fact. One meets with a purple reflection as a normal condition in some of the orange species, as in *C. Heldreichi*, *Aurora*, &c.; in others it is only met with occasionally, as in the case of certain specimens of *C. Herla*, *Myrmidone*, or of *Edusa*.

The Larvæ are smooth and covered with a slight down, slightly tapering at the extremities. All the known European species are green, with lighter lateral stripes. Their food-plants are mostly various low-growing species of *Leguminosæ.*

The Pupæ are straight, and terminated by a point anteriorly.

Obs.—This is one of the most interesting genera of Butterflies, not only on account of the extreme beauty of the species, but also of their habits and geographical distribution. The genus is spread over a very wide area, but is not found in the hotter parts of the earth, its place being apparently occupied in intertropical regions by the genera *Callidryas* and *Terias,* some species of the latter genus bearing a striking resemblance to those of the present. Many species of the genus *Colias* are confined to the northern regions; some are found exclusively within the Arctic circle—a curious variety of *Colias Hecla,* having been taken in 1876 by Capt. Feilden and Mr. Hart at the surprising latitudes of from 78° to 83° N.! For an account of the insects collected by these Arctic explorers in 1876, I must refer my readers to Mr. M'Lachlan's highly interesting report, which appeared in the 'Journal of the Linnean Society,' No. 74, May 23rd, 1878.* Besides *Colias,* three other genera of butterflies were found to be represented in these regions: "Thirty-five specimens of gaily-coloured butterflies," writes Mr. M'Lachlan, "were procured, belonging to certainly five distinct species. It may safely be asserted that there are desert regions in the tropics that would not furnish an equal number." The other genera found were *Argynnis, Lycæna,* and *Polyommatus.* It is now a generally received opinion, and one supported by actual geological discovery, that the climate of the Arctic regions was, immediately previous to the Glacial Epoch, not only milder than it is at present, but actually warm, and supported a luxuriant vegetation, traces of large trees, both deciduous and evergreen, having been proved to exist in these high latitudes.† Now it is believed by many zoologists that, during the Pliocene period, a

---

* "Report on the Insecta (including Arachnida), collected by Capt. Feilden and Mr. Hart between the parallels of 78° and 83° N. latitude, during the recent Arctic Expedition." By Robert M'Lachlan, F.R.S., F.L.S., &c.

† *Vide* Wallace, on 'The Geographical Distribution of Animals,' vol. i., pt. i., p. 70. London, 1876.

gradual cooling down took place, so as in time to reduce the
climate of these regions from a state which may be compared to
that of the warmer temperate areas of the present earth to a
condition which culminated in the glacial period, and that during
this age a special circumpolar fauna was established, some
survivors of which still inhabit their former abode, having
gradually regained their ground during post-glacial times.

Perhaps the above remarks may not be considered out of
place, whilst we have under consideration the genus *Colias*, of
which thirteen species occur in Europe, and out of this number at
least five are to be found within the Polar Circle. Two species of
the genus are found in Britain, the "Clouded Yellows," *Colias Edusa*
and *C. Hyale.* They are both very uncertain in their appearance,
occurring during some seasons in considerable numbers, after
having been scarce or altogether unnoticed for several years
previously. The year 1877 will always be remembered by British
entomologists on account of the vast numbers of *Colias Edusa* that
appeared throughout the country.

1. **C. Palæno**, Linn. Faun. Suec. p. 272 (1761); Syst. Nat. i. 2,
p. 764; Boisd. Sp. Gén. 2, p. 645.—*Europomene*,
Esp. 42, 1, 2; Hüb. 434-5.—*Philomene*, Hüb. 602-3,
740-1.

Expands from 1·75 to 2·70 in. Wings of the male pale yellow,
with a black border, dusted over, and sometimes finely veined,
with pale yellow; the fore wings have a small black discoidal spot,
sometimes this is altogether wanting; the hind wings have a very
faint and hardly visible pale spot touching the discoidal cell.
Under side:—Fore wings deeper yellow than above; the black
border of the upper side shows through, giving a darker shade to
the hind margin; discoidal spot small and nearly annular. Hind
wings yellow greenish, at the base, with a small pearly discoidal
spot. The head, antennæ and other appendages, as well as the
prothorax and legs, are red; the meso- and meta-thoracic segments
are black, as well as the abdomen, and are covered with white
downy hairs. The margins of all the wings are fringed with rosy

H

red, except at the anterior part of the hind margins of the hind wings, where the fringe is pale yellow. The female has the wings pale greenish white, instead of yellow; the black border is less sharply defined internally than in the male. Pl. XI., 1.

TIMES OF APPEARANCE.—July and August.

HABITAT.—Scandinavia, Russia, Lapland, the Alps of Switzerland, the Tyrol, and the Pyrenees. It is said also to occur on some of the Rhenish hills, and even to have been taken in Belgium, in the hilly districts of Namur and Luxembourg. It is essentially a mountain species, and may be seen very frequently in August on the passes of Switzerland. Besides these European localities it inhabits Northern Asia, and, like many other alpine species, may also be circumpolar, though in Kirby's Catalogue there is a query as to its inhabiting Arctic America.

LARVA green, dotted with black, and having stripes of yellow.

The female of this species is, like most of those of the genus, dimorphic, but it departs from the general rule in having the normal form white, as described above. There is, however, a yellow form, which is found in the higher alps, and may possibly represent the original appearance of the insect; this is the

VARIETY

a. **Werdandi**, H.-S. Schmet. Eur. i. 41. It resembles the common form of the female, but has the ground colour the same as the male, and bears traces of yellow spots upon the black border of the fore wings. Pl. XI., 2.

b. **Lapponica**, Stgr. This variety has the ground colour lighter than the type. The female has the black border of the fore wings traversed by five or six whitish lines running from within to the hind margin. It occurs in Lapland.

2. **C. Pelidne**, Boisd. Icon. 8, 1-3; Sp. Gén. i. 644; H.-S. Schmet. Eur. i. f. 35, 36, 43, 44.—*Anthyale*, Stand. Cat. p. 5.—Sed *Anthyale*, Hüb. alia est species.

Smaller than the preceding, having an expanse of from 1·50 to 1·75 in. The colour of the male is greenish yellow, and the hind margins of all the wings have a very narrow black border veined with yellow; the discoidal black spot is very small. The under side of the hind wings is bluish green, with a rose-coloured discoidal spot. The female is white, with the tips of the fore wings dusky; hind wings without any black border, and on the under side of a yellower green than in the male. Fringes of all the wings in both sexes rosy. Head, antennæ, &c., as in the last species. Pl. XI., 3.

HABITAT.—The Polar regions, chiefly of North America. I have specimens which are said to have been taken in Lapland, and have seen others from the same locality. I think, therefore, that it is circumpolar, and may properly be placed on the list of European Butterflies; though in Staudinger's Catalogue *C. Anthyale* has a * placed against it, as not occurring in Europe.

It is perfectly distinct from *C. Palæno,* and is not an Arctic var. of *C. Philodice,* Godt.

3. **C. Nastes,** B. Ic. 8, 4, 5; H.-S. 37, 38, 401-2.

Expands from 1·50 to 1·75 in. The male has the wings dark yellowish green. The fore wings have a black discoidal spot, and some lightish spots along the hind margin. Hind wings somewhat lighter, with a very indistinct orange discoidal spot. Fringes, head and antennæ, red. Under side:—Fore wings greenish white, dusky green at the tips; the discoidal spot indistinct. Hind wings dark green, with a paler hind margin; discoidal spot small and reddish, with a pearly white centre. The female is rather lighter, and with more distinct yellowish spots than the male. Occasionally this species is very dusky in appearance, and is sometimes shot with violet. Pl. XI., 4.

HABITAT.—The Polar regions. In Europe, the North of Lapland. It is chiefly met with in Labrador, and in the American Polar regions generally.

4. **C. Werdandi,** Zett. Ins. Lap. 908.—*Colias Nastes* var. *Werdandi,* Stgr. Cat. p. 5.

Expands from 1·75 to 2 in. Wings of male greenish white; the hind wings deeply tinted with green towards the base. Fore wings with the nervures black and strongly defined; hind margin with a black border, dentated along its basal edge; the discoidal cell has at its outer extremity a black linear mark. Hind wings with a very narrow black border, and hardly any trace of a discoidal spot. Female whiter than the male. Fore wings with a broad black border, on which are placed, parallel to the hind margin, five or six conspicuous white spots. Discoidal spot lozenge-shaped or oval, and white in the centre. The hind wings have the nervures defined as on the fore wings, a very narrow black border and light orange discoidal spot. Under side:—The fore wings are nearly white in both sexes, greenish yellow along the costa and hind margin; discoidal spot small and narrow. Hind wings dark bluish green at the base, four or five lighter square spots parallel to the hind margin; discoidal spot pearly white, and surrounded by a red ring. Fringes of all the wings rose-coloured, as well as the antennæ, head, prothorax and legs. Posterior thoracic segments and abdomen black, covered with fine white hairs. Pl. XII., 1.

HABITAT.—Lapland and Sweden, frequenting mountains.

Appears to be quite distinct from *C. Nastes.*

5. **C. Phicomone,** Esp. Schmet. i. 2, t. 5, 6; Hüb. Eur. Schmet. i. 436, 437; H.-S. Schmet. Eur. i. 399, 400.

Expands from 1·75 to 2·18 in. The male has the wings of a dusky greenish colour; hind margins black, spotted with light greenish yellow; fore wings yellowish towards their inner margins, and with a distinct black discoidal spot. Hind wings with a very pale orange discoidal spot. Under side:—Fore wings greenish white, yellow at the tip; discoidal spot white in the centre. Hind wings greenish yellow, with large lighter yellow spots along the hind margin; discoidal spot pearly, and surrounded by a red ring; on the costa a red spot. The female has the wings nearly white, with the base very dusky, and the hind marginal spots very large, the hind wings showing hardly any traces of a black border. Fringes of all the wings, in both sexes, rosy red; also the antennæ, prothorax, and legs as in the allied species.

TIMES OF APPEARANCE.—July and August.

HABITAT.—Mountain pastures in Europe and Siberia; in

Europe confined to the Pyrenees, the Tyrol, and Switzerland. In the latter country it is common, but always occurs in considerably elevated places. I have usually seen it at an elevation of not less than 5000 feet. Pl. XII., 2.

LARVA green, with a white stripe on each side, on which are yellow spots dotted with black. It feeds on *Leguminosæ*, chiefly of the genus *Vicia*, in June and July.

6. **C. Hyale**, Linn. Syst. Nat. x. 469; F. S. 272; O. I. 2, 181; H. S. 33, 34; Frr. 547.—*Palæno*, Esp. 4, 2; Hüb. 438-9.

Expands on an average 1·75 in., sometimes reaching 2 in. Wings sulphur-yellow. Fore wings with a black discoidal spot, and a black hind-marginal band enclosing a row of conspicuous yellow spots, and ending rather abruptly, so as not to reach the inner margin; as it approaches the latter it gradually becomes narrower, being widest at the costa. Hind wings yellow, blackish at the base; faint traces of a black band are seen on the anterior part of the hind margin, and there is a large discoidal spot of a bright orange colour. The female exhibits the dimorphism so common in this genus, often having the wings nearly white, instead of yellow. Under side:—Fore wings yellow, darker at the apices; a row of five or six black spots runs parallel to the hind margin; the discoidal spot is black. Hind wings deeper yellow, with a large pearly discoidal spot surrounded by a dull red ring, and having a smaller spot similar in character placed above it. At the base of the wing is a dull red mark, which is not found in the preceding species. Parallel to the hind margin is a row of dull red crescentic spots, with their concave sides outwards, and a conspicuous narrow reddish spot on the costa. Head, prothorax, antennæ, and legs dull red; fringes of hind margins pink. Pl. XII., 3.

TIMES OF APPEARANCE.—From July to the end of September; and again in the spring, after hybernation.

HABITAT.—The greater part of Europe, but not the northern regions; the extra-tropical part of Asia and North Africa. On the Continent it is a very common insect, and may be seen throughout the summer in Germany, Switzerland, &c., frequenting clover fields. It occurs in England, but seems to be confined to the more southern parts of the island, and to prefer chalky districts; it is very capricious in its appearance, and does not occur in Scotland or Ireland.

LARVA.—Cylindrical, dark green, sprinkled over with black dots; there is a narrow yellow and white lateral stripe. Feeds on *Leguminosæ*, principally *Trifolium*, the eggs being laid in the spring by hybernated females.

The PUPA is green, with a brownish yellow lateral stripe. Pl. XVI.

VARIETY.

a. **Sareptensis,** Staud. This variety resembles the male, but has the ground colour of the wings of a deeper yellow, and the hind marginal border is broader and darker, and more nearly approaches the anal angle of the fore wing. It occurs in South Russia, and is probably a hybrid between the present species and the next, *C. Erate.* The closely allied species of this genus occasionally intermingle, the result being hybrid forms, as in the present case.

7. C. Erate, Esp. 119, 3.—*Neriene,* Fr. 301, 1, 2; H.-S. 30, 32.

Expands from 1·50 to 1·75 in. All the wings are pale yellow, but brighter than in the last species; black at the base. Fore wings of the male with a tolerably broad hind marginal border, which is black powdered with grey, sometimes with light veins, broadest at the costa, and reaching quite to the inner margin; its internal edge is less sharply defined than in the succeeding species, and more as in *C. Hyale.* There is a distinct black discoidal spot. Hind wings with a distinct black border, well marked, but not very wide. In the centre of the wing is a bright orange spot. The female has the wings of the same colour as the male, but the borders of all the wings have large yellow spots. Under side almost exactly as in *C. Hyale,* but the discoidal spot of the fore wings is white in the centre, and the hind wings have a greener tint at the base than in the last species. Fringes of all the wings, head, antennæ, prothorax and legs, rosy. Pl. XII., 4.

TIME OF APPEARANCE.—August.

HABITAT.—South Russia, Siberia, and Northern India. It appears to be a very local species; found in hilly districts. LARVA and PUPA unknown.

VARIETIES.

a. **Pallida**, Stgr. Cat. p. 5 (1871). This is the pale form of the female. It is generally rather larger than the type, has the wings nearly white, and square light spots on the hind margin of the hind wings. It very much resembles *C. Edusa* var. *Helice*, and *C. Myrmidone* var. *alba;* but in these the discoidal spot of the fore wings has not a white centre beneath, and the black margin of the upper side is thicker towards the inner margin than in this variety. Pl. XII., 5.

b. Var. **Helichta**, Ld., is supposed by some to be a hybrid between *C. Erate* and *Edusa.* It occurs in Armenia and Persia, and perhaps South Russia.

8. **C. Hecla**, Lefev. Ann. Soc. Ent. Fr. 1836, p. 383, pl. 9; B. 3-6; Kirby, Man. of Eur. Butt. p. 17.—*Boothii*, Boisduval, Gén. Ind. Méth. p. 7, n. 40; H.-S. 459-60. —Sed *Boothii*, Curtis, alia est species.

Expands from 1·50 to 1·75 in. The male has the wings bright orange, rather dusky towards the base. Fore wings with a narrow but well-defined black hind-marginal border, generally divided by fine orange veins. The discoidal spot is long and narrow, and the costa, nearly throughout its whole length, is greenish yellow. Hind wings with a narrow black border, and a very bright reddish orange discoidal spot; the inner margin is greenish yellow. The female has the fore wings bright orange, with the nervures black and very distinct; the hind marginal border is black, and much broader than in the male, and is marked with seven or eight roundish and oblong yellow spots. The discoidal spot is large and black, and lozenge-shaped. The costa is yellow along its inner half. Hind wings yellow, with a dark greenish cast; the black hind-marginal border marked with five or six large light yellow spots; the discoidal spot large, and bright orange. The wings of both sexes are often shot with violet. Under side :—Fore wings orange, dull green along the hind margin, shading off into bluish at the apex; parallel to the hind margin is a row of small black spots; the discoidal spot is light in its centre. Hind wings dark bluish green, darkest at the base, and becoming yellower towards the hind margin, which has a row of faint yellow oblong spots; discoidal spot silvery, very small, and surrounded by a red ring. Fringes of all the wings broad, and deep rose-coloured. Head, antennæ, and legs of the latter colour. Thorax and abdomen black, clothed with white hairs. Pl. XIII., 1.

TIMES OF APPEARANCE.—July and August.

HABITAT.—The Polar regions of Europe and America. It seems to occur as far north as any Arctic expedition has yet penetrated. In Europe it is found not uncommonly in the north of Lapland.*

LARVA and PUPA unknown.

VARIETY.

a. **Glacialis**, M‘Lachlan, Journ. Linn. Soc. vol. xiv. p. 108. I have examined the specimens alluded to above, as having been taken in the high north by Capt. Fielden and Mr. Hart in 1876. They are now in the British Museum cabinet, and are of great interest. They differ from the type in being of a dull brownish yellow, instead of that brilliant colour that we meet with in the specimens from Lapland, and are altogether more dusky. They seem, however, to be merely varieties of *C. Hecla*, and are American, and not European.

OBS.—*C. Boothii* of Curtis is wholly distinct from *C. Hecla*, Lefev., and only occurs in Arctic America. It is smaller, greener, and wants a distinct hind-marginal border.

9. **C. Chrysotheme**, Esp. 65, 3, 4; Hüb. 426-8; Boisd. Ic. 9, 3, 4; Fr. 301, 3, 4.

Expands from 1·50 to 1·75 in. The male has all the wings pale orange, brightest in the centre, but pale, and occasionally greenish towards the margins. The fore wings have rather a broad black border, nearly the same width at the anal angle as at

* This species was once thought to inhabit Iceland, but recent observers state that the fauna of that island is entirely wanting in butterflies.

the costa, this border is crossed by very distinct yellow veins; the discoidal spot is black, with a reddish centre. The hind wings are dusky at the base, and light yellow along the inner margin; along the hind margin is a narrow but distinct black border, divided by yellow lines; the discoidal spot is orange. Female somewhat larger than the male. The black borders of all the wings broader, and without stripes, but enclosing a row of large spots of a greenish yellow colour. The hind wings are greener than in the male, the orange discoidal spot being very conspicuous. Under side :—Fore wings light orange at the base, greenish along the hind margins, parallel to which is a row of black spots; discoidal spot silvery, surrounded with black. Hind wings greenish yellow, with a silvery discoidal spot, surrounded by a dull red ring, and having a smaller spot of a similar character placed above it; at the base is a dull red mark, and another on the costa; an indistinct row of spots of the same colour runs parallel to the hind margin. Fringes of wings, head, antennæ and legs, red; thorax and abdomen black above, light yellow beneath. Pl. XIII., 2.

TIME OF APPEARANCE.—August.

HABITAT.—South-Eastern Europe, including Hungary, but not Greece; also Asia Minor, Armenia and Siberia; inhabiting meadows.

LARVA unknown.

OBS.—Stephens (Ill. Brit. Ent. Haust. i. p. ii.), in 1827, described a small variety of *C. Edusa* under the name of *Chrysotheme*, and several writers of that period followed him; *C. Chrysotheme* has never been taken in England, nor indeed in Western Europe. Such errors as this often crept into the works of the very foremost entomologists in the early part of the century, and arose, no doubt, from the great difficulty of becoming acquainted with continental species. (*Vide* p. 63.)

10. **C. Thisoa**, Mén. Cat. Raiso. p. 244; Boisd. Sp. Gen. 636; Stgr. Stett. Ent. Zeit. 1866, p. 49.—*Myrmidone* var., Ld. Ann. Soc. Belg. xiii. pp. 20, 21.

Expands 1·75 to 2 in. The male has all the wings of a brilliant orange, much brighter and clearer than *C. Hecla*. The fore wings

I

have a broad black border, reaching from the costa to the hind margin, and having nearly the same width throughout its whole course,—this is distinctly veined with yellow along the nervures; the black discoidal spot is very small, but not elongated as in the last species. The hind wings are greenish yellow at the base and inner margin; the black border is narrow and indistinctly veined with yellow. The head, antennæ and palpi are dull red. Under side:—Fore wings orange, with a white central discoidal spot; the hind margin greenish yellow, with three black spots. Hind wings greenish yellow, with a pearly discoidal spot, and some reddish marks at the base and parallel to hind margins.

The female, figured in this work from a specimen kindly lent me by Mr. R. D. Godman, is brilliant orange, with very broad black borders spotted with yellow, the hind wings being very dark.

The specimens in Mr. Godman's collection, though from Persia, almost exactly resemble Ménétriés' figure of a specimen taken in the Caucasus at an elevation of 8000 feet ('Enumeratio Corpor Animal. Musei. Petrop.' i., 1855, pl. 6).

M. Lederer describes and figures C. Thisoa ('Annales de la Société Entomologique de Belgique,' xiii.). His figures greatly resemble C. Myrmidone, of which species he considers it merely a variety. The male, however, presents an anatomical difference that Lederer appears to have overlooked. I allude to the absence of an ovate space immediately beneath the costal nervure of the hind wings on their upper surface. This structure is found constantly in the males of C. Myrmidone, Edusa, and other species. Its absence in the present species seems to place C. Thisoa much nearer to C. Hecla or Chrysotheme than to any other species.

Boisduval (Gen. et Ind. Meth.) includes C. Thisoa among those species which have a "glandular sac" on the hind wings of the male, though at the time when that catalogue was written he had not seen a specimen, and probably had only Ménétriés' representation of the female to which he could refer. Pl. XIII., 3.

Time of Appearance.—July.

Habitat.—The Caucasus and Trans-Caucasian Mountain regions; Siberia and the Altai.

Larva undescribed.

11. C. Myrmidone, Esp. 65, 1, 2; Hüb. 432-3; Dup. i. 14; H.-S. 393-4.

Expands from 1·50 to 1·90 in. The male has all the wings of a very brilliant orange, with hind marginal black borders, which are rather narrow, especially on the hind wings; these borders are finely dotted with yellow scales, very rarely those on the fore wings are, and have fine yellow veins running across them towards the apex of the wing. The fore wings have a black discoidal spot. The hind wings have their inner margins greenish yellow, and have likewise an orange discoidal spot, which is not very conspicuous, being very nearly of the same colour as the wings. The female is somewhat larger than the male; the fore wings have a larger discoidal spot, and the hind-marginal border, which is wider and has a row of yellow spots, similar in colour to those in *C. Chrysotheme*, female, but having the following arrangement:—Beginning at the costa there are five spots, always well defined; the two nearest the costa very narrow, the other three rounder; below this row and nearer to the anal angle are two or three more spots of the same colour, situated well within the black border, and not touching its edge. The hind wings have a more yellowish tint than those of the male, but have not the greenish cast that we see in *C. Chrysotheme*, female; the hind marginal band has a row of large square yellow spots, reaching quite to its inner edge, and sometimes obliterating it. Under side as in the allied species, but the discoidal spot of the fore wings has a light centre; that of the hind wings is very large and pearly. Head, antennæ, legs, &c., as in the other species. Pl. XIII., 4.

Times of Appearance.—May and August, the insect being double-brooded.

Habitat.—South and Eastern Germany, Hungary, Turkey, Russia, Asia Minor, the southern part of Siberia, Turkestan, and Northern India; frequenting meadows. It does not occur in the north or west of Europe.

Larva green, with the dorsal vessel showing through as a darker stripe, laterally there is a light green streak; the surface is covered with short black hairs growing from minute tubercular elevations. It is said to feed on *Cistus*, and thus departs from the general rule of this genus, which is to select plants of the order *Leguminosæ.*

TIMES OF APPEARANCE.—March and October.

VARIETY.

a. ♀ . **Alba**, Staudinger, Catal. p. 6. This is the white form of the female, and corresponds to the var. *Helice* of *C. Edusa*, from which it differs in the lighter colour of the hind wings, the narrowness of the black borders, and the arrangement of the spots on the fore wing; it is also smaller than var. *Helice*. From *C. Erate* var. *Pallida* it is more difficult to distinguish it; it differs, however, in having the black border of the fore wing much wider where it touches the inner margin than it is in the latter. Pl. XIII., 5.

OBS. I.—*C. Myrmidone* is very close to *C. Edusa*, but at the same time is perfectly distinct from it. The description I have given above is made after examination of a considerable number of specimens, and I believe I have not mentioned any characters as distinctive that are not constant. The female often greatly resembles the female of *Chrysotheme*, but the greenish colour of the hind wings in the latter, as well as the decided light mark in the centre of the discoidal spot on the under side of the fore wings, will assist in the distinction.

OBS. II.—The male of *C. Myrmidone*, as before mentioned, is provided with an ovate light-coloured patch immediately beneath the costal nervure of the hind wings, and close to the base. This is generally of a lighter colour than the surrounding portion of the wing, but varies, sometimes being even black. The males of other species have it in common with the present, such as *C. Edusa*, *Aurora*, *Heldreichi*, and *Aurorina*. Under a low power it will be seen to differ from the rest of the wing in having densely packed scales, not laid down smoothly and regularly like the ordinary wing-scales, but more elevated, and producing a granular appearance. Under a higher power these scales will be seen to be ovate in shape, with the margins entire, and without any notches or indentations. At the base is a small foot-stalk, as in the wing-scales. Boisduval called this structure a "glandular saccule." As far as my observations go I am bound to say that I am unable to detect anything very special in the tissue of the part; whether these slightly altered wing-scales can assist in any secretive or other function that is performed by glands, I cannot even conjecture. In any case it is a constant feature, and may therefore be made use of for the separation of species.

12. **C. Edusa,** Fab. Mant. p. 23, n. 240 (1787); Ochs. Schmet. Europ. 1, 2, p. 173; Boisd. Sp. Eur. i. p. 638.— *Electra,* Lewin, Ins. Brit. i. 31 (1795).—*Helena,* II.-S. 206-7 (var.).—*Hyale,* Wein. Verz. p. 165, n. 3 (1776); Esper, Sch. i. 1 (1777); Hüb. Eur. Sch. i. 329.—Sed *Hyale,* Linn. S. N. (1767), alia est spec.

Expands from 1·75 to 2·25 in. All the wings bright orange yellow, but without the red tint that is seen in *C. Myrmidone.* The male has on the fore wings a broad black hind-marginal border, veined with orange; the discoidal spot is conspicuous and black. Hind wings orange in the centre, but dusky, and covered with light hairs at the base, with a black hind-marginal border veined with orange; discoidal spot orange. The female has a broad black border to all the wings, without orange veins, and spotted with yellow, much as in *C. Myrmidone,* but is much more variable, hardly any two specimens being alike; the spot, however, which appears in *C. Myrmidone,* as the third from the costa, is often wanting in this species. The yellow spots on the hind wings are smaller and less square; the black border is wider. The under side has the usual discoidal black spot on the fore wings, but this never has a light centre. The hind wings are yellow, with the pearly spot bordered with reddish brown, as in the allied species. Head, thorax, legs, antennæ, &c., as in the last species. Pl. XIV., 1.

TIMES OF APPEARANCE.—May and June, and from August until November.

HABITAT.—The greater part of Europe, but not the northern portion, Western Asia, and North Africa; frequenting plains, meadows, and downs. As a British insect it is taken in England

and Scotland, but not in Ireland. In this country it is commoner than our other species of *Colias*, *C. Hyale*, but, like it, is very capricious and fitful in its appearance, sometimes being scarce or local for years together, and then, as in 1877, appearing in great numbers throughout the country.

LARVA.—Cylindrical, dark green, with a lateral narrow whitish stripe; spiracles yellowish. "The head is covered with minute warts, and each wart emits a short hair; the segmental divisions are indistinct and transversely wrinkled, the wrinkles dividing each segment into narrow sections, each section composed of a series of minute warts, and every wart emitting a short hair."—Newman, 'British Butterflies,' p. 145.

The PUPA is pale yellowish green, with a lateral yellow stripe on the abdominal segment; the wing-cases are of a deeper colour. It is pointed anteriorly, and is attached in an upright position on the stem of the food-plant. Pl. XVI.

TIMES OF APPEARANCE.—June to the middle of July.

FOOD-PLANTS.—Various species of *Trifolium*, chiefly the common clover, on the leaves of which the eggs are laid in the spring.

It is still an open question whether *Colias Edusa* is double-brooded in England, though the observations of Mr. Edward A. Fitch have gone far to prove that it is. For most valuable information on the subject of the life-history of this species, the reader is referred to a paper by the above-named gentleman in the 'Entomologist,' vol. xi., No. 178, March, 1878. In it he has collected the numerous accounts of its capture and observations made upon it during the year 1877, when it was so abnormally abundant. The paper is accompanied by a coloured plate, showing various aberrant forms of great interest. *Colias Edusa* certainly does hybernate as an imago; I have twice taken worn females at the beginning of June.

VARIETIES.

a. ♀. **Helice**, Hübn. 440, 441 (1798); *Pap. Edusa Alba*, Haw. (1802). This is the white form of the female. The markings have the arrangement as in the ordinary orange form, but the ground colour of the wings is nearly white, the hind wings being more or less suffused with dull green, and having the discoidal spot pale orange. Specimens sometimes occur which are intermediate between the ordinary form and the present. It bears some resemblance to (1) *C. Phicomone*, female; (2) *C. Hyale*, female, the light form; (3) *C. Erate* var. *pallida;* (4) *C. Myrmidone* var. *alba.* These, however, may be distinguished as follows :—The first has no black border to the hind wings ; in the second and third the black border of the fore wings does not touch the hind margin internally as completely as in *Helice* ; in the fourth the hind wings are much lighter, the spots on their black border are larger and squarer, and those on that of the fore wings more numerous and defined. The var. *Helice* is found most commonly in the South of Europe ; but some collectors, in 1877, record having taken it in the proportion of one to twenty in this country. Pl. XIV., 2.

b. **Chrysotheme,** Steph. Ill. Brit. Ent. Haust. 1, p. 11 (1827).

c. **Myrmidone,** Westwood & Humphreys, Brit. Butt. p. 130 (1841). Small specimens of *C. Edusa* were confounded with *C. Chrysotheme,* which, however, is a perfectly distinct insect, as will be gathered from the description in the present work. The *C. Myrmidone* of Westwood and Humphreys is merely a strongly-coloured variety of *C. Edusa.* When the variable character of *C. Edusa* is considered, it is not at all surprising that it should have been confounded with species so closely allied, at a period when it was difficult to obtain a knowledge of the nature and habits of these.

13. **C. Heldreichi,** Staud. Stett. Ent. Zeit. 1862, p. 257 ; Mill. Icon. vol. i. p. 335, 1865.—*C. Aurorina* var. b. *Heldreichi,* Staud. Cat. p. 6, 1871.

Expands from 2·06 to 2·18 in. The male has the wings of a deep orange-colour, speckled over with black scales, and strongly shot with rich purple. All the wings have a black border, which is about as wide as that of *C. Edusa,* and veined with orange. The fore wings have a black discoidal spot, and the hind wings a bright orange one. The female is somewhat larger than the male, of a brilliant reddish orange, slightly shot with violet. All the wings have a black border, spotted with yellow, almost exactly as in *C. Edusa.* The fore wings are greenish on the costa, and have a black discoidal spot. The hind wings are greenish at the base and inner margin, and have an orange central spot. Under side :—In the male this is much the same as that of *Edusa*, but the colour is more of a greenish yellow, and the marginal spots and other markings are brown. In the female the fore wings are orange-yellow, with the costa and hind margin greenish; discoidal spot with a pearly white centre. Hind wings greyish green, with a white discoidal spot. The head, thorax and abdomen are black above and clothed with light hairs, beneath they are greenish yellow; antennæ reddish. Legs light pink. Fringes of all the wings light pink, mixed with yellow. Pl. XIV. 3.

TIME OF APPEARANCE.—July.

HABITAT.—The mountains of Northern Greece.

LARVA undescribed.

Most authors consider this form to be a local variety of the Asiatic *Colias Aurorina*, H.-S., a notice of which will be found amongst the extra European species.

Genus 7.—**GONEPTERYX**, Leach. Edinb. Encycl. ix. p. 128 (1815); Doubld. Gen. Diur. Lep. p. 69 (1847).

ANTEOS, Hüb. Verz. bek Schmet. p. 99 (1816).

RHODOCERA, Boisd. Amer. Lep. p. 70 (1833); Sp. Gen. 1836.

All the wings with an angular projection; without any dark border; on each wing at the edge of the discoidal cell an orange spot, which has a slightly silvery centre beneath. Antennæ red, and curved downward. Body downy.

Besides the European species there are three or four others very closely allied; they inhabit various parts of Tropical and sub-Tropical America. The two European forms figured in this work are, by many entomologists, considered as varieties of one species.

The species forming this genus were separated from *Colias*,

Fab., by Dr. Leach to form the genus *Gonepteryx*, which certainly has the preference over Boisduval's *Rhodocera*, so generally adopted by continental entomologists, including Staudinger.

The Larvæ have the same shape as those of *Colias*, but the Pupæ differ in having the wing-cases very large and forming a conspicuous bulging mass. This form of pupa is seen in some of the allied exotic genera.

1. **G. Rhamni**, Linn. Faun. Suec. p. 272 (1761); Syst. Nat. 1, 2, p. 765. Esp. Schmet. t. 1, t. 4, f. 4. Hübn. I. f. 442-44. Boisd. Sp. Gen. 1, p. 602.

Expands from 2·25 to 2·50 in. The male has all the wings with a sharply projecting angle and of a bright greenish yellow colour, with an orange discoidal spot on each wing. Under side yellower; the hind wings have the discoidal spot pearly in the centre. The female is somewhat larger than the male, and of a lighter colour, being whitish, but otherwise resembling it. Head and antennæ dull red; body black, covered with white silky hairs; legs light yellow. Pl. XIV., 4.

TIMES OF APPEARANCE.—From the end of July till October, and in the spring after hybernation; thus the butterfly is on the wing during the greater part of the year.

HABITAT.—The greater part of Europe, excepting the Polar Regions. As a British insect it is generally common, but is local in Ireland, and very rare in Scotland.

LARVA cylindrical, or rather slightly thicker in the middle segments than at the extremities; it is of a dull apple-green, covered with minute excrescences from each of which grows a minute white hair. A lateral white waved stripe runs through the whole length of the body. Pl. XVI., 4.

PUPA bright green, pointed at both ends, thicker in the dorsal thoracic region, the wing-cases forming a considerable round-shaped projection; the cephalic and thoracic parts are shaded with purplish brown.

Food-plants, *Rhamnus frangula* and *R. catharticus.* The eggs are laid by hybernated females in April; the larva emerges early

K

in May, and is full grown by the end of June, the imago appearing towards the end of July at the earliest. Pl. XVI., 4.

VARIETY.

a. **Farinosa,** Z. Is. 1847, p. 5. This is larger than the type, and has the body and inner margin of the hind wings covered more thickly with down.

HABITAT.—The South and East of Europe and Western Asia.

2. **G. Cleopatra,** Linn. Syst. Nat. xii. 765. Esp. 48, 1. Hübn. 445-6. *G. Rhamni,* var. *Cleopatra,* Boisd. Gen. et Ind. Méth. p. 6.

Expands from 2·12 to 2·50 in. Differs slightly in outline from the last species, the angular projections of the wings being less acute, and the wings themselves broader. The male has the ground colour of the wings of a slightly deeper yellow than in *G. Rhamni,* and the fore wings have more than two-thirds of their area occupied by a blotch of brilliant orange, which does not appear on the under surface. The hind wings have an orange discoidal spot. The female greatly resembles that of *G. Rhamni,* differing only in shape and in having the hind wings tinged towards the base with rusty brown. Pl. XIV., 5.

TIMES OF APPEARANCE.—The spring and late summer or autumn.

HABITAT.—The South of Europe, North Africa, and Asia Minor.

LARVA.—Similar in shape to that of *G. Rhamni,* dull bluish green, rather darker on the dorsal aspect; it has the lateral streak narrower than in *G. Rhamni.* Pl. XVI., 5.

PUPA shaped like that of the last species, but browner in colour.

The larva appears in May and June on *Rhamnus alpinus;* the pupa in July.

Some entomologists consider this a variety of the last, but it is difficult to see upon what grounds.

OTHER SPECIES OF *PIERIDÆ* INCLUDED BY STAUDINGER.

Genus *APORIA*, Hüb.

A. Hippia, Brem. Lep. Ost. Sib. p. 7 (1864). *Crataegoides*, Luc. Ann. Soc. Fr. 1865.—Slightly smaller than *A. Crataegi*. The hind wings are straighter and more oblong. The fore wings have the nervures shaded with black; beneath, the apices are tinted with yellowish. The hind wings resemble those of *Crataegi* above, but beneath they have the nervures very much extended, or shaded with black; at the base is a patch of bright orange-yellow. Habitat, E. Siberia and China.

Genus *PIERIS*, Schrk.

P. Melete, Mén. En. p. 114, x. 1, 2.—About the size of *Pieris Brassicæ*. The male has all the wings white, and is very like the female of *Brassicæ*, having the apices dusky and two dusky spots on the fore wings; these, however, are very pale. The female is tinged with yellow, especially in the hind wings; the nervures are strongly shaded with black, the apices and black spots of the fore wings are very dark. The hind wings are white beneath in the male; yellow in the female, with a bright orange-yellow patch at the base. Habitat, Amoorland.

P. Frigida, Scud. Proc. Bost. viii.—Has size and form of *P. Napi*, but the wings are much paler, without dark spots or apices; the nervures are tolerably distinct, and are faintly marked with yellow on the under side of the hind wings. Habitat, Labrador. As Dr. Staudinger includes this insect in his Catalogue I mention it in this place. It seems to be a northern variety of the North American *P. Oleracea*, B.

P. Cheiranthi, Hüb. 647, 8.—The size and shape of *P. Brassicæ*, but the fore wings have a very large black tip extending half-way along the costa and two-thirds of the hind margin. The black spots in the female coalesce so as to form a large black patch extending from the centre of the wing to the inner margin. The hind wings are yellow, with the marginal spot larger than in

Brassicæ. Habitat, the Canary Islands. It is probably a local form of *P. Brassicæ.*

Genus *EUCHLOË*, Hüb.

E. Falloui, Allard, Ann. Soc. Ent. Fr., 1867, p. 318.—M. Allard gures this species; he describes it as being found in Algeria at the same time and in the same localities as *Glauce*, which it very much resembles, but is smaller and more strongly marked. The under side of the hind wings is striped with green and white. I have not seen a specimen.

E. Charlonia, Vouzel, Ann. Soc. Fr., 1842, 197 (the male only described).—Expands 1·50 in. Wings rounded, of a uniform sulphur-yellow; the fore wings with the apex and discoidal spot black; hind wings without markings. Under side :—Hind wings mottled with grey and yellow spots. Head and thorax black; antennæ and abdomen yellow. Times of appearance, February and March. Habitat, Algeria and Asia Minor.

E. Levaillantii, Luc. Ann. Soc. Fr., 1847, pl. L. *Charlonia* var. Staud. Cat. p. 4.—Smaller than the last, expanding 1·25 in. The fore wings are pointed at the apices, and the dark markings are paler than in *Charlonia;* the antennæ and abdomen greyish black. Time of appearance, the beginning of February. Habitat, Algeria. "It flies in dry sandy places where coarse grass grows" (G. Allard).

E. Eupheno, Linn. Syst. Nat. *Douei*, Pier. Ann. Soc. Fr., 1832. *Eupheno*, var. *Douei*, H.G. *Belia?* Linn. Syst. Nat. xii. 761 (female). —The size and shape of *E. Euphenoides.* This insect has been already alluded to on p. 42. Both sexes almost exactly resemble *E. Euphenoides* above, but beneath the hind wings have quite a different pattern. The markings are rusty red, instead of green, and disposed in three bands; one short, one starting from the costa, the second crossing the wing from the costa to the inner margin, and the third a short one starting from the inner margin. There are no white markings as in the under side of *E. Euphenoides.* Habitat, Algeria. It has been said to occur in the South of Spain and the Balearic Islands, and this seems very probable. Not having met with any authentic records of its capture in Europe, I do not figure it in the present work. Times of appearance, January to March.

Genus *ZEGRIS*, Rambur.

Z. Fausti, Christop. Hor. Ent. Soc. Ross. xii. pl. v. About the size and shape of *Z. Eupheme*, but the apices of the fore wings are more red than orange, and the hind wings have the under side greener. It is found in Siberia.

Genus *CALLOSUNE*, Doubl.

C. Nouna, Luc. Expl. Alg. Zool. iii. 50.—A small white species with black and red apices to the fore wings; hind wings without mottling beneath. Habitat, Algeria.

Genus *LEUCOPHASIA*, Steph.

L. Sinapis v. *Amurensis*, Mén. Schrk. p. 15.—This variety is larger than the type and has longer and narrower wings. It inhabits the south-east of Siberia and the Amur, probably also Japan.

Genus *IDMAIS*, Boisd.

I. Fausta, Olivier, Voyage en Syrie, pl. 33.—Expands, male, 1·50 in., the female slightly larger. All the wings are of a very delicate flesh-colour. The fore wings have a black discoidal spot and a row of black spots along the hind margin; there is also a black streak running downwards from the costa. The hind wings have no markings, except a row of small black spots along the hind margin; these are faint in the male, but more apparent in the female, in which all the markings and the ground colour are much darker than in the other sex. The head, thorax, and abdomen are black above, but, with the legs, white beneath. The antennæ are slender and black. The species belongs to a genus that is principally African: it occurs in Syria, and on that account is included by Staudinger in his Catalogue.

Callidryas Pyranthe, L.—This species, belonging to the well-known tropical genus, is also included by Staudinger on account of its reputed occurrence in Syria.

Genus *COLIAS*, F.

C. Melinos, Ev. Bull. Mosc., 1847, 4, 8, 624-7.—This species seems, from Herrich-Schäffer's figure, very greatly to resemble

C. Werdandi, but is somewhat larger. It is probably a southern form of that insect. Habitat, S.E. Siberia and the Amur.

C. Sagartia, Ld. Hor., 1869, t. iv. 1, 2.—Expands from 2 to 2·37 in. The male has all the wings of a bluish green colour and bordered with black. The fore wings are tinted with yellow towards the costa and have yellow spots on the black border, on which also the nervures show as yellow stripes. The discoidal spot is large and black; the hind wings have a row of yellow spots internal to the black border, and an orange discoidal spot. The head, prothorax, and antennæ are faintly red, as well as the marginal fringes; they are sometimes, however, tinged entirely with green. The female is pale greenish white, and has very large spots of the same colour on the marginal borders; the discoidal spot of the hind wings is bright orange. The under side is very much like that of *Phicomone*, but paler. This species is undoubtedly the handsomest of all the light-coloured Coliades, and is found in July and August in those regions which lie to the south and east of the Caspian Sea.

C. Boothii, Curtis, ' Voyage in search of N.W. Passage' (Lond. 1835).—Expands 1·50 in. The fore wings are orange-yellow, shading off along the costa and hind margin into light green; the discoidal spot is small and roundish; the marginal black band is very narrow. The hind wings are greenish, with an orange discoidal spot and a very narrow hind-marginal black band. The fringes of all the wings are red. This species is found in Arctic America in the district known as Boothia Felix, between 70° and 75° N. lat. It is quite distinct from *C. Hecla*, with which it has been confounded. The extremely narrow unveined black border and the greenish tint of the wings are never found in *C. Hecla*, and the discoidal spot in the male of the latter species is never rounded.

C. Aurorina, H.-S. 453.—Expands from 2·25 to 2·50 in. The male has all the wings of a dull reddish orange, with black border, like that of *Edusa*, with distinct orange veins. The hind wings are greenish in the inner margin and have a very large and bright orange discoidal spot. The female has the border spotted as in *Edusa*, and the ground colour of the wings is purer than in the male, owing to the absence of the minute black scales with which the wings of the latter are covered. Both sexes have purple reflections. Times of appearance, July and August. Habitat, the Trans-Caucasus and Armenia, frequenting mountains, at great elevations, as high as 1200 ft. Larva cylindrical in shape, rather pointed at the extremities; in colour it is reddish grey above and flesh-coloured beneath; on each side of every segment is a yellow spot surrounded by black, and below this there are two longitudinal black stripes. It feeds on *Astragalus Argyrothamnus*.

Var. a. *Libanotica*, Led. Wein. Mts., 1858, 140.—Somewhat larger than the type. The male is of a duller orange and has the black borders without orange veins, excepting one or two near the apices of the fore wings. The borders are speckled minutely with yellow scales. The female greatly resembles that of *C. Heldreichi*. This variety inhabits the mountains of the chain of Lebanon and the Anti-Libanus.

C. Aurora, Esper, 83, 3. Hüb. 544. Dup. 1, 6, 45.—Expands 2·37 to 2·50 in. All the wings are very brilliant orange, with a purple reflection; the nervures of all the wings are dusky and nearly black; the hind margins of the male have a narrow black border, which is veined with yellow on the fore wings near the apex. The discoidal spot of the fore wings is long and narrow. The hind wings have a narrow unveined border and an internal row of scarcely perceptible light spots; the discoidal spot is nearly of the ground colour; the inner margin and costa are greenish, and immediately below the costal nervure is a conspicuous "glandular space" of a light ochre colour. The female has the border of all the wings spotted with yellow and unveined, and is somewhat larger than the male; the white form of this sex seems to be frequent in this species as well as in *Aurorina*. Habitat, Siberia, principally the south-eastern portion and the Amur.

Genus *GONEPTERYX*, Leach.

G. Rhamni var. *Aspasia*, Mén. Bull. Acad. Pet. xvii.—This variety is found in the Amur and Japan; it has the wing narrower and more pointed than the type. Staudinger considers it a distinct species.

G. Cleobule, Hüb. Zutr. Ex. Schmet. p. 455. *Rhodocera Rhamni*, var. B., Boisd.—This is most probably a variety of *G. Cleopatra*.

The male has the fore wings entirely suffused with orange. The angles of the wings are about equal to those of *G. Cleopatra*. The female of the latter and of the present form are similar. Habitat, the Canaries.

NORTH AMERICAN SPECIES OF *PIERIDÆ* ALLIED TO THOSE OF EUROPE AND NORTH ASIA.

Picris Oleracea, Harr.—Expands 1·7 in. A white species found throughout the United States. It is very close to *P. Rapæ*, but is less strongly marked with black above and without any yellow beneath.

Picris Virginensis, Edw.—Very close to the last, but darker, being powdered with greyish brown scales. Habitat, the Southern States.

Picris Vernalis, Edw.—This species, figured in the 'Butterflies of North America,' appears to be very close to our *P. Callidice*, but is smaller and lighter.

Picris Beckeri, Edw., and *Picris Protodice*, Boisd., from the Southern States, Colorado, and California are very like the Russian *P. Chloridice*, but somewhat larger.

Zegris Olympia, Edw. Butt. N. Am. ii. t. 1-4.—About the size of *Z. Eupheme*. The upper surface of the wings is white; the fore wings have a large grey patch at the apex partly replaced by white; the costal margin is slightly speckled with black. The hind wings have a few black scales at the outer angle and a small wedge-shaped black spot on the costa. The under side is white, with the apices of the fore wings greenish. Hind wings caused by bands of yellow and green on a grey ground. Habitat, Texas (Edwards).

Colias Philodice, Godt. Enc. Meth. ix. p. 100.—This is the commonest and most widely distributed North American *Colias;* it greatly approaches our *C. Palæno;* both sexes are yellow. The male has a narrow black border, veined with yellow. The female has the border spotted with yellow. The discoidal spot of the fore wings is distinct. This is the species that was erroneously described as British by the older entomologists under the name of *C. Europome*.

Colias Scudderi, Edw. Butt. N. Am. p. 60.—This species seems to be identical with *C. Pelidne*, and is probably a southern form of that species, and rather larger, expanding from 1·80 to 2 in. Habitat, Colorado, Upper Arkansas Valley.

Colias Interior, Scudd., found in Canada, appears to be likewise a form of the same insect.

Colias Behrii, Edw. Proc. Ent. Soc. Philad. 1866.—This species is taken in California on the Yo Semite Mountains, at an elevation of 10,000 ft. It appears to me identical with *C. Nastes*.

Colias Kewaydin, Edw. Butt. N. Am. 49.—Expands from 1·60 to 1·80 in. Very close to *C. Chrysotheme*. Habitat, New Orleans and Illinois.

Colias Christina, Edw. Proc. Ent. Soc. Phil. 1863.—Expands from 2·10 to 2·50 in. Not unlike *C. Hecla*, but greener, being somewhat nearer to *C. Boothii*, of which it may possibly be a southern form. "Taken at the portage of Slave River by Mrs. Ross in 1862."

Colias Eurytheme, Boisd. Ann. Soc. Fr. 1852.—Expands from 2 to 2·50 in. Has very much the appearance of *Edusa*, but is without the glandular space found in the male of that species. Its colour is lighter than *Edusa*, and more like that of *Chrysotheme*, near to which is the true position of the species. Habitat, the fields about New York in the spring and autumn (Edwards).

Colias Meadii, Edw. Trans. Am. Ent. Soc. 1871.—Intermediate between *Edusa* and *Myrmidone*, but smaller than either, expanding from 1·70 to 1·90 in. The border is not veined and the discoidal spot of the fore wings very small. Habitat, Colorado.

The study of the Nearctic forms of the genus *Colias* is very interesting; most of the species are very close to those of the Palæarctic region, and many seem to be absolutely identical, though they have received different names. Figures of many of them will be found in Mr. Edwards's magnificent work 'The Butterflies of North America.'

Fam. 3.—LYCÆNIDÆ, Stephens.

CHARACTERS.—Small-sized butterflies, with the anterior pair of legs perfectly formed in both sexes. Antennæ straight, with the clubs elongated and not curved. Palpi fully developed. Hind margins of fore wings without angular projections; inner margins of hind wings never concave, but forming a shallow groove to receive the abdomen.

The larvæ are onisciform or woodlouse-shaped, being short and thick, the middle segments having a larger diameter than those near the extremity. The head is small and retractile.

The pupæ are short, thick and rather rounded, entirely without points or angular projections. They are attached by the tail and girt by a belt round the middle, as in the last two families.

Next to *Nymphalidæ*, this is the most extensive of butterflies; it contains more than 1200 species, and is found in every region of the earth. The "blues," "coppers," and "hair-streaks," which represent this family in our own country and in Europe generally, are remarkable for their extreme beauty, though small size; the Oriental and Neotropical regions, however, are where the *Lycænidæ* chiefly abound, and are most magnificent in colouring and design; some of those from South America, for instance, may be fairly said to constitute some of the most beautiful objects in Nature.

Five genera are represented in Europe—THECLA, LÆOSOPIS, THESTOR, POLYOMMATUS, and LYCÆNA. The total number of species does not amount to more than seventy.

Every description of country furnishes examples. Woods, downs, fields, roadsides, and marshes, all have their inhabitants belonging to this family; some are peculiar to mountainous districts, and others are content with the cold sunshine of the Arctic Regions.

Their flight is generally fast and jerky, seldom continuous, and they most frequently settle on low-growing flowers; some species, however, especially those of the genus *Thecla*, are fond of flying about the branches of trees, often in groups; their habit may be observed in our *Thecla Quercus, T. Betulæ, T. W.-album, Lycæna Argiolus,* &c.

By far the greater number of species exhibit on their under sides a pattern composed of small dark spots, often ocellated, a peculiarity which has gained for them the name of "Argus" butterflies both in this and other countries.

The larvæ of the genus *Thecla* are mostly arboreal feeders, the leaves of various forest-trees forming the food-plants. Those of the remaining European genera feed upon various low-growing plants, the orders *Papilionaceæ* and *Leguminosæ* being the most frequently used; the larvæ of *Polyommatus* seem partial to various species of *Rumex,* as will be seen hereafter. Some species of *Lycæna* feed on the pods of *Leguminosæ,* as in the case of *L. Bætica* and *L. Iolas;* others on the flowers of various plants.

Genus 1.—THECLA, Fabr. Ill. Mag. vi. 286 (1807); Leach
(1815); Boisd. (1833); Westwood (1852).
CUPIDO, Sect. C., Schrank. (1801).

Fore wings nearly triangular in shape, without angular projections; hind wings generally with a notch or tail-like prolongation near the anal angle. The colour of the upper surface of the wings is usually dark brown; the under side lighter, with white streaks either formed by continuous lines or broken up into a chain of dots.

The larvæ are mostly of a green colour, with lateral and dorsal stripes of a brighter shade.

1. T. Betulæ, Linn. Syst. Nat. x. 482; Faun. Suec. 282;
Esper. 19, 1; Hüb. 383-5.

Expands 1·25 to 1·50 in. The ground colour of the wings in both sexes is dark brown; all the wings have the fringes whitish brown. The head, thorax, and abdomen are black above, but with the legs and palpi are white beneath; the antennæ are black, ringed with white. The male has a faintly black oblong discoidal spot on the fore wings, and external to it a light but inconspicuous patch; the hind wings have the tail orange, and a small orange patch at the anal angle. The female differs from the male in having a large and bright orange patch on the fore wings external to the discoidal spot, crossed by two or three black veins, and occupying nearly a fourth of the area of the wings. The under side is nearly the same in both sexes; the ground colour is reddish brown, brightest in the female; the fore wings have an elongated discoidal spot, and, external to this, reaching from the costa, a long tapering streak of a darker colour, coming to a point as it approaches the inner margin and bounded externally by a white line; the hind wings have a patch of bright reddish brown running from the costa to the inner margin, bounded internally by an indistinct white line and externally by a very distinct wavy line of the same colour; the hind margins are reddish orange.

TIME OF APPEARANCE.—From the end of July to the middle of September.

HABITAT. — Central Europe, including Germany, the Low Countries, Switzerland, and the Tyrol; also France, Scandinavia, Britain, and South Russia. Besides being European, it occurs in Armenia and Central Asia as far as the Amur.

LARVA, when full grown, is apple-green; the segments are very definitely divided, and each segment has four longitudinal white stripes, two dorsal and two lateral, and, besides these, several oblique pale lines. The head is brown, and very much smaller than the segment immediately posterior to it.

PUPA.—Pale brown and smooth, and (according to Newman's observations in 'British Butterflies,' p. 115) not attached by silken threads *; in this manner resembling the pupa of *T. Quercus*, which also does not attach itself by the head and tail, and is by some authors placed with the present species in another genus— *Zephyrus*, Dalm.

"The egg is a depressed sphere, and white."—Newman. It is laid in September on the twigs of birch and blackthorn, the caterpillar emerging in the spring and being full-fed by June.

2. **T. Spini**, Wien. Verz. 186, n. 5; Hübn. i. f. 376, 377.—
Papilio Lynceus, Esper. i. 1, p. 356.

Expands 1·18 to 1·40 in. The wings are brown on the upper surface. The hind wings distinctly tailed. The male has one or

---

* See also Westwood, Proc. Ent. Soc., 1866, p. xxxiv.

two small orange spots near the anal angle of the hind wings. The female is larger and lighter in colour than the male, and has a large indistinct lighter patch on the fore wings, and a row of orange spots on the hind margin of the hind wings, extending from the anal angle.

The under side is brownish grey; the fore wings have a distinct white line beginning on the costa at a point which is distant from the hind margin by a space equal to a fourth of the width of the wing. The hind wings have a white line running from about the middle of the costa to the inner margin, taking near the anal angle an upward and then downward direction, so as to give it somewhat of a W-shape. Along the hind margin is a row of orange spots, and near the anal angle and filling up the whole distance between that point and the tail is a patch of light blue.

Time of Appearance.—May and June.

Habitat.—All Central and Southern Europe, but not Great Britain; Northern and Eastern Asia. It frequents open places in woods and roadsides.

Larva.—Light green, with two yellowish streaks on the sides; on each segment is a row of oblique lines of a darker green than the ground colour; on the dorsal surface, which has a dark streak, are some pink dots.

Time of Appearance.—June.

Food-plants.—*Prunus Spinosa* and *Cratægus Oxyacantha.*

2. T. W-album, Knoch. Btr. ii. p. 85; Hüb. 380-1.

Expands 0.93 to 1·25 in. Hind margin of hind wings scalloped, and with a small slender tail. Wings dark blackish brown, the nervures showing a little darker. Fore wings quite without any pattern. The hind wings have an orange spot at the anal angle; the tail is tipped with white. Under side:—The colour of all the wings is greyish brown; the fore wings have a narrow white line, somewhat wavy, reaching from the costa to the inner margin, and placed closer to the hind margin than to the base of the wing. The hind wings have a white W-shaped line beginning at the costa and ending at the inner margin; along the hind margin is a row of brilliant orange crescents, which are largest near the anal angle and decrease in size as they approach the costa.

TIME OF APPEARANCE.—July and August.

HABITAT.—Central Europe, including Britain; the South of Europe, excepting Spain and Portugal; Siberia and the Amur. It frequents commons, wood clearings, and roadsides where there is plenty of bushy vegetation; bramble-bushes seem to be chiefly affected by this species. As a British insect it is not generally common, but during some seasons it occurs in great abundance in certain localities. Its extraordinary appearance at Ripley, in Surrey, in 1827, recorded by Stephens, is an historical fact well known to every British lepidopterist.

LARVA onisciform, light green, with a yellowish brown dorsal stripe; every segment has two oblique light yellow lines on each side; the head is black and retractile. It feeds on the leaves of the common elm (*Ulmus campestris*), from whence it may be beaten at the end of May and the beginning of June.

The PUPA is attached to a twig by a belt of silk, and also by the tail.

"The eggs are laid on the twigs of the elm and wych elm in July and August, and are shaped something like an orange, but are more depressed at the crown; they are of a whitish or putty colour, and remain firmly glued to the rind of the twigs throughout the winter. The full-fed caterpillar rests on the surface of the leaves."—Newman, Brit. Butt. 108.

4. **T. Ilicis,** Esp. i. p. 353; Hüb. 378-9; O. I. 2, 105.—*Linceus,* Fab. Mant. 69, i. 3, 279.

Expands 1·16 to 1·33 in. The male has all the wings dark brown, much less black than those of the last species; at the anal angle of the hind wings is a very small orange spot; the hind margin has a slight tail. The female is lighter brown than the male, and is shot with a greenish bronze much in the same way as *Epinephele Janira;* the fore wings have a large dull orange patch divided into four by the nervures, which are dark brown. The under side is lighter brown than the upper surface; all the wings have a chain of small white spots of a crescentic shape running parallel to the hind margins, but some distance internal to them; these white spots are faint on the fore, and more strongly marked on the hind, wings; between this row of white spots and the hind margin is a row of dull orange spots, only seen on the hind wings; outside this is a narrow white line.

TIMES OF APPEARANCE.—May, June, and September.

HABITAT.—South-Eastern and Central Europe (but not Great Britain), Scandinavia, Asia Minor, and Armenia. It frequents woods and bushy places.

LARVÆ onisciform, of a pale green colour, with yellow streaks on the back and sides. The head is black and retractile; legs black. It may be found in April and August on the leaves of the oak.

VARIETIES.

a. **Cerri** (female), Hüb. 863-6. Has the orange patch on the fore wings larger than in the common form of the female.

b. **Esculi** (*recté Æsculi*), Hüb. 559-60. Smaller than the type. Fore wings with the white chain of spots nearly or entirely wanting.

HABITAT.—South-Western Europe. This variety of the species replaces the typical form in Spain and the South of France.

5. T. Acaciæ, Fab. Mant. 69; E. S. 299; Hüb. 743.

Expands 0·87 to 1 in. The wings are of a uniform brown in both sexes; there is no orange blotch on the fore wing. The hind wings have a short tail, tipped with white; between this and the anal angle there are one or two small hind-marginal orange spots. The under side has the wings of a lighter colour than above; the fore wings have an indistinct row of whitish spots shaded on their inner sides; the hind wings have a similar row, forming a white line which extends from the costa to the inner margin just above the anal angle; the hind margin has a row of orange crescentic spots.

TIME OF APPEARANCE.—June.

HABITAT.—South of Europe; principally South of France, Italy, Spain, and South Russia; also Asia Minor. It frequents dry elevated meadows.

Larva.—Undescribed. Herrich-Schäffer* says that it feeds on sloe, but does not mention it further; and I am unable to find any account of its habits or appearance in other works.

6. T. Pruni, Linn. Faun. Suec. p. 283 (1761); Esper Schmet. i. 1, p. 19, 3, 39, 1; Hüb. Eur. Schmet. i. 386, 887.

Expands 0·87 to 1·12 in. Wings brownish black. The fore wings in the female and sometimes in the male exhibit faint traces of a brownish orange band running parallel to the hind margin. The hind wings have a short tail something like that of *T. Spini*, and parallel to the hind margin is a row of well-defined semilunar spots of an orange-brown colour, decreasing as they approach the costa. The colour of the under side is paler than that above. The fore wings have a bluish white interrupted line running from the costa to the inner margin. The hind wings have a similar line which does not assume a W-shape. The hind margin has an orange band bordered with bluish white, and having a row of black spots, one being placed in each inter-neural space. Fringes black. Antennæ black, ringed with white.

Time of Appearance.—June.

Habitat.—Woods in Central Europe (including England), Northern and Central Italy, and Dalmatia. It also occurs in the Altai.

Larva.—Similar in shape to those of the allied species. "Green, darker on the back; two rows of long yellow spots on the back, and a row on each side above the legs; six long oblique yellow stripes on each side" (Stainton, from Duponchel). "The chrysalis is attached by a belt and by the anal extremity; it is obese, blunt-headed, and hump-backed; and has a medio-dorsal series of five rather conspicuous warts or tubercles" (Newman, from Hübner). The larva feeds on the leaves of *Prunus Spinosa*, on the twigs of which the eggs are laid in the summer and remain all the winter, the larva appearing in May.

---

* Europ. Schmett. p. 136.

7. **T. Quercus**, Linn. Faun. Suec. p. 283 (1761); Syst. Nat. 1, 2, p. 788; Esp. Schmet. i. 1, t. 19, f. 2 (1777); Hüb. Eur. Schmet. i. f. 368-70.

Expands 1·16 to 1·59 in. Wings blackish brown. Hind wings with a short but well-defined tail. The male has all the wings suffused with dull violet, excepting along the margins. The female has the purple decoration confined to a triangular patch of very brilliant bluish violet, extending from the base of the fore wings along their inner margins. The under side is ashy grey; fore wings with a whitish streak running parallel to the hind margin. The hind wings with a similar streak running across their centre; near the anal angle are two orange spots, the external one having a black centre. Pl. XVIII., 2.

Times of Appearance.—July and August.

Habitat.—The whole of Central and South Europe, and the Trans-Caucasian Provinces, frequenting oak woods; as a British insect, commoner in England than in Scotland or Ireland.

Larva shaped much like others of the genus; it is reddish brown, often with green tinge, and has two rows of oblique stripes of black colour on the dorsal surface; it feeds on oak-leaves, the eggs being laid on the twigs of the tree in July.

The Pupa is brown, without angles, and according to the most accurate observations does not attach itself by the tail or by a silken belt; thus resembling the pupa of *T. Betulæ*, with which species the present one is placed in the genus *Zephyrus* by many modern entomologists, as before mentioned.

VARIETY.

Bellus, Gerh. Pl. 4, 2; Hüb. 621. Has a yellow spot in the centre of the fore wings.

Habitat.—Hungary and Austria.

8. **T. Rubi**, Linn. Syst. Nat. x. 483; Esp. 21, 2; Hüb. 364-8.

Expands 1 to 1·75 in. All the wings are brown, with a reddish gloss. The hind margins of the hind wings are slightly

M

dentate, but without any tail. The under side is bright green, the hind wings having a row of white spots. Pl. XVIII., 3.

TIMES OF APPEARANCE.—April, May, and August.

HABITAT.—Europe, Western Asia, and North Africa. (The North-American *T. Dumetorum*, Boisd., seems to be a variety of this species.) In Great Britain it is somewhat local, but common where it occurs, and of wide distribution; it frequents woods and heaths.

LARVA.—Shaped as in the other species, of a green colour, with lateral yellowish white stripes and oblique yellow lines, and a whitish dorsal streak.

The PUPA is dark brown, without angles, with a silken belt attaching it by the middle, and with an anal attachment. Mr. Joseph Greene mentions that he has found this pupa under moss on an old tree-trunk.

The larva feeds on the flower-buds of the bramble and on various Papilionaceous plants. The eggs are laid early in June; the larva is full grown by the beginning of July.

Thecla Rubi is a double-brooded insect in the more southern parts of its area of distribution; it is, I believe, only single-brooded in the north, as in Scotland, Scandinavia, North-Western Asia, &c.

Genus 2. --**LÆOSOPIS**, Rambur, Cat. Lep. Andal. i. p. 33 (1857).

AUROTIS, Dalman; Kirby, Man. Eur. B. p. 87.

The eyes in this genus are not hairy, as in *Thecla*. The hind margins of the hind wings are not scalloped near the anal angle, and the under side does not exhibit streaks of light colour, but has rows of black spots along the hind margins.

This genus contains but one species, unless we include *Lycæna Ledereri* of Herrich-Schäffer, which has tailed hind wings.

1. **L. Roboris**, Esper. 103-5.; O. I. 2, 95.—*Evippus*, Hüb. Vög. & Schmet. t. 56.

Expands 1·25 to 1·40 in. The male has the fore wings of a dull purple, like that in the male of *Thecla Quercus;* the costæ and hind margins are dark brown, the latter very broadly so. The hind wings are dark brown, with their bases purple, and a row of hind-marginal purple spots. The female is brown, with three purple streaks at the base of the fore wings of a brighter colour than in the male; the hind wings are brown, with a row of blue spots on the hind margin. The under side is brownish grey. The fore wings have a row of faint yellow and black hind-marginal spots. The hind wings have a hind-marginal row of lilac-blue spots, internal to this a faint orange interrupted band, and inside this again a row of black spots surmounted with white angular marks. The fringes are white. The clubs of the antennæ black, tipped with yellow. The head, thorax, and abdomen black. Pl. XVIII., 4.

TIMES OF APPEARANCE.—May and June.

HABITAT.—Woods in the South of France and Andalusia; also Botzen in the Tyrol (Kirby). It has not been found in any non-European localities.

LARVA.—"Dull brown, with a black dorsal streak bordered with obscure yellow markings" (V. G.) "On oak?" (Kirby, Man. Eur. Butterflies).

OBS.—Dr. Staudinger includes this species in the genus *Thecla*, but I have thought it best to separate it, as was done by Rambur and others. Its characters are certainly marked enough to entitle it to represent a separate genus.

Genus 3.—**THESTOR**, Hüb. Verz. Bek. Schmet. (1816); Led. Wein. Ent. Mon. i. p. 26.

TOMARES, Ramb. (1858).

CHRYSOPHANUS, Westwood (1852).

Small butterflies, brown in colour, with reddish or orange markings above, but never with a metallic lustre. The under sides are spotted with black. The hind wings are rounded, with their margins entire, and without tails. The eyes are hairy; the palpi short. The antennæ have long and thickish clubs.

This genus approaches very close to the next, but has characters sufficient to separate it.

There are but four known species, and these have their habitats confined to the shores of the Mediterranean and Black

Seas. Three of these species may be said to occur in Europe; the fourth is exclusively North-African.

1. T. Ballus, Fab. Mant. Ins. ii. p. 80 (1787); Hüb. Eur. Schmett. i. f. 360-61 (1798).

Expands 1 to 1·25 in. The male is brown. The hind wings have one or two indistinct orange spots at the anal angle. The female has the fore wings bright orange, with broad dark brown hind marginal and costal borders; the hind wings are dark brown, with large orange hind-marginal blotches. The under sides of both sexes are similar. The fore wings are orange, with a coppery tinge, with the costa and apex brown; they are covered with numerous black spots, which are disposed in three rows. The hind wings are green, with a bronzy lustre, and thinly covered with white spots; hind margins reddish brown. Head and pro-thorax shiny green, the rest of the body black; clubs of antennæ long and black, their shafts ringed with white. Pl. XVIII., 5.

TIME OF APPEARANCE.—March.

HABITAT.—The South of France, Spain, and North Africa, on the shores of the Mediterranean.

LARVA.—" Yellowish white, with a row of reddish dorsal spots, bordered with brownish red, and bisected by a blue line; oblique streaks and lateral line violet-red. Head brown; first and second segments washed with reddish. On *Lotus Hispidus* " (Kirby, from Guenée.)

2. T. Nogelii (recté Nogelli), H.-S. 529-32; Frr. 574, 1, 2.

Expands 1 to 1·12 in. The male has all the wings dark brown, nearly black; the marginal fringes are white. The under side is grey. The fore wings have an orange discoidal spot; along the hind margin is a row of orange spots, and internal to this a row of larger ones bordered with black. Hind wings grey, with three rows of orange spots, bordered with black. The female has the wings dark brown; the fore wings have a bright orange patch in the centre. The hind wings have an orange patch on the hind margin, near the anal angle. The under side resembles that of the male, but the fore wings have the ground colour orange, the costa and hind margin being grey. Pl. XVIII., 6.

Times of Appearance.—April and May.

Habitat.—The North-East of Turkey in Europe and Asia Minor, frequenting hills and mountains.

Larva and Pupa.—Unknown.

3. **T. Callimachus,** Ev. Bull. Mosc. 1848, iii. 208; Nord. Bull. Mosc. 1851, ii. 123-5.—*Hafis*, Koll. Ins. Pers. p. 10.— *Epiphania*, B. Ann. S. Fr. 1848, p. xxix.; H.-S. 438-41 (1850).

Expands 1 to 1·25 in. The male has the fore wings orange, the tip and hind margin being black, and the costa grey; there is a fine black line along the median nervure. The hind wings are brownish black, a large orange patch reaching from the anal angle along two-thirds of the hind margin. The female has the wings of a lighter orange, and the hind wings have the orange patches larger than the male. Under side:—Fore wings orange, hind margin and costa grey; towards the centre, near the costa, is a group of black spots, and a row of black spots runs parallel to the hind margin. The hind wings grey, with an orange discoidal spot; two rows of orange spots bordered with black run parallel to the hind margin. Pl. XVIII., 7.

Times of Appearance.—April and May.

Habitat.—Mountains in South-East Russia, Armenia, and Persia.

Larva.—The following notice of the Larva is translated from Millière, Ann. Soc. Lin. Lyon. xxv. pp. 8 and 9:—" The shape is flat below, convex above, with the sixteen feet quite visible, although short. The head is small and brown; the first segment is covered by a narrow scaly plate of a uniform brown colour, with a fine white border. The colour of the caterpillar is of a testaceous ochre, and on the dorsal region there is a fine brown vascular line continued from the fourth to the ninth segments. Accompanying the dorsal line, these segments present a double chevron of a dull red colour. Each segment is covered above with numerous hairs, very short and close. The sides and the ventral surface have the same colour as the feet. Above the lateral line are placed the stigmata, which are black and very small, not well seen without the aid of a glass. The larva lives on the pods of *Astragulus Physodes*, L.; it probably has the same habits as that of *L. Bœtica.*"

Genus 4.—**POLYOMMATUS**, Latr. Hist. Nat. Crust. et
Insect. (1805); Boisd. Gen. Ind. Meth. (1840).
CHRYSOPHANUS, Hübn. Verz. Bek. Schm. p. 72 (1816).

Small butterflies with the hind wings generally denticulated on the hind margin, near the anal angle. The colour of the wings is usually brilliant orange-red, with a metallic coppery lustre; sometimes it is dark brown, and occasionally the wings are shot with violet. The upper surface has generally rows of black spots, though in some species the males have the wings unspotted. Beneath, the colour is grey, varying from blue to warm brown in tint, always with numerous black spots which are without ocelli. The females are larger than the males, and usually more spotted with black. The eyes are not hairy, as in *Thestor;* the antennæ are long, with an elongated club.

The Larvæ have the shape of those of the *Lycænidæ* generally —being woodlouse-shaped, with small and retractile head. Their food-plants are usually, but not exclusively, different species of *Rumex.*

The butterflies frequent meadows, woods, and marshes; and some species are found only in mountainous districts.

Though this genus is represented in almost every part of the world, the more temperate regions are where the species most abound; hence it is that a large proportion of these are found in Europe.

1. **P. Virgaureæ**, Linn. Faun. Suec. p. 285; Esp. Schmett.
22, 2; Hüb. 349-51; Frr. 115.

Expands 1·12 to 1·25 in. The male has all the wings of a brilliant golden copper, without spots, and with a narrow black hind-marginal border. The hind wings are dentated near the anal angle, and have their inner margins brownish black. The female has the wings orange-brown, with a coppery lustre, and marked with numerous black spots. Under side :—Fore wings orange, with black spots. Hind wings light brown, with a row of white spots, forming an irregular band across the centre; between this and the base are a few black spots; along the hind margin is an orange band. Pl. XIX., 1.

TIMES OF APPEARANCE.—May and August.

HABITAT.—The whole of Europe (except Britain, the southern part of Spain and Portugal, and South Russia). It also inhabits Asia Minor and a considerable part of Siberia. Some authors describe it as having been formerly taken in Britain, but no authentic British specimens exist. It is generally found in woods and meadows.

LARVA downy, of a dull green colour; every segment is streaked laterally with dark green, and on the dorsal aspect there is a line of pale yellow. It feeds on *Solidago Virgaurea* and various species of *Rumex* in June and September.

VARIETIES.

a. **Oranula,** Frr. Beitr. v. 455. A Boreal form of the insect occurring in Lapland; it differs from the type in being smaller.

b. **Miegii,** Vogel, 1857, n. Schm. Zeit. Ges. Isis, p. 201, t. vi. In this variety the male is spotted above, and the female is paler in colour than in the type.

HABITAT.—Spain and the Pyrenees; Armenia.

c. **Zermattensis** (female), Fallou, Ann. Soc. Ent. Fr. 1865, p. 101, pl. 2, 3. This variety of the female, which is found at Zermatt and other places in Switzerland, is very dark in appearance, the ground colour of the wings being dark brown, instead of orange, as in the type.

2. P. **Ottomanus,** Lefebrve, Guérin. Mag. 1830, p. 19; Boisd. Ic. 10, 4, 5; Dup. i. 9, 1, 2; H.-S. 236-9.—*Legeri,* Frr. Beitr. 133, i., iii.

Expands 1 to 1·25 in. The male has all the wings of a most brilliant copper-colour. The fore wings have narrow discoidal spot, and two more between this and the base; external to and midway between it and the hind margin are four or five small spots; all the above-mentioned spots are very indistinct; the hind margin narrowly black throughout its whole length. The hind wings are deeply emarginate near the anal angle, unspotted, but with a hind-marginal black border; the inner margin is brown. The female has all the wings spotted with black; the fore wings have a row of black spots parallel to the hind margin, and internal to this another row of eight, arranged irregularly; in the discoidal cell are three spots. The hind wings are brownish, with a hind-marginal orange band; there is one black discoidal spot, and two rows of black spots running parallel to each other and to the hind margin. Under side:—Fore wings orange, with black spots, surrounded by grey rings. Hind wings brownish grey, with black spots, surrounded by light rings; hind margin bluish grey, with a row of bright red almost scarlet spots, forming a band; in the female there is a row of large light spots internal to these. Pl. XIX., 2.

TIMES OF APPEARANCE.—March and June.

HABITAT.—Greece, Turkey, and Asia Minor.

LARVA.—Unknown.

3. **P. Thetis**, Klug. Symb. Phys. t. 40, 17, 18; H.-S. 643.— *Ignitus*, H.-S. 332.

Expands 1·19 to 1·25 in. All the wings are very brilliant copper in the male. The fore wings are unspotted, and with the apices narrowly deep black. Hind wings with a short tail, and a hind-marginal row of black spots. The female has the tail on the hind wings much longer than the male, and all the wings have black spots above. Under side:—Fore wings yellowish grey, spotted with black; hind wings ashy grey, with indistinct black spots, and a hind-marginal row of orange ones. Pl. XX., 2.

TIME OF APPEARANCE.—June.

HABITAT.—Greece and Asia Minor, frequenting mountainous places.

LARVA.—Unknown.

4. **P. Ochimus**, H.-S. 523-6 (1851).—*Phaëton*, Frr. 571, 2 (1852).—*Kersteinii*, Gerh. pl. 9, 4 A—C.

Expands 1·25 to 1·30 in. The male has all the wings shining golden copper; the hind margins have a narrow black border. The fore wings have an indistinct discoidal spot, and three others near the apex. The hind wings are without spots above; they are emarginate near the anal angle. The female has two discoidal spots on the fore wing, and two rows running parallel to the hind margin. The hind wings have two discoidal spots, one oblong, the other small and round; three rows of black spots run parallel to the hind margin. Under side:—Fore wings light orange, the costa and hind margin grey; there are three discoidal spots, a double hind-marginal row, and another irregular row internal to this. Hind wings grey, with a double row of black spots along the hind margin, enclosing a faint orange band; there are three basal spots, an oblong discoidal spot, and between the basal and discoidal spots three others; besides these, there is another row midway between the discoidal cell and the hind margin. Pl. XX., 3.

TIMES OF APPEARANCE.— ————?

HABITAT. — Turkey, Asia Minor, Armenia, and Persia; frequenting mountains.

5. **P. Thersamon**, Esp. 89, 6; Godt. ii. 22, 7, 8; Frr. 109, 3, 4.—*Xanthe*, Hüb. 346 (female).

Expands 1·10 to 1·15 in. The male has the wings copper. The fore wings unspotted above, but showing the spots of the under side through the wing; hind margin with a narrow black border. Hind wings blackish at the base, and with a hind-marginal row of black spots. All the wings are more or less strongly shot with violet. The female has the fore wings lighter, and marked with numerous black spots. Hind wings brownish copper, with a lighter hind-marginal band, and numerous black spots. The wings in the female are not shot with violet. Under side:— Fore wings reddish yellow, with black spots. Hind wings warm grey, with black spots, and a hind-marginal orange band. Pl. XIX., 3.

TIME OF APPEARANCE.—July.

HABITAT.—South-Eastern Europe, North and Central Italy, Asia Minor, and Persia.
LARVA.—Unknown.

6. P. **Dispar**, Haw. Lep. Brit. p. 40; Dup. i. 13, 3-6; Boisd. Ic. 10, 1-3.—*Hippothoë*, Lewin, Insects, pl 40; Donovan, Brit. Ins. pl. 117; Esp. 38, 1, 2 (*Hippothoë*, Linn. est nomen speciei sequentis).

Expands 1·62 to 2 in. The male is brilliant copper-red; all the wings have a narrow black hind-marginal border. The fore wings have two spots in the discoidal cell, the outer one being elongated, and the inner one being merely a black dot; about midway between the outer spot and the hind margin, and parallel to the latter, is an indistinct row of dark spots; the hind wings possess a similar row, and also an elongated discoidal spot. The female is larger than the male; the fore wings have a broader hind-marginal band, and parallel to this a row of seven large black spots; there are three spots in the discoidal cell. The hind wings are copper, with three rows of black spots, generally clouded over with dusky brown, except along the hind margin, which has a broad copper band; sometimes the hind wings are rather broadly veined with copper. The under side is similar in both sexes. The fore wings are bright orange-red, with a grey hind-marginal border. In the discoidal cell there are three black spots, surrounded by grey rings, and parallel to the hind margin is a row of spots similar to those just mentioned, and seven in number. The hind wings are pale grey, strongly tinged with light blue towards the base, and with a broad and very distinct hind-marginal orange border enclosing a double row of black spots; at the base are two black spots, and one near the centre of the costa; there are three spots, two small and round, and one elongated, in the discoidal cell, and then one midway between the latter and the inner margin; beside these, there is an irregular row of ten large spots running parallel to the hind margin, and these, as well as all the other spots, are surrounded by light rings. Pl. XIX., 4.

TIME OF APPEARANCE.—From the end of June to the end of August.

Habitat.— ————?

This insect, known in England as "the large copper," once inhabited the fen-districts of Cambridgeshire and Huntingdonshire, but it has not been seen for many years, the last capture having been made in 1848; we are therefore forced to conclude that it is extinct in England. As regards its further habitat, I cannot find any certain account of its ever having been taken anywhere but in Britain, though there have been rumours of its occurrence in the Pontine marshes, near Rome; also in Egypt.

All the continental specimens of *Dispar* I have seen belong to the variety *Rutilus*, to be described presently, and are so distinct that there ought not to be any confusion between them and the true typical form once taken in England.

Larva.—Green in colour, with a darker dorsal stripe. Its food-plant was the great water dock (*Rumex hydrolapathum*), upon which the eggs were laid in August, the larva hybernating and becoming full-fed in the June following.

VARIETY.

Rutilus, Wernb. Btr. i. p. 394.—*Hippothoë*, Hüb. 352-4; O. I. 2, 83. This is smaller and less brightly coloured than *Dispar*, the spots on the under side are much smaller, and the colour of the under side is ashy grey, with very little tinge of blue. The most distinctive feature, however, is the narrowness of the orange band on the under side of the hind wings, near the hind margin. I have examined a great number of specimens of *Rutilus*, and also of *Dispar*, with the object of fixing upon some constant character by which they may be differentiated, and have never seen a specimen of *Rutilus* with the hind-marginal band so broad and so well defined as it always appears in *Dispar*. I am therefore inclined to look upon this character as diagnostic.*

Times of Appearance.—July and August.

Habitat.—This form of the species is distributed throughout France, Germany, and the South-East of Europe. It also inhabits Asia Minor, Armenia, and the Altai, inhabiting moist meadows. In France and Germany it is very local, but commoner in the more eastern parts of its range.

* I am indebted to Mr. Howard Vaughan for first suggesting to me this method of distinction.

LARVA green, with a paler lateral stripe. It feeds on various species of *Rumex*, and on *Polygonum bistorta.*

7. P. Hippothoe, Linn. Faun. Suec. ii. 274 (1761); Esp. 22, 3. —*Eurydice*, Rott. Naturf. vi. 28 (1775).—*Chryseis*, Hüb. 337-8; O. I. 2, 79; Stainton, Man. Brit. Butt. & Moth. vol. i.; Samouelle, Useful Compend. p. 249 (1819).

Expands 1 to 1·40 in. Male brilliant copper-red. Fore wings with a black hind-marginal border, and with a narrow black discoidal spot. Hind wings with the hind margins and bases brown, shot with violet. There is a narrow band of copper running parallel to the hind margin. The fore wings are often shot with violet, as well as the hind pairs. Female brown; fore wings lighter in the centre than at the hind margins or costa, and with several black spots; hind wings brown, with a narrow hind-marginal border of orange spots. Under side:—Similar in both sexes. The fore wings are orange, shaded with grey on the costa and hind margin, with numerous black spots surrounded by light rings. Hind wings brownish grey, with a hind-marginal orange band, and numerous black spots. Pl. XX., 4.

TIMES OF APPEARANCE.—June to September.

HABITAT. — North and Central Europe, and Siberia. It probably occurred in Great Britain at one time, but, like the foregoing species, is now extinct here.

LARVA.—" Green, with a darker dorsal stripe, and two paler lateral lines " (Freyer). Feeds on *Rumex hydrolapathum* and *R. acetosa* in June.

VARIETIES.

a. **Confluens**, Gerh. pl. 8, fig. 1 A—D. This is an aberrant form in which the spots on the under side are confluent.

b. **Eurybia**, O. I. 2, 81; Fr. 163-4.—*Euridice*, Hüb. 339-42. Usually smaller than the type. The male has the borders of the wings blacker and narrower; there are no violet reflections. The female is darker than the type. Under side dark grey, with very small spots, and without the orange band on the hind wings.

HABITAT.—The Alps of Europe, and the Altai. Pl. XX., 5.

c. **Stieberi**, Gerh. pl. 35, 1 A. Much smaller than the type. The female fulvous. This is the boreal form of the insect, being found in Lapland and the northern part of Scandinavia.

d. **Candens**, H.-S. 229-31; Fr. 571, 3, 4. Larger than the type. The male is a very brilliant copper-red, without violet reflections, and with an even black border, as in *Eurybia*, but wider. The female is fulvous.

HABITAT.—Turkey and Asia Minor.

8. **P. Alciphron**, Rott. Naturf. vi. p. 11 (1775). — *Hipponoë*, Esp. 78, 6 (1782); O. I. 2, 76.—*Hiere*, Fab. Mant. 80 (1787); Godt. ii. 23, 3, 4.—*Lampetie*, Hüb. 356-7.

Expands 1·16 to 1·25 in. The male has all the wings brownish violet, streaked with copper towards the bases, especially on the hind wings. The fore wings have two black discoidal spots; between the outer one of these and the hind margin is a row of five or six black spots. The hind wings have an elongated discoidal spot. The female has the wings brown, without any violet colour; they are spotted as in the male, and the hind wings have an orange band running parallel to the hind margin. Under side :—Grey, inclining to orange, especially on the fore wings, and more so in the male than in the female; with numerous black spots, and on the hind wings an orange hind-marginal border. In this species, as well as in several others of the genus, the shape of the wings differs slightly in the two sexes; the fore wings of the male have their hind margins nearly straight, whilst in the female they are more or less convex. Pl. XXI., 1.

TIME OF APPEARANCE.—July.

HABITAT. — Woods in the East of Europe, Asia Minor, Armenia, Persia, and the Altai. It is found more sparingly in Central Europe and Russia.

LARVA.—" Dull green, darker on the back and sides; stigmata blackish; head brownish (V. G.); on *Rumex acetosa* " (Kirby).

TIMES OF APPEARANCE.—April and May.

9. **P. Gordius**, Sulz. Gesch. Ins. t. 18, f. **7, 8** (1776); Esp.
i. 1, t. 30, f. 3 *a*, *b* (1778); Hüb. Eur. Schmet. 1,
f. 343-5; Godt. Enc. Méth. ix. p. 665 (1823).—
Alciphron, var. *a. Gordius*, Staudinger, Cat. p. 8.

Expands 1.16 to 1·40 in. Very close to *Alciphron*, but the
wings are copper in both sexes. The male has a tinge of violet,
which is slight when compared with the deep violet colour of the
last species. The female is light copper, without any violet tinge.
Both sexes have well-marked black or dark violet spots on all the
wings. The under side is very similar to that of *P. Alciphron*, but
somewhat warmer in colour, and the black spots are larger.
Pl. XXI., 2.

TIMES OF APPEARANCE.—June to August.

HABITAT. — Valleys in Switzerland and the Tyrol, South-
Western Europe, and Sicily.

LARVA and PUPA.—Unknown.

10. **P. Dorilis**, Hufn. Berl. M. ii. 68 (1766).—*Circe*, Schiff. S. V.
p. 181 (1766); Hüb. 334-6.—*Xanthe*, F. Mant. 81
(1787).—*Phocas*, Esp. 35, 1, 2.

Expands 0·90 to 1·08 in. The male has all the wings very
dark brown, with black spots arranged as in the last species,
except that the inner-marginal spot on the fore wings is wanting.
All the wings have a marginal row of faint orange spots generally
present. The female is larger than the male. The fore wings are
copper-coloured, with black spots arranged as in the last species,
except that there are two rows of spots running parallel to the hind
margin. The hind wings, which, though slightly emarginate
towards the anal angle in both sexes, are more so in the female,
are dark brown, with a hind-marginal band of the colour of the
fore wings, and enclosing a row of black spots. The under side is
grey, tinged with yellow in the male, and with orange in the
female; all the wings are spotted with black, and have orange
hind-marginal bands. Pl. XXI., 3.

TIMES OF APPEARANCE.—The greater part of the year, from
early spring to late autumn.

HABITAT.—Meadows, wood-sides, commons, &c., throughout Central and Southern Europe. (It is absent from Great Britain, Andalusia, and Sicily.) It also occurs in Asia Minor, Armenia, and the Altai.

The LARVA, which is light green, with lighter spots and reddish brown short hairs or bristles, feeds on *Rumex acetosa*. There are three or four broods in the year.

VARIETIES.

a. **Subalpina,** Spr. Stett. E. Zeit. 1851, p. 339. The male is without any orange bands above, and the under side of both sexes is entirely without any red or orange tinge.

HABITAT.—Switzerland.

b. **Hypoxanthe,** Kirby, Man. Eur. Butt. p. 91. Mr. Kirby describes under this name two specimens from the Polish Ukraine. They seem to differ from typical specimens of *Dorilis* in being somewhat larger and more strongly marked, and with a much more decided orange tinge beneath. Though these specimens do not appear to be specifically distinct from *P. Dorilis*, I think the name *Hypoxanthe* may justly stand as that of a variety, and not merely as a synonym.

11. P. Phlæas, Linn. F. S. 285; Esp. 22, 1; Hüb. 362-3.

Expands 1 to 1·25 in. Fore wings shining copper, with a hind-marginal band of dark brown; there are two square black spots in the discoidal cell, and an irregular row parallel to the hind margin. The fore wings of the male are darker than those of the female, and the black spots are larger. Hind wings rather strongly emarginate near the anal angle, dark brown, with a faintly black linear discoidal spot; parallel to the hind margin is a broad copper band, spotted with black, and above this there often appears the traces of a row of blue spots, generally four in number. Under side:—Fore wings bright orange; hind margins brownish grey; the wings are spotted with black, as above. Hind wings brownish grey, with a few indistinct darker spots, and a faint reddish orange hind-marginal band. Pl. XXI., 4.

TIME OF APPEARANCE.—The butterfly is on the wing through-out the greater part of the year, there being several broods.

HABITAT.—The whole of Europe, the greater part of Asia, North America, the United States, and perhaps Canada. It inhabits fields, road-sides, heaths, &c., and is a common British butterfly.

LARVA.—"The colour of the head dingy green, with a few dark brown markings; of the body, opaque apple-green, the warts being white and the bristles sienna-brown; in some specimens the green is interrupted by three stripes of a delicate purplish pink, one of them medio-dorsal, the others marginal" (Newman).

PUPA.—Dirty white, speckled with black or dark brown.

The Larva feeds on various species of *Rumex*.

VARIETIES.

a. Schmidtii, Gerh. pl. 10, 3, 1, B.—*Phlœas*, var. Esp. 60, 5; Hüb. 636 - 7. In this form all those portions of the wing that are normally copper-coloured are brilliant shining white. It is found most commonly in the southern districts of the territory inhabited by *P. Phlœas*, but it occasionally occurs in the more northern parts; has been several times taken in Britain, but is very rare.* Pl. XXII., 1.

b. Eleus, F. Suppl. E. S. 430. — *Turcicus*, Gerh. pl. 5, 5. Larger and much darker than the type. The hind wings are more deeply emarginate near the anal angle, so as to exhibit short tails. This variety occurs in the South of Europe and Asia Minor in the summer and autumn.†

12. **P. Helle,** Wien. Verz. 181, n. 4 (1776); Hüb. 331 - 3.— *Amphidamas*, Esp. 58, 4 (1779). — *Hille*, Fab. Mant. (1787).—*Xanthe*, Lang, Verz. Schmett. (1789).

Expands 0·75 to 0·90 in. The fore wings are copper-coloured, the hind margin dark brown or black; parallel to the hind margin there is a row of evenly-placed black spots; in the female there are

---

* The specimen from which the figure was drawn was kindly lent to me by Mr. J. T. Carrington; it was taken in Perthshire.

† I have seen a dark specimen from the New Forest, which is apparently identical with *Eleus*, Fab.; it was taken last summer (1881) by Mr. Clark, of London Fields.

two such rows; in addition to these there are two or three spots in the discoidal cell and one near the inner margin. The hind wings are dark brown, with a hind-marginal band of copper; above this there are sometimes faint blue spots. All the wings in the male, but not in the female, are shot with brilliant bluish violet on their upper surface. Under side:—Fore wings bright orange, spotted with black. Hind wings brownish grey, spotted with black; hind margins with an orange border, internal to which are two rows of black spots, and between these a row of white ones. Pl. XXI., 5.

TIMES OF APPEARANCE.—May to August.

HABITAT.—Northern and Central Europe (but not the North-West of Europe). It is found in moist meadows. In Switzerland it occurs in the mountains. It also inhabits Armenia, Siberia, and the Amur.

LARVA.—Yellowish green, with a darker green dorsal line, and lateral stripes. The head is yellow. It feeds on various species of *Rumex*, and on *Polygonum bistorta*. Pl. XXVIII., 4.

Genus 5.—**LYCÆNA**, Fabr. Ill. Mag. vi. p. 285 (1807).

POLYOMMATUS, Latr. Hist. Nat. Crust. et Ins. xiv. 116.
CUPIDO, Sect. B, Schrank, Faun. Boica.
ARGUS, Boisd. Lec. Lep. Amer. Sept. 118 (1833).

Small Butterflies, with the antennæ slender, and with their clubs elongated; the wings seldom or never emarginate, the posterior pair rounded, and in a few species furnished with a slender tail-like prolongation from the hind margin. In most of the species the upper surface of the wings is blue in the male, and more or less brown in the female; yet we occasionally meet with species that are blue in both sexes, and with others that have both males and females of a brown colour.

The under sides have always a ground colour of grey, sometimes inclining to blue and sometimes to brown. In one or two species they are striped with darker colour, but in most cases they are spotted with black. The pattern of the under side when complete may be said to exhibit the following characters:—1. On each wing is a black discoidal spot, generally elongated. 2. A central row of spots, sometimes irregular, sometimes quite even, and parallel to the hind margin. 3. A double hind-marginal row of black spots, often enclosing an orange band. 4. One or more spots placed between the base of the wing and the discoidal spot. In a few species the hind-marginal spots of the hind wings are studded with metallic silvery points. Some species exhibit all these markings, as for instance, the "common blue" (*L. Icarus*), the "chalk blue" (*L. Corydon*), &c. In others the orange bands are wanting between the hind-marginal rows, as in *L. Arion*. In many there is no hind-marginal row, as in *L. Argiolus, Semiargus*, &c. The basal spots are often wanting, as in *L. Medon, Argiolus Minima*, &c.

The black spots are generally surrounded by light or white rings; sometimes they are wholly wanting, and their places supplied by white spots or patches, as in *L. Artaxerxes, Pheretes*, &c.

Variations very frequently occur. That most commonly noticed is where the normally brown female of a species shows nearly as much blue on the upper surface as the male. This is often seen in the female of *L. Icarus*, and is very marked in the variety *Ceronus* of *L. Adonis*. On the under sides the spots are very liable to enlargement or coalescence, or to be elongated into dashes or streaks, often producing a very remarkable appearance. Sometimes the normally present basal or discoidal spots are "conspicuous by their absence." Almost every extensive series of blues in a collection of butterflies will show a variety of these differences.

Locality also has a great influence on the species, Boreal and Alpine forms especially differing from their types, both in size, colour, and times of appearance.

Hermaphrodism is occasionally seen in species of this genus, and when it occurs is very marked, on account of the dissimilarity of the sexes.

The size of the European species varies from rather more than an inch and a half in the largest to less than half an inch in the smallest.

The genus *Lycæna* is a very large one, containing more than 320 species, and, like the family LYCÆNIDÆ, is represented in every zoological region of the world; it is universally distributed throughout Europe, which contains forty-eight species at least.

The Larvæ have the onisciform shape common to the LYCÆNIDÆ; they feed on a variety of low plants, generally preferring *Leguminosæ* and *Papilionaceæ*.

1. **L. Bœtica**, Linn. Syst. Nat. xii. 789 ; Esp. 27, 3 a, b, 91, 3 ; Hüb. 373-5,

Expands 1·10 to 1·33 in. The male has all the wings violet-blue ; in the female they are brown and blue only towards the base. The fore wings have the hind margins narrowly brownish black. The hind wings have faint traces of a hind-marginal orange band, and near the anal angle are one or two black spots; attached to the hind margin is a slender tail. Under side :—Brownish grey, with numerous white streaks, and white bands running parallel to the hind margin. At the anal angle of the hind wings are two light orange spots, and below these two black ones, spotted at their lower part with metallic silvery green. Pl. XXII., 2.

TIMES OF APPEARANCE.—August, September, and October.

HABITAT. — The South of Europe (excepting the Eastern portion), Central and South-West France. It is rare in Switzerland, Germany, and Belgium ; and is very occasionally taken in England on the South Coast, the first recorded capture being at Brighton in 1859 ; since that time it has been taken several times in this country. It also occurs in Western Asia and Persia, and in the more temperate parts of Africa (North and South), being found plentifully at the Cape.

LARVA.—Green or olive, or sometimes reddish brown, with a dark dorsal stripe. The spiracles are yellow, and below there is a white lateral stripe ; above the spiracles on each segment is an oblique line, paler than the ground colour.

PUPA.—Reddish yellow, dotted with brown, and with black spiracles. Pl. XXVIII., 5.

The Larva feeds in the pods of the common pea, also on *Colutea arborescens*, and on various leguminous plants, devouring the seeds. The eggs are laid in the autumn on the twigs of the plants, the newly-emerged Larva entering the young pods in the following summer; when it is fully grown it undergoes its pupation on the stems or in the leaves.

2. L. Telicanus, Lang, Verz. p. 47 (1789); Hüb. 371-2, 553-4; Frr. 86.

Expands 1·0 to 1·10 in. The hind wings have a short tail. The male has all the wings blue, with a tinge of violet. The fore wings are unspotted; the hind wings have two small black spots near the anal angle. The female has all the wings brown, their bases being shot with brilliant lilac-blue; the fore wings have two rows of rather large but indistinct dark spots converging towards the inner margin. The under side somewhat resembles that of the last species, being light brown, with numerous white streaks and bands; these, however, are very much more waved and indistinct than in *L. Bætica.* Along the hind margin of the hind wings, near to the anal angle, there are placed three black spots in rings of brilliant silvery green, surmounted by faint traces of orange. Pl. XXII., 3.

TIMES OF APPEARANCE.— July and August (in the North). Throughout the summer in the South.

HABITAT.—The South of Europe, North Africa, and Western Asia; in short, those countries that approach the shores of the Mediterranean. Its northern range is chiefly confined to Germany, in which country it is found sparingly in gardens and meadows.

LARVA.—Something like that of *Bætica,* but has the ground colour reddish purple and the lateral streaks darker. It feeds on the flowers of the purple loose-strife (*Lythrum salicaria*). Pl. XXIX., 1.

TIMES OF APPEARANCE.—August and September.

3. L. Balcanica, Frr. 421, 1, 2 (mens. Feb. 1844);— *Psittacus,* H.-S. 220-3 (*Fin.* 1844).

Expands 0·60 to 0·80 in. Hind wings with a short tail. Male violet-blue; fore wings with three or four oblong rectangular dark spots; hind wings with two small black spots near the anal angle. Female brown, shot with violet-blue at the base of the wings; fore wings with two parallel rows of dark spots; hind wings with faint dark markings, and some whitish lines or streaks along the hind margin. Under side :—White, slightly shaded with grey. Parallel to the hind margins of all the wings is a row of well-defined black spots, those on the hind wings near the anal angle being spotted with silver. Internal to this is a narrow dark band continued for the entire breadth of fore and hind wings. Inside this band the basal areas of the wings are more shaded with grey, and marked with numerous well-defined dark brown or black spots. Pl. XXII., 4.

TIMES OF APPEARANCE.—July and August.

HABITAT.—Turkey, Western Asia, and Persia. It is very closely allied to the African *L. Theophrastus,* F.

4. L. **Argiades,** Pall. Reis. i. p. 472 (1771).—*Tiresias,* Rott. Naturf. vi. 23 (1775); Esp. 34, 1, 2.—*Amyntas,* S. V. (1776); Hüb. 322-4.

Expands 0·80 to 1·12. in. Hind wings with a short tail. The male has all the wings violet-blue, with a narrow brown hind-marginal border; the fore wings are without spots; the hind wings have two or three small brown spots along the hind margin, near the anal angle. The female is brown, slightly tinged with violet-blue at the bases; the hind wings, besides having two or three black spots near the anal angle, show faint traces of an orange band. The fringes in both sexes are white. Under side:—Greyish white, tinged with blue at the base; all the wings have a faint hind-marginal orange band. The fore wings have a narrow linear discoidal spot, and a row of seven black spots parallel to the hind margin. The hind wings have a row of black spots enclosed by the orange band, and an irregular row running across the central area of the wings, besides two placed near the base. Pl. XXII., 5.

TIMES OF APPEARANCE.—From May to the end of August.

HABITAT.—Central and Southern Europe (except Britain and Spain), North-Western Asia, the South of Siberia, and the Amur. It frequents meadows, and is generally a common insect. There are two or three broods in the year; the individuals of the spring brood are smaller than those which appear later in the season, and to these the name *Polysperchon*, Berg., is generally given.—Expands 0·70 to 0·80 in. Besides this seasonal variety, there is an aberration found at the same time and in the same localities as the typical form: this is the

VARIETY.

Coretas, Och. i. 2, 60. Differs from the type in the absence of the orange spots on the under side.

LARVA.—Pale green, with a darker dorsal stripe, dark lateral streaks, and light brown and white spots.

TIMES OF APPEARANCE.—June, and again in autumn and early spring, after hybernation. It feeds on trefoil and other *Leguminosæ.*

5. L. Fischeri, Ev. Bull. Mosc. (1843) iii. 537; II.-S. 218-9; Frr. 440, 2.

Expands 0.90 to 1 in. Hind wings with very minute tails. All the wings are dark brown; the fore wings have a small dark discoidal spot; on the hind wings is a row of faint light spots placed parallel to the hind margin. The fringes of all the wings are white. Under side :—The ground colour is light grey. The fore wings have a double row of narrow black spots placed parallel to the hind margin; internal to this is an irregular row of larger spots, that near the anal angle being the largest, and besides these there is a black discoidal spot surrounded like the others by a light ring. There are no basal spots. The hind wings have a double row of hind-marginal spots enclosing a faint orange band, the black spots near the anal angle being spotted with silvery green; between this row and the narrow linear discoidal spot is an irregular row of black spots in light rings; besides these, there are basal spots four in number. Pl. XXIII., 6.

TIME OF APPEARANCE.—July.

HABITAT. — The Steppes of South-Eastern Russia; dry meadows in the Ural Mountains and the Altai. Like most other exclusively Russian species, rare and difficult to obtain.

LARVA.—Unknown.

6. L. Trochilus, Frr. 440; H.-S. 224-6.

Expands 0·60 to 0·80 in. All the wings are brown, without any tinge of blue. The fore wings are unspotted; the hind wings have a hind-marginal row of black spots; three or four of these that are nearest the anal angle are surrounded by conspicuous light spots of an orange colour, producing a short band. Under side :—Light grey. The fore wings have a double row of darker spots surrounded by light rings parallel to the hind margin; discoidal spot black, with a light ring, and between this and the hind margin is a row of six similar spots. The hind wings have four small basal and one costal spot all black, with light rings. The central area of the wing is marked with grey and white spots, and beginning at the anal angle and extending for two-thirds of the hind margin is a bright orange band, enclosing four deep black round spots, surrounded by rings of shining greenish silver. The fringes of all the wings are white. Pl. XXII., 7.

TIMES OF APPEARANCE.—July and August.

HABITAT.—Turkey, Greece, and Dalmatia; also Asia Minor, Syria, and the North of Persia. It also occurs in South Africa.

LARVA.—Unknown.

7. L. Ægon, Schiff. S. V. (1776); Hüb. 313-5; O. I. 2, 57; Frr. 175.—*Argryrotoxus*, Bgstr. Nom. ii. p. 77 (1779).

Expands 0·90 to 1·10 in. The male has all the wings deep blue, rather inclining to violet, with a narrow brownish black hind-marginal border. The hind wings have some faint brown spots along the hind margin. The fringes of all the wings are *broadly* white. The female is brown, with faint traces of an orange band along the hind margin of the fore wings; the hind wings have a more distinct orange hind-marginal band, most conspicuous towards the anal angle, and enclosing four or five black spots. Under side :—Grey; in the female with a tinge of brown; the hind margins of all the wings with a double row of black spots, enclosing an orange band; internal to these an irregular central row of black spots, surrounded by white rings. The fore wings have a round discoidal spot similarly surrounded by white, and no basal spots ; the hind wings have four spots between the central row and the base, two of them nearly touching the costa. The bases of all the wings are strongly tinged with blue in the male, more faintly so in the female. The outermost spots of lower hind-marginal row are minutely studded with metallic silvery blue. The anterior tibiæ are furnished with short spines.   Pl. XXIII., 1.

TIMES OF APPEARANCE.—May to August.

HABITAT. — The whole of Europe (excepting the Eastern portion), Asia Minor, Armenia, Persia, and Eastern Siberia.   In England it is generally distributed, but local, frequenting heaths, open places in woods, &c.   In Scotland and Ireland it seems to be rare.

LARVA.—" Bright yellowish green, with the dorsal stripe blackish brown, edged with whitish from the beginning of the third to the end of the tenth segment. . . . . The subdorsal stripe is visible from the beginning of the third to the end of the eleventh segment as a greenish yellow line running between two green ones, darker than the ground colour.   At the bottom of the sides, along the lateral ridge, commencing on the third segment and continued round the anal extremity, is a whitish line.   Between the dorsal and subdorsal stripes, on all the segments from the third to the tenth, both inclusive, are faintly paler oblique lines of yellow-green, *viz.*, one on each segment sloping downwards and backwards. The warts on the twelfth segment are very often suddenly projected considerably, and then a circle of fine short hairs is visible on their extremities.   The surface of the body is also clothed with similar hairs. . . . . The chrysalis is about five lines long, smooth but without polish, the tip of the head slightly projecting, the thorax rounded, the body plump, curving on the back outwards and backwards towards the tip, which is hidden in the caterpillar-skin ; the wing-cases are prominent, and long in proportion ; it is of a dull green tint, with a dark brown dorsal line of arrow-head marks." —Buckler, Entom. Month. Mag., p. 241, March, 1869 (also quoted more fully by Newman, Brit. Butt., p. 119).   The Larva feeds on vetches, trefoil, and other *Leguminosæ.*

Bella, H.-S. 227-8, i. 127. Has the wings of a lighter blue than the type; the hind wings have a more conspicuous row of hind-marginal black spots.

HABITAT.—Turkey (?), Asia Minor, Armenia, and Persia.

L. Argus, Linn. Syst. Nat. x. 483; Esp. 20, 3; Hüb. 316-8; Och. i. 2, 52; Frr. 169.

Expands 1 to 1·10 in. The male is dark blue, nearly the same colour as *L. Ægon,* which it altogether greatly resembles; the hind-marginal border, however, is narrower and more defined, the hind-marginal spots of the hind wings are more distinct, and the white marginal fringes are very narrow—not broad, as in *Ægon.* The female on the upper surface almost entirely resembles the female of *L. Ægon,* but the orange hind-marginal spots are rather more distinct. Under side :— Ground colour uniform brownish grey in both sexes; the central row of spots on the fore wings is more even, the last spot but one from the inner margin being more in a line with the rest, and not, as it were, thrust inwards, as in *L. Ægon;* the silvery spots on the hind margin are much more distinct than in *Ægon.* The anterior tibiæ are not provided with spines. Pl. XXIII., 2.

TIMES OF APPEARANCE.—June and July.

HABITAT.—The greater part of Europe (not, however, in Great Britain); Asia Minor, and Armenia. It is a more local insect than the last, though it is found in the same kind of localities.

LARVA.—According to Guenée, dark green, with a red dorsal line, and an oblique reddish streak on each segment, bordered with whitish. Feeds on *Genista, Melilotus,* &c. Pl. XXIX.

a. **Argyronomon** (female ab.), Berg. Nom. ii. p. 76. This is merely an aberration of the female in which the wings are suffused with blue.

P

b. **Ægidion,** Meissner, Natur. Anz. Allg. Schw. Ges. (1818). This is a Boreal and Alpine form, which differs from the type in being smaller, and having the wings deep violet in the male.

HABITAT.—Alpine valleys in Switzerland; also Scandinavia and Lapland.

c. **Hypochiona,** Rbr. Cat. S. Andal. p. 35. Larger than the type; the ground colour of the under side is very light, being nearly white.

HABITAT.—Andalusia and Greece. (? Asia Minor, *Staudinger*).

9. **L. Optilete,** Knoch. Btr. i. p. 76, T. V. 5, 6 (1781); Esper. 79, 4, 5 (1782); Fab. Mant. 74; Hüb. 310-2; Frr. 451, 2, 3, 656.

Expands 0·90 to 1·13 in. The male has all the wings of rich dark purplish blue, unspotted, with a narrow brown border along the entire length of the hind margins. The female is brown, generally dark blue or purple towards the base; the hind wings have two or three orange spots towards the anal angle. Under side:— Brownish grey; fore wings with a crescentic discoidal spot, a central row of six, and a double hind-marginal row without any orange; the basal spots are absent. Hind wings slightly blue at the base, with three basal spots, an elongated discoidal, and an irregular central row; the hind margin has a double row of black spots, three or four pairs enclosing each a spot of bright orange, two or three of the external row nearest the anal angle having a spot of shiny blue or silver. Pl. XXIII., 3.

TIMES OF APPEARANCE.—June and July.

HABITAT.—Europe and Siberia, inhabiting peaty or boggy places and mountain-sides. Its range in Europe is limited to Germany, Scandinavia, Russia, and the Alps of Switzerland. It does not occur in Western or North-Western Europe.

LARVA.—According to Freyer, pale green, with short reddish hairs, the spiracles marked with white, and the head dark brown or blackish. Its food-plant is *Vaccinium oxyococcus*, on which it feeds in September and October, and again in the spring after hybernation.

a. **Cyparissus**, Hüb. 654-7. This variety is smaller than the type, and paler in colour beneath. It is found only in Lapland.

10. L. **Zephyrus**, Friv. Imr. ii. (1835); H.-S. 208-11.

Expands 0·90 to 1 in. The male has all the wings violet-blue; the hind margins have a narrow but distinct black border; on the hind margin of the hind wings, near the anal angle, are one or two very small black spots. The fringes of all the wings are white. The female has all the wings brown, with traces of an orange band on the hind margin, most distinct near the anal angle. Under side:—The ground is of a greyish fawn-colour. All the wings have a well-marked double row of black spots parallel to the hind margin, enclosing a distinct orange band; they have also a central row of black spots somewhat irregularly placed, and surrounded by light rings; internal to this, on all the wings, is the discoidal spot; the fore wings have no basal spot, but on the hind wings there are four; and the base itself is slightly tinged with blue. Pl. XXIII., 4.

TIMES OF APPEARANCE.—June and July.

HABITAT.—Mountains in Greece, Turkey, Asia Minor, and Armenia.

LARVA.—Unknown.

a. **Hesperica**, Rambur, Faun. And. p. 270, pl. x. 1-4; H.-S. 4-15 (male), 349-50 (female). Differs from the type in the colour of the male, which is light greenish blue, instead of being nearly violet. It is found in Andalusia. Pl. XXIII., 5.

11. L. **Pylaon**, Fisch. Nouv. Mém. Mosc. ii. 357; H.-S. 333-4, 339-42.

Expands 0·90 to 1 in. The male has all the wings light purplish blue, with a narrow black hind-marginal border. Near the anal angle of the hind wings there are two or three black dots or spots sometimes marked with orange. The female is dark brown, with hind-marginal orange spots, most distinct near the anal angles. Under side:—Light grey; all the wings with a marginal row of large orange spots enclosed in a double row of black ones; discoidal spots conspicuous; fore wings with no basal spots, but with a central row of eyes; hind wings with a central row of eyes and three basal spots. The fringes are white. Pl. XXIII., 6.

TIME OF APPEARANCE.—May.

HABITAT.—The Steppes of South Russia.

12. **L. Bavius**, Evers. Nouv. Mém. Mosc. (1832) 349, t. 19, 3, 4; H.-S. 10, 11, 357-60.

Expands 1·16 to 1·36 in. The male has all the wings bluish purple, with hind-marginal black borders; near the anal angle of the hind wings are three or four black spots surrounded by orange. The female is dark brown, with a hind-marginal row of orange spots on the hind wings. Under-side:—Fore wings with the spots much larger than the hind wings; there is a hind-marginal row of black spots, and the central row is much curved; the discoidal spot is well marked, and there are one or two basal spots. The hind wings have a hind-marginal row of orange spots enclosed by a double row of black ones; central row wavy; discoidal and four basal spots well marked. Pl. XXIII., 7 (from H.-S.)

TIME OF APPEARANCE.— ————?

HABITAT.—South Russia, Asia Minor, and Syria; a rare and not well-known species.

LARVA.—Unknown.

13. **L. Orion**, Pallas, Reis. t. i. p. 471 (1771).—*Telephii*, Esp. 41, 2 (1779).—*Battus*, Hüb. 328-30, 801-2 (1793).— *Bathus*, Fab. Mant. 76 (1787).

Expands 0·85 to 1·12 in. The male has all the wings dark brown, covered with purple scales, excepting along a broad band parallel to the hind margins. Fore wings with a black discoidal spot. Hind wings with a hind-marginal row of black spots in light blue rings. Fringes of all the wings spotted with black and white. The female resembles the male, but has no purple scales. Under side :—The ground colour is light grey, nearly white. The fore wings have a row of conspicuous black spots on the marginal fringe ; internal to this are two parallel rows of black spots, those composing the inner row being much larger than the external ones ; the spots of the central row are large and black, placed somewhat irregularly ; internal to this are a large discoidal and two basal spots. Hind wings with a ciliary row of black spots, as in the fore wings ; internal to this are two parallel rows of large black spots enclosing a light orange band ; the central row is conspicuous, and internal to it are a discoidal and six basal spots. Pl. XXIV., 1.

TIMES OF APPEARANCE.—May to the end of July.

HABITAT.—Central and Southern Europe (excepting Britain, and the South of Spain and Portugal), Asia Minor, Armenia, the South of Siberia, and the Amur. It frequents rocky or stony places.

The LARVA is bluish green, with a violet dorsal line ; it feeds on *Sedum telephium*, and perhaps other species of *Sedum*.

TIME OF APPEARANCE.—July.

14. **L. Baton,** Berg. Nom. t. 60, 6-8, ii. p. 18 (1779).—*Amphion,* Esp. 53, 1 (1780).—*Hylus,* Fab. Mant. 75 (1787).— *Hylas,* S. V. p. 185 ; Hüb. 325-7 (sed *Hylas* Linn. Syst. Nat. x. alius erat *Papilio*).

Expands 0·75 to 0·85 in. The male has all the wings pale lilac-blue ; the marginal fringes are black, spotted with white ; both pairs have a discoidal spot, which is black and elongated. The hind wings have a row of black dots along the hind margin. The female is pale brown, having only the basal halves of the wings blue ; the hind margins show very faint traces of an orange band. Under side :—Pale grey, the bases being faintly blue ; the fore wings have a single hind-marginal row of black spots, the central row is wavy, and there is one basal spot internal to the discoidal ; the hind wings have a double hind-marginal row of black spots, enclosing a light orange band ; internal to this a central row, an elongated discoidal, and three basal spots.

Times of Appearance.—May to August. Pl. XXIV., 2.

Habitat.—Central and South-Eastern Europe (not in Britain), and Western Asia, being found in dry sandy or chalky places.

VARIETY.

a. **Panoptes**, Hüb. 6, 670-3; Mill. Ic. pl. 85, 1. Rather smaller than the type, the wings being of a duller blue, and the ocellated spots of the under side smaller and closer. It is found in Spain and the South of France.

Larva.—Millière thus describes the larva of *Panoptes:*— "It is green, inclining to olive; the head is black and retractile. There is a broad dorsal stripe of pink, bordered on each side by a narrow stripe of light yellow. In a line with the stigmata is a brilliant white stripe, the stigmata themselves being whitish. The ventral surface is green, but duller than the back and sides; the legs (*pattes écailleuses*) are brown; the claspers (*pattes membraneuses*) are green. It feeds on *Thymus vulgaris.* The egg is hatched when the thyme is in full flower, and its growth is rapid, the pupa being formed by the first week in June. The pupa is ovoid, short, and slightly tapering at the extremities; its surface is smooth, clay-coloured, the wing-cases being tinged with green." Millière describes it as occurring at Hyères, Cannes, and other places on the Mediterranean.

15. L. **Panope**, Eversmann, Bull. Mosc. 1851, ii. 619.

I regret my inability to give a figure of this species, which is described by Eversmann as inhabiting the Steppes of the Lower Ural. It appears to be a great rarity. I have never seen a specimen, and am not aware that any figure exists. The following is the original Latin description, and a translation from Eversmann's paper in the Moscow Bulletin referred to above:—

"L. Alis supra fuscis, disco cæruleo, ciliis albo nigroque variis; subtus fusco canis; lunula discoidali nigra, serie flexuosa externa sericoque submarginali duplici punctorum nigrorum; posticis præterea punctis basalibus tribus."

"The wings are blackish brown above, with the base and disc violet-blue, slightly silvery, merging insensibly into the black margin. Fringe black and white. Beneath, the wings are ashy grey, slightly brownish; they have a black discoidal lunule, a waved central row of black spots, and a double submarginal row. The fore wings have no basal spots; there are three or four basal spots on the hind wings; these spots, except those in the sub-marginal bands, are large and surrounded by white circles. The female differs little from the male, only the wings are duller beneath, being smoky grey."

16. L. Lysimon, Hüb. 534-5; H.-S. 28, 29.

Expands 0·67 to 0·87 in., being probably smaller than any other European butterfly. The male has all the wings brown, with bluish violet; the fore wings have a small black discoidal lunule and a row of indistinct spots. The female is brown, slightly blue at the base. Under side:—Light grey; fore wings with black discoidal and basal spots, a double hind-marginal and a central row. Hind wings similarly marked, but with three basal spots; and the bases of the wings slightly tinged with blue. Fringes of all the wings brown. Pl. XXIV, 3.

Time of Appearance.—July.

Habitat. — The South of Europe, Western Asia, China, Africa, and a part of Australasia. Thus it is one of the most widely distributed of butterflies. In Europe it is confined to the South of France and Spain, principally the southern part of the latter, such as Andalusia and Granada, frequenting meadows.

Larva.—Unknown.

17. L. Rhymnus, Eversmann, Nouv. Mém. Mosc. 1832, 350, t. 19, 1, 2; H.-S. 22, 23.

Expands 0·75 to 1 in. All the wings are brown, the bases greenish blue. Under side:—Dark brown; all the wings with a faint hind-marginal orange band; internal to this a row of white spots, and a waved central row of similar spots; the fore wings have a white discoidal and are without basal spots; the hind wings have, besides the white discoidal, two or three white basal spots or streaks. Pl. XXIV., 4 (from H.-S.)

Times of Appearance.—May and June.

Habitat.—South Russia and the Altai. A rare species, frequenting mountains.

Larva.—Unknown.

18. L. Psylorita, Frr. 469, 3, 4, v. p. 146; H.-S. 328-31.

Expands 0·75 to 1 in. All the wings are pale brown, with an indistinct orange hind-marginal band, the discoidal spots absent. Under side :—Very light grey, nearly white, with marginal and central rows of small black spots, which with the discoidal and basal spots are very contracted and indistinct; between the hind-marginal rows is a very faint yellowish orange band. Pl. XXIV., 5 (from H.-S.)

Time of Appearance.—June.

Habitat.—Mountains in Candia. This remarkable insular species has as yet only been found in the above island, receiving its name from the Psiloriti Mountains (the ancient range of Mount Ida), on which it occurs at a considerable elevation. It seems to me possible that it is a local variety of some other form, perhaps of *L. Astrarche.*

Larva.—Unknown.

19. L. Pheretes, Hüb. Text. p. 45; O. I. 2, 25.—*Atys*, Hüb. 495-6, 548-9; Esp. 118, 4, 5 (*Atys*, Cramer, 1782 alius erat *Papilio*).

Expands 1·12 in. Fringes of all the wings white. The male has the wings deep rich violet-blue, with a narrow black border; discoidal spots absent. The female has all the wings uniformly brown in colour, without any markings. The under side is grey, slightly tinged with greenish blue at the base in both sexes. The fore wings have a central row of black spots, and an elongated discoidal surrounded with white. The hind wings have two rows of large white spots without ocelli. Pl. XXIV., 6.

Times of Appearance.—June and July.

Habitat.—Mountain-pastures at a great elevation in Switzerland and the Pyrenees.

It also occurs in Norway, Sweden, and Lapland. It is local, but, where it occurs, abundant. Besides being an European species, it is also found in South Siberia. Pl. XXIV., 6.

Larva.—Unknown.

20. **L. Orbitulus**, Prun. Lep. Ped. p. 75 (1798); Esp. 112, 4 (1800); Hüb. 840 ab.; Frr. 421, 3, 4.—*Meleager*, Hüb. 522-3.

Expands 0·87 to 1 in. The fringes of all the wings are white. The male has the wings pale silvery grey, mixed with brown; hind margins brown; all the wings have a small black discoidal spot; the hind pair have a row of faint brown spots along the hind margin. The female is uniform brown, with a black discoidal spot on the fore wings. Under side pale grey; fore wings with large black spots surrounded by white rings, arranged in a central row, discoidal and basal, and besides these a faint double hind marginal row. Hind wings with a triangular white discoidal spot, one ocellated basal, and a central row of four or five, sometimes white and sometimes ocellated; there are faint traces of a hind-marginal row and orange band. Pl. XXIV., 7.

Times of Appearance.—June and July.

Habitat.—The higher Alps of Switzerland and the Tyrol; abundant, but local.

Larva and Pupa.—Unknown.

VARIETIES.

a. **Pyrenaica**, Boisd. Gen. p. 11; H.-S. 483-5. Larger than the type, and with white spots on the under side.

Habitat.—Pyrenees.

b. **Aquilo**, Boisd. Ic. 12, 7, 8; H.-S. 24-5, 343-4; Dup. i. 47, 6, 7.—*Franklinii*, Curtis, Descr. App. Nar. p. 68. Smaller than the type. Both sexes greyish, lighter towards the hind margins; discoidal spots surrounded by light rings.

Habitat.—The Polar Regions. This insect was taken by Capt. Feilden at lat. 81° 45′. I do not know whether it has ever

been taken in Europe, though it is said to occur in the North of
Lapland. Most of the specimens I have seen are from Labrador.
As it also occurs in Siberia it is probably circum-polar; I have
therefore figured it on Pl. XXIV., 8.

c. **Dardanus,** Frr. 419, 2, 3; H.-S. 240-3. A small and light-
coloured variety found in Alpine regions of Asia Minor and Armenia,
also in Andalusia (Sierra Nevada).

21. **L. Astrarche,** Bgstr. Nom. iii. p. 4, t. 49, 7, 8.—*Medon*, Esp.
　　　32, t. 55, 7.—*Agestis,* Hüb. 303-5; Frr. 235, 1.—*Idas,*
　　　Lewin, Ins. p. 82, t. 39, 1, 2 (1795).

Expands 1 to 1·20 in. Marginal fringes white. The wings
are brown in both sexes, with a hind marginal band of bright
orange spots. Fore wings with a black discoidal spot. The orange
bands are broader and more conspicuous in the female than in the
male; there is no tinge of blue at the base. Under side brownish
grey, untinged with blue. All the wings have the usual hind
marginal row of spots and orange band, and one irregular central
row of black spots; basal spot absent from fore wings; hind
wings with three basal spots; all these spots are conspicuously
surrounded by white rings. Pl. XXIV., 9.

TIMES OF APPEARANCE.— April to October, there being a
succession of broods in the year. The individuals of the spring
brood are larger and lighter than those of the later ones.

HABITAT.—Dry sunny meadows and hill-sides throughout
Europe (excepting the Polar Regions); Northern and Western
Asia, as far as the Himalayas, and in North Africa. As a British
insect it is commonest in the South of England, where it is double-
brooded.

LARVA.—Pale green, with a brownish purple medio-dorsal
stripe and faint pale lateral stripes; each segment has two small
wart-like eminences with projecting white bristles. The ventral
surface is pale green, with whitish bristles. The claspers are
semi-transparent and pale yellow in colour; the legs are spotted
with black. The larva when full grown is about half an inch in

length, and has the usual *Lycæna* shape. Its food-plant is the stork-bill (*Erodium cicutarium*).

The PUPA has the usual *Lycæna* form, pale yellow in colour, with a green tinge, with a dorsal stripe of reddish purple. It is spun up among the dry leaves of *Erodium* and *Artemisia*.

VARIETIES.

a. **Allous**, Hüb. 990. This variety has the wings entirely dark brown above in both sexes, without any trace of an orange band. It is found as a varietal form of the summer brood in South, Central and Southern Europe and North Africa.

b. **Salmacis**, Steph. This and the following form are insular varieties peculiar to the British Isles. *Salmacis*, which has only been found in the North of England, is intermediate between the type and the var. *Artaxerxes.* The male has no orange band on the fore wing, and the black spots on the under side of the wings are very small. The female has a white discoidal spot on the fore wings. It is single-brooded.

c. **Artaxerxes**, Fab. E. S. 297; Lew. Ins. pl. 39, 8, 9; Haw. Lep. Brit. p. 47. The male is often without any orange hind-marginal bands above. In the female they are generally distinct. Both sexes have a white discoidal spot on the fore wings. Under side entirely without black spots, the white ones only remaining. Pl. XXIV., 10.

HABITAT.—Scotland, as far as Aberdeenshire.

LARVA.—Pale green, with a darker dorsal line, and pink lateral stripe. It feeds on *Helianthemum vulgare.*

Artaxerxes is single-brooded, appearing at the end of June; the larva in May.

22. L. Anteros, Frr. 265, 1; iii. p. 101; H.-S. 16, 17.

Expands 0·60 to 1 in. Fringe white, narrowly spotted with brown. The male has all the wings light greyish blue, brown along the hind margins; the fore wings generally have a black discoidal spot, though often this is absent, or is very narrow; the hind wings have a hind-marginal row of black spots, and show faint traces of an orange band.

The female is brown, the hind wings having a hind-marginal orange band and a row of black spots. The under side is light brown, and the spots and bands are arranged very much the same as in *L. Astrarche*, but the fore wings have a small black basal spot. Pl. XXV., 1.

TIMES OF APPEARANCE.—May to July.

HABITAT.—Greece, Turkey, Asia Minor, and Syria.

LARVA.—Unknown.

23. **L. Eros**, O. i. 2, 42; Dup. i. 12, 5, 6; B. Ic. 14, 4-6; H.-S. 212-3.—*Tithonus*, Hüb. 555-6.

Expands 1 to 1·12 in. in the typical form. Fringe of all the wings white. The male has all the wings shining light blue above, without discoidal spots, with a well-defined dark brown hind-marginal border, and on the hind margin a row of dark spots. The female is brown; all the wings with a light orange hind-marginal band and black spots; fore wings with a black discoidal spot. Under side pale grey in the male, brownish grey in the female, with the usual orange bands and rows of ocellated spots; fore wings with two basal spots; the bases are tinged with pale blue. Pl. XXV., 2.

TIMES OF APPEARANCE.—June, July, and August.

HABITAT.—Mountain pastures in Switzerland, the Pyrenees, and the Altai. Not very common.

LARVA.—Not described.

VARIETY.

a. **Eroides**, Friv. Imm. 1835; H.-S. 12, 13.—*Interos*, Frr. 386, 3, 4.—*Eceros*, Dup.—*Boisduvalii*, H.-S. 7-9. Much larger than the type, expanding 1·37 in. The wings in the male are of a much deeper and more brilliant blue, and the markings on the under side, especially the orange bands, are much more defined. Pl. XXV., 3.

HABITAT.—The plains of North-East Germany, South Russia, and Asia Minor.

24. **L. Icarus**, Rott. Naturf. vi. p. 21 (1775); Esper, 32, 4; Frr. 651, 2, 3.—*Alexis*, S. V. p. 184; Hüb. 392-4; O. i. 2, 38. —*Thetis*, Esp. 32, 2 ♀.

Expands 0·75 to 1·14 in. Fringes of all the wings white, without spots. The male has all the wings deep lilac-blue, with a narrow black border; there is no discoidal spot on the fore wings. The female is brown, with an orange hind-marginal border on all the wings, the hind wings having a row of black spots; the bases of all the wings are blue, which colour sometimes suffuses the entire wings. Under side pale grey in the male; pale brown in the female. There are the usual spots and orange bands, the fore wings having two basal spots; the hind wings have an elongated white spot towards the middle of the hind margin. The base of all the wings is strongly tinged with blue in both sexes. Pl. XXV., 4.

TIMES OF APPEARANCE.—All through the fine season, from April to October, there being several broods.

HABITAT.—Europe, Northern and Western Asia and North Africa, frequenting meadows, roadsides, pastures, &c. In Great Britain it is the commonest of the genus, and in many places one of our commonest butterflies (generally known as the "Common Blue").

LARVA.—Shaped like those of its congeners, green or olive, with the head black; there is a dorsal stripe of a darker shade than the ground colour, a lateral stripe of lightish green, and on each segment three lateral stripes inclined obliquely from before backwards. The larva feeds on various low-growing *Leguminosæ*, especially on *Ononis spinosa*.

PUPA.—Dull green, tinged with brown on the wing-cases.

VARIETY.

a. **Icarinus**, Scriba, Journ. Ent. p. 216 (1795).—*Thersites*, Boisd.; *Alexis*, var. H.-S. 46.—*Alexius*, Frr. 676, 1, 2. This variety differs from the type in having no basal spot on the under side of the hind wings. It is common on many parts of the Continent, especially in Switzerland. Occasionally it occurs in Britain. Pl. XXV., 5.

25. **L. Eumedon,** Esp. 52, 2 (1780); Hüb. 301-2; O. i. 2, 48; Frr. 235, 2, 3.—*Chiron,* Rott. Naturf. vi. p. 27 (1775), sed *Chiron,* L., alius erat papilio.

Expands 1 to 1·16 in. The male has all the wings very dark brown, the fore wings with a black discoidal spot. The female is somewhat lighter in colour, and has a row of faint orange spots on the hind margin of the hind wings. The fringes of all the wings white. Under side:—Brownish grey; the fore wings have a row of faint orange spots along the hind margin, and internal to this a straight row of round black spots in white rings. The discoidal spot is oval and black, with a white ring: there are no basal spots. The hind wings have the same arrangement of spots as the fore wing, and in addition to these there is one basal spot, black and surrounded by a white ring. The base of the wing is dusted with shining greenish blue, and from the discoidal spot to the orange hind-marginal band there runs a rather broad white line. Pl. XXV., 6.

TIMES OF APPEARANCE.—July and August.

HABITAT.—South-Eastern and Central Europe (not Britain), the southern parts of Scandinavia, mountains in Asia Minor and the Altai. It frequents moist pastures, generally at rather a considerable elevation.

LARVA.—Unknown.

26. **L. Idas,** Rambur, Faun. And. p. 266, pl. 10, 5-7.

Expands 0·87 to 1 in. All the wings are brown, with a white marginal fringe; the fore wings have a black discoidal spot. In the female there is a faint trace of an orange hind-marginal band on the hind wings. The under side much resembles that of *Astrarche,* but the orange hind-marginal spots are very pale and small, those of the fore wings being almost obscured by white; the central row of the fore wings is very irregular, the two spots nearest the costa being considerably internal to the rest. The shape of the wings is much longer and narrower than in *Astrarche.* Pl. XXVI., 1.

TIME OF APPEARANCE.—June.

LYC*ÆNA*.

LYCÆNA. **119**

HABITAT.—Mountains in Andalusia.
LARVA.—Unknown.

27. **L. Amanda** (-dus), Schn. N. Mag. iv., p. 248 (1792); Hüb.
283-5, 752-4.—*Icarius*, Esper, 99, 4; O. i. 2, 37;
B. Ic. 12, 1-3.

Expands 1 to 1·25 in. Fringes of all the wings white. The
male has the wings blue and somewhat glossy; the fore wings
have a broad indistinctly defined black border and a small oblong
black discoidal spot; the hind wings have a narrow but defined
black border. The nervures of all the wings become rather broad
and black as they enter the hind-marginal border. The female is
dark brown; fore wings with a very indistinct discoidal spot; hind
wings with three or four orange spots towards the anal angle.
Under side:—Pale grey; brownish in the female. Fore wings
without basal spots; discoidal and central row black, with white
rings; hind margin with faint traces of an orange band. Hind
wings bluish at the base, with three basal spots; discoidal and
central row as in the fore wings; the hind margins have a well-
defined double row of black spots enclosing a pale orange band.
Pl. XXVI., 2.

TIMES OF APPEARANCE.—June to August.

HABITAT.—Mountain meadows in Eastern and Central Europe
(except Britain), the South of Scandinavia, Greece, Asia Minor,
the Altai and the Amur.

LARVA.—Unknown.

28. **L Escheri**, Hüb. 799, 800 (1819); Dup. i. 11, 3-6.—*Agestor*,
Godt. Enc. Méth. p. 690.

Expands 1 to 1·30 in. Fringes of all the wings white. The
male has the wings blue, tinged with lilac, much resembling
L. Icarus in colour, but rather brighter; all the wings have a very
narrow black hind-marginal border. The female is brown, slightly
tinged with blue at the base; the fore wings have a black discoidal
spot and an orange hind-marginal band indistinctly defined on its
inner edge. The hind wings have a sharply hind-marginal band.
Under side very much as in *L. Icarus*, var. *Icarinus* (there being no

basal spots). The ground colour is, however, lighter, and the black spots are very large and defined. Pl. XXVI., 3.

TIMES OF APPEARANCE.—May to July.

HABITAT.—Mountains in the South of Europe, including Spain and Portugal, the South of France, Switzerland, North Italy, and the Balkans. A local species always found in rather elevated positions, as on the Simplon and other Swiss Passes.

LARVA.—Unknown.

29. L. **Bellargus**, Rott. Naturf. vi. p. 25 (1775); Esp. 32, 3, 55.
　　—*Adonis*, S. V., p. 184 (1776); Hüb. 290-300; Frr.
　　487; O. i. 2, 33.

Expands 1·16 to 1·50 in. Fringes of all the wings black and white. All the wings in the male are brilliant glossy blue, the hind margins have an exceedingly narrow black border, and on the hind wings there is generally a hind-marginal row of black dots. The female is brown, generally streaked with blue at the bases; the fore wings have a black discoidal spot, and a rather faintly defined orange hind-marginal band; the hind wings have the orange band distinct, and enclosing a row of black spots. Under side grey; brownish in the female. The spots arranged very much as in *Icarus*, but those on the fore wings are rather larger. The discoidal spot on the hind wings is generally white, and rarely with a black centre; there is a triangular white blotch placed between the central row and the hind margin, about mid-way between the costa and the anal angle. The bases are slightly tinged with blue. Pl. XXVI., 4.

TIMES OF APPEARANCE.—From May to middle of September, being double-brooded.

HABITAT.—Central and Southern Europe, North Africa and Western Asia. Its preference for chalky or limestone districts is well known to British entomologists.

LARVA.—Green, with a lateral yellow stripe, and a double dorsal row of fulvous spots (Freyer). It feeds on various *Leguminosæ*, such as *Lotus, Hippocrepis, Vicia*, &c.

OBS.—I regret to have to use Rottenburg's name of *Bellargus* for this species, instead of that of *Adonis*, by which it has been known in this country for so many years.* Yet as this alteration in nomenclature has been adopted not only by Staudinger, but by entomologists generally, and applies equally to other species, there is nothing left but submission.

VARIETIES.

a. **Cinnus**, Hüb. 830-1. An aberration, figured by Hübner, in which the spots on the under side of the hind wings are not ocellated.

b. **Ceronus**, Esp. 90, 2 (1784); Hüb. 297.—? *Thetis*, Rott. Naturf. vi. 24. A very beautiful variety of the female, in which the ground colour, instead of being brown, is blue, as in the male. Hübner considered as belonging to the male *Ceronus* those specimens of that sex which have a row of black dots on the hind wings. *Ceronus* seems to occur throughout Europe, but most commonly in the south. Pl. XXVI., 5.

30. **L. Corydon,** Poda. Mus. Græc. p. 77 (1761); Esp. 33-4, 79, 1; Hüb. 286-8; O. i. 2, 28; Frr. 223.—*Tiphys*, Esp. 51, 4.

Expands 1·25 to 1·50 in. Fringes of all the wings black and white. The male has the wings pale silvery blue. Fore wings with a broad black hind-marginal border; that of the hind wings narrower, but accompanied by a row of black dots. The female is brown, with a black discoidal spot on the fore wings; all the wings with the usual hind-marginal orange band, which is pale in colour. Under side:—I am unable to see any difference between this species and the last as regards the arrangement of the spots. The male, however, is much lighter than the male of *Bellargus* in the ground colour, and the spots are smaller, and more apt to be replaced by white ones. I know of no character in the design or colouring of the wings by which the female of this species may, with any certainty, be distinguished from that of the last. Pl. XXVI., 6.

* The reader will, perhaps, be reminded of Crabbe's allusion to "*Adonis* blue," in the poem of the ' Borough,' published in 1810.

TIMES OF APPEARANCE.—May to September.

HABITAT.—South and Central Europe, including Britain, and Western Asia. It is generally found in open places and on hillsides, where the soil is chalk or limestone.

LARVA dull green, with dorsal lines and oblique lateral stripes of a bright yellow colour. The head is dark brown or black. It feeds on *Papilionaceæ* in May and June.

The PUPA is pale greenish brown, and is shaped like those of the genus generally.

VARIETIES.

a. **Syngrapha**, female, Keff. Stett. Ent. Zeit., 1851, 308; ab. fœm., maris colore, B. Gen., p. 12. This is a variety of the female in which the wings are more or less entirely suffused with blue, as in the male. It thus corresponds to the variety *Ceronus* of the last species. Pl. XXVI., 7.

HABITAT.—The Pyrenees.

b. **Appenina**, L. Is., 1847, p. 148.

HABITAT.—Italy.

c. **Hispana**, H.-S. 500 - 1.—*Arragonensis*, Gerh.

HABITAT.—Northern and Central Spain.

Both these forms are local mountain varieties, and differ principally in being much paler than the type, between which and the following they form transitions.

d. **Albicans**, H.-S. 494-5; Mill. i. 159, pl. 4, 2. In this remarkable variety the wings of the male are entirely dull white, without any trace of blue. Pl. XXVI., 8.

HABITAT.—Andalusia.

31. **L. Hylas**, Esp. 45, 3 (1777).—*Dorylas*, Hüb. 289 - 91 (1793).

Fringes of all the wings white. Male bright blue, with a narrow brown hind-marginal border; along the hind margin of the hind wings is a row of not very distinct brown spots. The female, above, very closely resembles that of *L. Icarus*, but the wings are darker brown, the orange band on the fore wings less distinct, and the white marginal fringe broader. Beneath, the wings somewhat resemble those of *L. Icarus* var. *Icarinus* (the fore wings having no basal spot). It differs, however, in having the orange bands paler and less distinct, in the greater size of the black spots on the fore wings, and in having the discoidal spot of the hind wings white, without any, or with a very minute, black central dot. Pl. XXX., 1.

TIMES OF APPEARANCE.—May to August.

HABITAT. — Hill-sides and grassy places in Southern and Central Europe; also mountains in Asia Minor. Some believe it to be a British species, occurring in the South of England, and its non-occurrence in England is queried by Staudinger accordingly. Considering, however, its close resemblance to *L. Icarus* var. *Icarinus*, it seems to me very likely that specimens of that insect have been confounded with it.

LARVA.—Dark green, with a dorsal line still darker in colour, and lateral yellow streaks; the head is black. There are two broods in the year, the larva appearing in May and August, and feeding on the flowers of *Melilotus officinalis*.

VARIETY.

a. **Nivescens**, Kef. Stett. Ent. Zeit., 1851, 309. Differs from the type in having the wings dull white. This variety, as will be seen, is analogous to the white variety of *Corydon;* and like it is a Spanish form, inhabiting hilly limestone districts in Andalusia and Catalonia. Pl. XXX., 2.

32. **L. Meleager**, Esper. 45, 2 (1777); Fab. Mant. 71.— *Daphnis*, S. v. p. 182; Hüb. 280-2; O. i. 2, 26.

Expands 1·16 to 1·35 in. Hind margin of hind wings dentated, most markedly so in the female. Fringes of all the wings white. The male is blue, rather lighter than *Bellargus;* the costa of the fore wings is white along its extreme edge, and all the wings have a narrow black border internal to the hind-marginal fringe. The female bright violet-blue; the nervures black; the fore wings are marked with a broad band of brown on the costa; all the wings have a black discoidal spot surrounded by a whitish ring, and a broad well-defined hind-marginal band dark brown in colour, and marked with lighter brown spots; internal to these bands are some bluish white lunules. The under side much resembles that of *L. Corydon*, but there are no black spots on the fore wings, and the hind-marginal orange bands are very faint. Pl. XXX., 3.

TIMES OF APPEARANCE.—May to July.

HABITAT.—Southern and Eastern Germany, Switzerland, the South of France, Italy, the South-east of Europe, and Western Asia. It is a local species, frequenting flowery meadows, open places in woods, &c.

LARVA.—Unknown.

VARIETY.

a. **Stevenii**, Tr. x. 1, 66; Frr. 427, 2; H.-S. 244-5. The male has a distinct black margin, and the female is brown, without any blue colour, and rather smaller than the typical form. This variety appears in mountainous districts: it is found in the Tyrol, Switzerland, South Russia, and Greece.

33. L. Admetus, Esp. 82, 2, 3; Hüb. 307-9; O. i. 2, 50; H.-S. 488-9.

Expands 1·08 to 1·58 in. The fringes of the fore wings are brown; those of the hind wings lighter. The fore wings of this and the following species are thickly clothed with brown hair-like scales towards the base of the wings along the course of the nervures. The wings in both sexes are uniformly dark brown; the female sometimes shows traces of an orange hind-marginal band. Under side:—Pale brownish grey; fore wings with an elongated discoidal spot, a central row of black spots in white rings, and a double hind-marginal row of brown spots; there are no basal spots. Hind wings with markings similar to those of the fore wings, and in addition two black basal spots in white rings; in the female there is a slight tinge of orange between the hind-marginal rows. Pl. XXX., 4.

TIME OF APPEARANCE.—June.

HABITAT.—South-Eastern Europe and Asia Minor. A very local species.

LARVA.—Unknown.

a. **Ripartii,** Frr. B. 133, 3, iii. p. 128.—*Rippertii,* B. Sc. 16, 4-6; Dup. i. 10, 1, 2. Slightly smaller than the typical form, and with a white basal streak on the under side of the hind wings. It very much resembles the female of the next species, but can be distinguished by the presence of a faint hind-marginal double row of spots, and by the greater distinctness of the basal streak.

TIMES OF APPEARANCE.—June and July.

HABITAT.—The southern slopes of the Alps; mountains in Bulgaria, Greece, and Asia Minor.

34. **L. Dolus,** Hüb. 793-6; Frr. B. 97, 3, 4; B. Ic. 15, 6-8; Dup. i. 10, 3, 4, p. 63.—*Lefebvrei,* Godt. Enc. Méth. ix. p. 695.

Expands 1·0 to 1·25 in. The fringes of all the wings greyish white. The male has the wings very pale greenish blue, nearly white towards the hind margins, but decidedly blue at the base; all the wings have a very narrow dark brown border. The fore wings are clothed with dark brown scales, as in the last species. The female is dark brown, much resembling the var. *Ripartii* of the last species, but the basal streak beneath is less distinct. Pl. XXX., 5.

TIMES OF APPEARANCE.—June to August.

HABITAT.—The South of France and Piedmont.

LARVA.—"Green, with slightly oblique yellowish dorsal streaks, separated by more conspicuous green lines; violet at the sides, and bounded by a yellowish line" (De Villiers and Guenée). Feeds on *Onobrychis sativa* in May.

a. **Menaclas,** Frr. 223, 2, 3, iii. p. 64 (1837).—*Epidolus,* B. Gen. p. 13 (1840); H.-S. 18, 19. Differs from the typical form in having the wings of the male dull white, being only tinged with blue at the base; the brown scales of the fore wings are very conspicuous. The hind wings, beneath, have a white basal streak, faint in the male, but conspicuous in the female.

Time of Appearance.—July.

Habitat.—Turkey and Asia Minor.

Obs.—Some entomologists are of opinion that the forms *Admetus, Ripartii, Dolus, Menaclas,* and the Asiatic *Hopfferi,* are all local races or varieties of the same species. It must be admitted that the definition of a true species is a very great difficulty at the present time, when zoologists are so divided in their opinions on this subject. In the case of the group of insects now under notice I have strictly followed Staudinger's division of the different forms.

36. **L. Damon,** S. v. p. 182 (1776); Hüb. 275-7; O. i. 2, 19.— *Biton,* Esp. 33, 5, 62, 4.

Expands 1·0 to 1·50 in. Fringes of all the wings white. The male has the wings pale blue, somewhat brighter than in *Corydon.* The fore wings have a rather broad dark brown hind-marginal band, somewhat indistinctly defined on its inner side; the hind wings have a similar band, but narrower and more defined; the nervures of all the wings become broad and dark as they approach the hind margin. The wings have not the thick patch of brown scales seen in the two preceding species. The female is uniformly brown. Under side brownish grey, greenish blue at the base. Fore wings with an elongated discoidal spot, and a central row of black spots in white rings. Hind wings with two small black spots in white rings near the costa, and four similar ones forming a central row; a very sharply defined white streak runs straight across the wing, nearly from the base to the hind margin. Pl. XXX., 6.

Times of Appearance.—June to August.

Habitat.—West Central and Southern Germany, Switzerland, the South of France, North Italy, Dalmatia, and the East of Russia. In Western Asia it assumes various varietal forms, which will be found enumerated under the extra-European *Lycænidæ.* It is a somewhat local species, but common where it occurs, frequenting open sunny places, generally on a chalky soil, in saintfoin fields.

Larva.—Green, with a darker dorsal stripe, and two pale

lateral ones. Feeds on saintfoin, *Onobrychis sativa*, in May and June. Pl. XXIX., fig. 3.

VARIETY.

a. **Damone,** Ev. Bull. M. 1841, i. 18; Frr. 386, 2. Rather larger than the type. The male is deeper blue, and with a narrower hind-marginal border. Both sexes have the white streak on the under side much narrower.

HABITAT.—The Ural Mountains.

37. **L. Donzelii,** B. Ic. 15, 1-3; Frr. 145, 2, 3; II.-S., 351-2.

Expands 0·81 to 1·25 in. Fringes white and grey. The male has all the wings greenish blue, with broad hind-marginal band of brown, indistinctly defined internally; the fore wings have a black elongated discoidal spot. The female is brown, with a black discoidal spot on the fore wings. Under side brownish grey. Fore wings with a rather large discoidal and a central row of smaller spots black, surrounded by white rings. Hind wings bluish at the base, with a white streak, which is more triangular in shape than that seen in *L. Damon;* besides the usual central row and discoidal spot there is a faint orange hind-marginal band enclosing a row of black spots.

TIMES OF APPEARANCE.—June and July.

HABITAT.—The Southern Alps, the Pyrenees, Scandinavia, Finland, South-East Russia, and the Altai; at a high elevation in the Alps and the South of France, but on the plains in its Boreal localities. Pl. XXX., 7.

LARVA.—Unknown.

38. **L. Argiolus,** Linn. Syst. Nat. x. 483; Faun. Suec. 284; O. i. 2, 17; Frr. 445, 3, 4.—*Acis,* Fab. Mant. 73; Hüb. 272-4.—*Cleobis,* Esper, 40, 3.

Expands 0·80 to 1·40 in. Fringes of the fore wings black and white; those of the hind wings white. Wings of the male clear, light blue; fore wings with a very narrow black hind-marginal border. The female has the outer half of the costa and all the hind

margin of the fore wings broadly brownish black; the hind wings
are similarly brownish black on the costa; sometimes the hind
margin is also dark brown, and always has a row of black dots.
Under side bluish grey; fore wings with an elongated black
discoidal spot, and a row of black spots running parallel to the
hind margin; hind wings blue at the base, with an irregular
central row, an elongated discoidal and two basal spots. The spots
in this species are all black, without white rings. Pl. XXXI., 1.

TIMES OF APPEARANCE.—April to August, being double-
brooded.

HABITAT.—The whole of Europe and North Asia (except the
Polar region), Western Asia, and North Africa.

LARVA.—Dark greenish grey, with a dark green dorsal line.
Feeds on the flowers of *Ilex*, *Hedera* and *Rhamnus* in June, and
again in the autumn.

39. L. Sebrus, B. Ic., Pl. 71, 1-3, p. 72; Tr. x. 1, 65; Frr. 451, 1.

Expands 0·75 to 1 in. Fringes white. The male is violet-
blue, with a very narrow and defined black border. The female is
dark brown, generally with traces of blue at the base. Under side
bluish grey. Discoidal spots slender and crescentic. Fore wings
without basal spots, and with central row evenly arranged and
parallel to the hind margin. The hind wings have two basal spots,
and an irregularly arranged central row. None of the spots are
ocellated. Pl. XXXI., 2.

TIMES OF APPEARANCE.—May to July.

HABITAT.—Mountains in Austria, Piedmont, the Valais, and
the South of France; the Balcans; mountains in Asia Minor and
Armenia, frequenting dry places at a considerable elevation.

LARVA.—Unknown.

40. L. Minima, Fuessl. Verz. p. 31 (1775); Esp. 34, 3.

Expands 0·75 to 1 in. Fringes of all the wings white. Wings
dark brown in both sexes. The male slightly tinged with light
blue at the bases. Under side pale grey. Fore wings with a

straight central row of black spots and an elongated discoidal spot. Hind wings tinged with blue at the base, with an irregular central row and two basal spots; all the spots are surrounded by light rings.

TIMES OF APPEARANCE.—May to August.

HABITAT.—The whole of Europe except the Polar Regions and the Southern parts of Spain and Portugal. Northern and Western Asia, and the Amur. In Britain it occurs in dry chalky places.

LARVA.—Green, with an orange dorsal stripe edged with pale yellow, a lateral yellowish line and an oblique streak. It feeds on several kinds of vetches, as *Anthyllis vulneraria, Coronilla varia, Astragalus,* &c., in June and August.

VARIETY ?

Lorquinii, H.-S. 442-4. This form is larger than the type, and has the wings of the male blue except on the hind margins. The female is dark brown, larger and blacker than *Alsus.* It is found in Andalusia and the South of France, and is very probably a distinct species. Pl. XXXI., 4.

41. L. Semiargus, Rott. Nat. vi. p. 20 (1775); Hbst. xi. p. 177.

—*Acis,* Schiff. S. v. p. 182 (1776); O. i. 2, 14; Frr. 451, 4.—*Argiolus,* Esp. 21, 1, 2; Fab. Mant. 73; Hüb. 269-71.

Expands 1·0 to 1·25 in. Fringes of all the wings white. The male is dull violet-blue, with a dark brown ill-defined hind marginal border. Female uniform brown; fore wings with a narrow discoidal spot. Under side:—Light brown. Fore wings with a curved central row and an elongated discoidal spot. Hind wings tinged with blue, with a somewhat irregular central row, a very narrow discoidal and one basal spot. The rings surrounding the spots are less white than in the last species. Pl. XXXI., 5.

TIMES OF APPEARANCE.—May to July.

HABITAT.—The whole of Europe, Northern and Western Asia, and the Amur. In Britain, rare and local, frequenting rough pastures in the South-Western and Midland Counties of England, and sometimes taken in Wales.

LARVA.—Undescribed !

S

VARIETIES.

a. **Bellis,** 398, 1, 2; H.-S.—232-4. Larger than the type, and with traces of a marginal row of red spots on the hind wings beneath.

HABITAT.—The South-East of Europe.

b. **Parnassia,** Stgr. Hor., 1870, p. 55. A smaller form of the last, found in the mountains of Northern Greece.

c. **Helena,** Stgr. Stett. Ent. Zeit., 1862, 265; Mill. Ic. pl. 39, 1, 3. In this variety or sub-species the female has a marginal orange band on the upper surface of the hind wings. Both sexes have an orange marginal band beneath on all the wings, that on the fore wings being faintly marked.

HABITAT.—The mountains of Southern Greece.

OBS.—Dr. Frey has named the small specimens occurring in the higher Alps of Switzerland, var. *Montana.* It does not differ from the type excepting in its smaller size.

42. **L. Cælestina,** Eversm. Bull. Mosc. 1843, iii. 535; H.-S. 335-8; Frr. 445, 1, 2.

Expands 1·0 to 1·20 in. Fringes of all the wings white. The male is bright blue, with a slight purplish tinge; all the wings have a narrow dark brown hind marginal border. Female uniform dark brown. Under side :—Slaty grey. Fore wings with an elongated discoidal and a central row of black spots small and uniform in size. Hind wings with a central row similar to those of the fore wings; discoidal spot wanting; there is a conspicuous patch of shiny blue scales at the base, without any tinge of green. Pl. XXXI., 6.

TIME OF APPEARANCE.—June.

HABITAT.—The South-East of Russia (principally Sarepta.)

LARVA.—Unknown.

43. **L. Cyllarus,** Rott. Nat. vi. p. 20 (1775); Esp. 33, 1, 2; O. i. 2; Frr. 271; Mill. Ic. pl. 108, 2.—*Damœtas,* Hüb. 206-8.

Expands 1·0 to 1·30 in. Fringes of all the wings white. The male is blue, brighter than the last species. The female is brown, slightly blue at the base. The under side somewhat resembles that of the last, but differs in the following particulars :—The black spots on the fore wings are much larger than those on the hind wings ; instead of being equal in size, the blue tinge at the base is strongly mixed with green, has an almost metallic brilliance, and extends over quite two-thirds of the wing area, often passing beyond the central row of spots. Pl. XXXI., 7.

Times of Appearance.—May to the end of August.

Habitat.—The greater part of Europe, though not in Britain, the Low Countries, or the Polar Regions, and probably not in Andalusia. It occurs throughout the greater part of Northern and Western Asia and the Amur. It generally frequents flowery meadows and open places in woods.

Larva.—Greenish yellow, with a bright reddish brown dorsal stripe and lateral oblique stripes of the same colour ; lateral stripe greenish. It feeds on several species of *Leguminosæ*, in June and July.

Pupa brown, the sides being lighter and the wing-cases darker. Pl. XXIX., 4.

44. L. Melanops, Boisd. Ind. p. 13 ; Frr. B. 97, 1, 2 ; Mill. Ic. 108, 1.—*Saportæ*, H.-G. 922-8 ; Marchandi, Gerh. 15, 1.

Expands 0·75 to 1·0 in. Fringes of all the wings white. The male has all the wings blue, with a tinge of purple, and with a narrow black hind marginal border. The female is dark brown, slightly tinged with blue at the base of the fore wings. The under side has the spots almost exactly as in *L. Cyllarus*, but the ground colour of the wings is darker grey, and the hind wings are only slightly dark blue at the base ; there is also a trace of a hind marginal band of black spots in addition to the central row.

Times of Appearance.—April and May. Pl. XXXIII., 1.

Habitat.—The South of France, Spain and Portugal, and North Africa, along the shores of the Mediterranean.

Larva, as figured by Millière, green, with a blackish dorsal stripe and oblique lines, and with two whitish lateral lines. It feeds on trefoil in May and June.

Pupa light brown, darker on the wing-cases. Pl. XXIX., 5.

a. **Marchandii,** Hüb. i. 996-7. Differs from the type in the spots on the under side of the hind wings not being ocellated.

45. L. **Iolas,** Ochs. Schmett. Enc. iv. 144; Boisd. Ic. t. 11, f. 1-3; Frr. 97.—*Iolaus,* Hüb. 879-82.

Expands 1·25 to 1·70 in., being the largest European species. Fringes of all the wings white. Both sexes have the wings violet-blue. The male has a narrow black hind marginal border, the hind wings having very faint dots along the hind marginal edge. The female has a very broad dark brown border extending along the costæ and hind margins of all the wings, and the hind wings have a row of rather large black-lined marginal spots; the nervures of all the wings are strongly marked in this sex. Under side :—Pale fawn-coloured grey ; all the wings have a narrow crescentic discoidal spot, and a central row of black eyes in light rings, that on the fore wings being parallel with the hind margin, that on the hind wing somewhat waved. There is a faintly marked hind marginal light band enclosing dark spots, more strongly marked in the female ; the base of the hind wings is slightly blue, and there are two basal spots. Pl. XXXII., 2.

TIMES OF APPEARANCE.—June and July.

HABITAT.—North-Eastern Germany, Hungary, the Balkan, the South of France, Catalonia, rarely in Switzerland. It also occurs in Asia Minor.

LARVA.—Olive-green, with a black dorsal and a broad pale green lateral stripe. It feeds on the pods of *Colutea arborescens.*

46. L. **Alcon,** Fab. Mant. 72; Hüb. 263-5; Boisd. Ic. 13, 1-3.— *Arcas,* Esp. 34, 4.—*Euphemus,* Godt. i. 11.

Expands 1·25 to 1·35 in. Fringes of all the wings white. The male is purplish blue, with a narrow brown hind marginal border. The female has the wings of the same colour but with a broad costal and hind marginal border. Under side :—Brownish grey. Fore wings with an elongated discoidal spot and a curved central row. Hind wings slightly blue at the base, with a narrow discoidal, two basal spots, and a wavy central row. All the spots are about equal in size, and, in addition to them, all the wings have a double hind marginal row of dark brown spots enclosing a faint light band. Pl. XXXII., 3.

TIMES OF APPEARANCE.—July and August.

HABITAT.—Central and Southern Europe, namely, Germany, the South-East of Holland, France, Piedmont, Switzerland, the South of Sweden, Central and South-Eastern Russia, Hungary and Bulgaria. It probably also occurs in Western and Central Asia, and is taken in the Altai; it is a local species, occurring on moorland meadows. Any accounts of its capture in Britain are no doubt entirely erroneous.

LARVA.—Unknown.

47. L. Euphemus, Hüb. 257-9; O. i. 2, 9; Boisd. Ic. 13, 4-8.— *Diomedes*, Rott. Nat. vi. p. 26 (sed *Diomedes*, L., Syst. Nat. 1758, al. erat. Pap.)

Expands 1·25 to 1·35 in. Fringes of all the wings white. Wings blue in both sexes, with a dark brownish black hind marginal border, broader in the female than in the male; all the wings have a narrow black discoidal spot and a central row of round black spots. Under side :—Brownish grey, with the spots arranged as on the upper side, but in addition to these all the wings have a hind marginal row of black spots, and the hind wings, which are very slightly blue at the base, have two basal spots. Pl. XXXII., 4.

TIME OF APPEARANCE.—July.

HABITAT.—The greater part of Central Europe (but not Great Britain), probably also Asia Minor, Armenia, and the Caucasus. It is a local species, frequenting marshy places.

LARVA.—Not as yet described by any author.

48. L. Arion, Linn. Syst. Nat. x. 483; Fab. Syst. 283; Esp. 20, 2; Hüb. 254-6; O. i. 2, 4.

Expands 1·25 to 1·50 in. Fringes of all the wings white. The wings are deep blue in both sexes, with a brown hind marginal border; the spots have the same arrangement as in the last species, but those on the fore wings are much larger and rounder, whilst those on the hind wings are fainter and smaller, and are often absent altogether in the male. Beneath, this species differs from the last in having a double row of hind marginal spots, and in having the bases of the hind wings more decidedly tinged with blue.

TIMES OF APPEARANCE.—From the end of May to the middle of July. Pl. XXXII., 5.

HABITAT.—The greater part of Europe, excluding the Polar regions and the South-West; it occurs also in Western Asia and in the Southern parts of Siberia, frequenting meadows and open places. In Britain it is very local, its chief localities being in the Midland and South-Western parts of England.

The LARVA has never been observed in its full-grown state; it has been described when fifteen days old as being of a dirty pink colour, the head brown and shiny, the dorsal line rust-colour. It was reared by Mr. Porritt, on wild thyme, on the flowers of which it feeds.

OBS.—Varieties of this and of the preceding species (*L. Euphemus*) occur occasionally in which the wings are strongly suffused with dark brown, almost obscuring the spots and giving the insect the appearance of a brown rather than a blue butterfly. The dark form of *L. Arion* is called var. *obscura* by the German and Swiss Entomologists, and occurs in Alpine districts of Switzerland and the Tyrol.

49. **L. Arcas,** Rott. Nat. vi. p. 25 (1775).—*Erebus,* Knoch. Btr. ii., t. 6, 6, 7 (1782); Fab. Mant. 72; Hüb. 260-2; Esp. 101, 1; O. i. 2, 10; B. Ic. 11, 4-6.

Expands 1·25 to 1·30 in. Fringes of all the wings brown. The male is very dark, irony blue, with broad brown hind marginal borders. Fore wings with a narrow black discoidal and a central row of elongated spots. Hind wings with similar spots, but those of the central row are smaller and rounder. The female is strongly suffused with dark brown. Under side uniformly brown, with no tinge of blue, the discoidal spots almost imperceptible; all the wings have a central row of round black spots enclosed in lightish rings. Pl. XXXII., 6.

TIMES OF APPEARANCE.—July and August.

HABITAT.—Germany, and the Alps of Switzerland and the South of France, also the Altai. A local species, frequenting marshy meadows.

LARVA.—Unknown.

———

OTHER SPECIES AND VARIETIES OF *LYCÆNIDÆ*, INCLUDED BY STAUDINGER, &c.

Genus *THECLA*.

T. Spini var. *Melantho*, Klüg. Symb. Phys. t. 40, 10, 11.—Has the under side paler and the tails of the hind wings longer than in the type. Habitat, Asia Minor and Persia.

T. Ilicis var. *Caudatula*, Z. Is. 1847, p. 6.—A variety with the tails longer than in the type. Habitat, Asia Minor.

T. Taxila, Brem. Lep. O. S. p. 26, 95.—Expands 1·40 to 1·60 in. Somewhat resembles *T. Quercus* in size and shape, but the wings are suffused, except along the hind margins, with a brilliant emerald-green. The under side is very like that of *T. Quercus*, but the orange spots near the anal angle are much larger. The female is brown above, very much resembling the male of *T. Betulæ*. Habitat, the Amur and Japan.

T. Ledereri, B. Ann. Soc. Fr. (1848; H.-S. 445-8 (1850); Frr. 572-4. *Lœosopis? Ledereri*, Kirby, Cat. 377.—Expands 1·30 in. Fringes white. Wings brown above. Hind wings with an orange hind marginal band and with short tails. Under side light brown. Fore wings with two parallel rows of black spots and a double black discoidal spot surrounded with white. The hind wings are bluish at the base, and along the hind margin have a row of orange and black spots; internal to this a row of white spots with black pupils, and two discoidal spots. Habitat, Armenia and the Trans-Caucasus.

T. Smaragdina, Brem. Lep. O. S. p. 25, iii. 5.—Much resembles *T. Taxila*, but is rather larger and more brilliant, and the markings on the under side are bolder. Habitat, the Amur and Japan.

T. Arata, Brem. Lep. O. S. p. 25, iii. 6.—Expands 1·20 in.

Wings dark brown; those of the male shot with greenish blue.
Hind wings with a short slender tail, and with a bright orange spot
at the anal angle. Under side striped with dark brown and white;
Hind wings with four black spots placed on a patch of orange at
the anal angle. Habitat, the Amur.

T. Myrtale, Klüg. Symb. Phys. t. 40, 15, 16.—Expands 0·9
to 1·9 in. Wings uniform brown and fringes grey. Hind wings
with a very rudimentary tail and, with a bright orange spot at the
anal angle. Under side olive green, darkest at the base. Hind
wings with short yellow band near the anal angle, enclosed by a
double row of crescentic black spots. (Descr. from Klüg's fig.)
Habitat, Syria.

*T. Sassanides, Kollar. Ins. Pers. p. 10.—" Alis supra furvis
unicoloribus, posticis caudatis; subtus omnibus dilute cinereis,
striga communi alba, intus fusco marginata, serie punctorum
nigrorum ocelliformium ante marginem externum; cauda longiore
nigra, apice alba. Expans. alar. 13 lin. Statura Th. Acaciæ, a
quâ tamen punctis marginalibus nigris in pagina inferiore strigaque
alba latiore valde differt." (The original description. I have seen
neither a figure nor a specimen of the insect).

T. Frivaldszkyi, Ld. z., b., V. 1855, p. 100, t. i. fig. 1.—
Expands about the same as *T. Rubi.* Hind wings somewhat
indented on the hind margins. The wings are steel-blue; the
fore wings having a black border extending along the costa and
hind margin, and very broad at the latter part. Hind wings
blue, black on the costa, with a row of wedge-shaped black spots
placed parallel to the hind margin. Under side chocolate-brown
streaked with light grey. Fringes of all the wings black and
white. Taken by Kindermann at Ustbuchtarnisk, on June 2nd,
shortly after the melting of the snow. (Desc. from Lederer's fig.)

T. Fusca, Brem. Bull. de l'Acad. 1861.—"Alæ anticæ maris
supra cærulescente violaceo micantes, subtus fusco-griseæ, maculis,
fasciis lunulisque marginalibus fuscis, albido-annulatis, 32 mm."
Habitat, the Amur (Mount Bureja). Time of appearance, July.
Probably to be referred to another genus—*Amblypodia.*

---

* In cases like the present, where I am unable to describe a species, either from
specimens or from figures, the original Latin diagnosis will be given verbatim.

Genus *THESTOR.*

T. Mauretanicus, Luc. Expl. Alg. p. 360, Pl. i. 3.—Very much resembles *T. Ballus* above, having about the same expanse; the female, however, brighter orange, having the orange patches nearly oval and bordered very narrowly with brown. Under side :— Fore wings almost as in *T. Ballus*, but the spots are more uniform in size and more closely placed. Hind wings brownish grey, with four faint bands or rows of brown spots running across the wing, and having a direction parallel to the hind margin. Habitat, Algeria and the North of Morocco.

Genus *POLYOMMATUS.*

P. Caspius, Ld. Hor. 1869, p. 77.—" Alis fuscis, posticis caudatulis anticis e basi longe igneo-violascentibus, punctis disci duabus nigris ; subtus omnibus dilute grisescentibus, multi ocellatis, posticarum fascia marginali rufescente obsoleta, 27 mm." Habitat, the shores of the Caspian.

P. Phœnicurus, Ld. Hor. 1871, t. i. 4, 5.—" Alis supra, mas violaceo-fuscis (fœm. fuscis), macula venæ transversæ nigra, fascia marginali rufa (in anticis maris evanescente), cordula posticarum in basi rufa. Subtus mas et fœm. exalbido canis, fascia marginali læte rufa, ocellis multis, vena transversa striolam nigram gerente. Ex. 25—28 mm." Habitat, the shores of the Caspian.

P. Lampon, Ld. Hor. 1871, t. i. 2, 3.—" Alis mas supra croceo-aureis, anticis apice limboque nigris ; posticis caudatulis, in margine nigro maculatis ; fœm. pallidioribus seriebus tribus macularum nigrarum ante-marginalibus, anticarum maculis disci duabus, posticarum una obsoleta nigris. Subtus mas et fœm. ut in Thersamone, sed pallidioribus, ocellis latius pallido-cinctis. 30 mm." Habitat Persia.

P. Asabinus, H.-S. 527-8; *Helius*, H.-S. vi. p. 33.—About the size of *Phlœas*. Male golden copper above, with a narrow black hind-marginal band and a row of black dots parallel to the hind margins. Discoidal spot black. Under side :—Fore wings with a black discoidal spot, internal to which are two more black spots, and external to it two rows of spots, one regular and the other

T

irregular. Hind wings grey-blue at the base, with a hind-marginal orange band and two rows of black spots ; internal to these a row of black spots in light rings ; there are five or six basal spots.

P. Thersamon, v. *Omphale*, Klug. Symb. Phys. t. 40, 12, 14. *Xanthe*, Hub. 347-8.—This variety is slightly smaller and brighter than the type, the hind wings having small tails. Probably a variety of the second brood. Habitat, Asia Minor.

P. Athamanthis, Ev. Bull. Mosc. 1854, iii. 180.—" P. alis supra nigris, fulvo-micantibus et nigro-punctatis; fascia subterminali fulva ; subtus cano-albis, nigro-punctatis; fascia subterminali fulva ; alis posticis caudatis." Very close to *Amphidamas*, but the coppery lustre of the wings is limited to a broad band running parallel to the hind margins. The ground colour of the under side is greyish white. Habitat, Western Siberia, the Steppes to the northward of the Sea of Aral.

Genus *CIGARITIS*, Lucas, Ann. Soc. Fr. ii. 98, 1850.

Butterflies having very much the aspect of those of the genus *Polyommatus;* the upper surface of the wings, however, have not the metallic coppery lustre of that genus. Beneath they are white, with various brown and silvery markings. The hind wings are strongly emarginate, and often tailed. All the known species occur in Africa, one in Syria also.

C. Acamas, Klug. Symb. Phys. t. 40, 7-9 (1834) ; *Epargyros*, Ev. Bull. Mosc. 1854, iii. 178.—Expands 1·25 to 1·35 in. Fulvous ; fore wings with a black stripe running from the costa downwards, and three black costal stripes. Hind wings with two hind marginal tails, with blackish stripes and spots arranged parallel to the hind margins. Under side white, beautifully spotted and banded with brown and silver. Habitat, Syria, Persia, and the Steppes to the South and South-East of the Ural. The female is larger and darker than the male.

C. Cilissa, Ld. Wien. Mts. 1861, p. 147.—On an average the size of *P. Phlæas*, the male being smaller than the female. The wings are brilliant copper-red, but without any metallic gloss, with a narrow hind-marginal border and square black spots ; the hind wings with two tails. Under side whitish, with square spots having a dull metallic lustre and black borders. Habitat, Syria, and the Southern parts of Asia Minor.

C. Siphax, Luc. Expl. Alg. p. 362.—Expands 1 in. The wings are bright reddish brown without metallic lustre, and with a black hind-marginal border and two narrow black interrupted lines running across them parallel to the hind margins. The tails of the hind wings are very minute. Under side :—Fore wings reddish brown, with a yellowish hind-marginal border enclosing a row of black spots; the principal area of the wings covered with large silvery spots bordered with black. Hind wings brownish white with silvery spots. Habitat, Algeria.

C. Zohra, Donz. Ann. S. Fr. 1847, p. 528.—(The female only described). About the size and shape of *P. Phlœas.* The anterior wings are fulvous, with three bands and a border of brownish black; the first two bands are very short, and do not pass beyond the medium nervure; the third is wavy and longer. The hind wings are brown, with a posterior fulvous band surmounted by another less regular but equally deep in colour. Under side :—Fore wings mottled with brown and white patches, especially along the costa. Hind wings golden brown, with several white spots arranged in lines; hind margin with two tails. Habitat, Algeria.

C. Massinissa, Luc. Expl. Alg. p. 364.—Rather larger than *C. Zohra.* The wings are fulvous above, with a black margin. The fore wings spotted with black above, and beneath rust-coloured spotted with grey and decorated with silver. The hind wings fulvous above, with black spots; beneath they are white, with reddish brown spots bordered with silver; hind marginal band greyish, with six black spots. Habitat, Algeria and Morocco.

Genus *LYCÆNA.*

L. Fortunata, Stgr. Berl. E. Z. 1870, p. 99.—Very close to *L. Telicanus,* and about the same size. Male violet-blue above; beneath it is brighter grey than *Telicanus,* and more strongly marked with brown. Female grey-brown above, slightly blue at the base. Habitat, the Canary Islands. Probably an insular form of *Telicanus.*

L. Theophrastus, Fab. E. S. 281 ; Luc. Expl. Alg. pl. 1-6. *Psittacus,* Allard. Ann. S. Fr. 1867, p. 313.—Very close to *L. Balkanica,* but somewhat larger and probably brighter, the dark spots on the fore wing being less distinct. Habitat, Algeria. The larva on *Liziphus vulgaris.* This species also inhabits South Africa.

L. Attilia, Brem. Lep. O. S. p. 24, t. ii. 3.—About the size of *L. Bætica.* Wings brown, with white fringes. Hind wings tailed. Under side bluish white. Fore wings with a square black discoidal spot, a transverse band across the centre of the wing, and a submarginal row of rather large black spots. Hind wings with similar markings, and near the anal angle two orange spots. Taken by Radde on July 1st, at Mount Bureja in Eastern Siberia.

L. Gamra, Ld. Z. b. v. 1855, p. 189.—Size of *Ægon.* Hind wings not tailed, violet-blue in the male, with a very narrow brown border ; brown in the female. Fore wings with an indistinct discoidal spot. Hind wings with a faint hind marginal row. Under side bluish grey, with hind-marginal rows of black spots surrounded by light orange ; the other spots, both on the fore and hind wings, coalesce so as to form bands. Habitat, Syria. In June and July.

L. Galba, Ld. Z. b. v. 1855, p. 190, i. 1, 4.—Larger than *L. Trochilus.* The male is blue, with a rather wide dark brown border and a faint discoidal spot. Female brown. The under side is light brownish grey, with spots arranged much as in *Trochilus,* but larger. Habitat, Syria.

P. Cyane, Ev. Bull. Mosc. 1837, i. 22 ; H.-S. 646-9.— Expands 1·30 to 1·50 in. Male bright blue with a narrow black hind marginal border, internal to which is a band of indistinct whitish spots. Female brown, strongly blue at the base. Fore wings with a row of whitish spots parallel to the hind margin. Hind wings with a bright orange hind-marginal band, internal to which is a row of white spots. Under side very similar to *Pylaon,* of which species some consider it a variety. Habitat, South Siberia and the Altai.

L. Cleobis, Brem. Bull. de l'Acad, 1867.—About the size of *L. Ægon.* Fringes broad and white ; the male has all the wings bluish white, with a rather broad black border, the nervures black ; the female, the wings are brown, powdered with bluish white.

Under side bluish or brownish white. Fore wings with a discoidal lunule, a central row of ocellated spots and a double row of hind-marginal spots enclosing an orange band. Hind wings with four basal spots and discoidal lunule, a central row and double marginal row enclosing an orange band; this row is often marked with silver as in *Ægon*. Habitat, Mount Bureja.

L. Elvira, Ev. Bull. Mosc. 1854, iii. 177.—" L. Alis supra cærulescenti-argenteis, nigro marginatis (mas), aut fuscis (fœm.); subtus cinerascenti-albis; punctis ordinariis nigris satis magnis; anticarum punctis basalibus nullis; posticarum punctis anguli analis argentea-pupillatis subtribus." Habitat, the Steppes of South-Western Siberia.

L. Hyrcana, Ld. Hor. 1869, p. 78.—Expands 0·75 to 1 in. Fringes white. Male brilliant violet-blue, faintly brown along the hind margins. Fore wings with an indistinct discoidal spot. Female uniform brown, all the wings with a hind marginal row of black spots. Fore wings with a black discoidal spot. Under side brownish grey, all the wings with a faint hind-marginal orange band. Fore wings with a discoidal and one row of ocellated spots external to it. Hind wings with a central row, discoidal and four basal spots, all ocellated. Habitat, Persia (Astrabad.)

L. Subsolana, Ev. Bull. Mosc. 1851, i. 620.—Expands 1·30 in. Male only described. " L. Alis superne basi cærulcis, externe nigro-fuscis, nervis nigris, ciliis albis; subtus albido-cinereis, lunula discoidali, serie flexuosa serieque submarginali duplici punctorum nigrorum, hac lunulis fulvis signata; alis posticis præterea punctis basalibus nigris quatuor punctisque nonnullis seriei limbalis externæ argenteo-notatis." Habitat, Eastern Siberia and the Altai.

L. Loewii, Z. Is. 1847, p. 9; H.-S. 4, 34-7.—Expands 1·10 to 1·20 in. Fringes of all the wings white. Male deep blue, with a narrow dark brown hind-marginal border. Fore wings paler, blue on the costa. Hind wings with a row of faint hind-marginal spots. Female greatly resembles that of *L. Argus*, but is larger. Under side much as in *Argus*, with silver centres to the hind-marginal row of hind wings. (H. S.)

L. Erschoffii, Ld. Hor. Ent. Ross. 1869, p. 80.—" Alis supra fuscis anticis (mas antice late cyaneo-pulveratis) strigula media nigra, ciliis apicalibus niveis; subtus omnibus fuscenti cinereis, multiocellatis, anticarum striga e maculis ocellaribus magnis atris post ocellum medium magnum; posticis sine rubidius, radio medio, longitudinali albo, 25 mm." Apparently easily distinguished from all other species by the brilliant violet costa of the male. Habitat, Persia.

L. Eurypilus, Frr. 573.—Size of *L. Icarus*. Wings dark brown, with white fringes. Hind wings with two or three reddish orange lunules. Under side brownish white, slightly greenish blue at the bases; all the wings with the usual ocellated spots and hind-marginal bands. Fore wings without basal spots. Habitat, Amasia.

L. Diodorus, Brem. Lep. Ost. Sib. p. 29.—About the size of *L. Icarus.* The wings are brown, shot with violet above with indistinctly marked discoidal and central spots. Under side dark-grey. Fore wings with a discoidal lunule, a whitish hind-marginal band, a central row of irregularly arranged square spots, and one large basal spot. Hind wings similarly marked, but with two spots between the central row and the discoidal near the costa, and with three basal spots. All the spots are conspicuously surrounded by white rings. Habitat, Eastern Siberia (Sea of Baikal.) End of June.

L. Panagea, H.-S. 490-3.—Expands 0·75 to 1 in. Somewhat close to *Lysimon*; the male is deep blue with very broad hind-marginal border, and with a black discoidal spot on the fore wings. Female brown, dusted with blue at the base. Under side slaty-grey, and bluish at the base. Fore wings with a hind-marginal row of black spots; internal to this three large black spots arranged in a triangle; above them and beneath them two smaller ones. Discoidal spot very small and narrow. Hind wings with a faint orange band, an irregular central row, narrow discoidal and two basal spots, all very small. Habitat, Asia Minor.

Var. *Anisophthalma*, Koll. Ms. Pers.—Has both sexes brown, the blue colour being absent from the male. Habitat, Persia.

L. Pheretiades, Ev. Bull. Mosc. 1843, iii. 536; H.-S. 650-1. Expands 1 in. Dull grey, dusted with blue at the base. Fore wings with a black discoidal spot. Fringes all white. Under side almost as in *Pheretes* (H.-S.)   Seems to be intermediate between *Pheretes* and *Orbitulus*.   Habitat, Eastern Siberia.

L. Orbitulus var. *Wosneseuskii*, Mén. En. ii. p. 95.

L. Isaurica, Stgr. Berl. e. z. 1870, p. 327.—A small species, somewhat resembling *L. Eros*, but somewhat more emerald-green. A broad black border extends along three-fourths of the hind margins, the hind wings having hind marginal black spots.   The under side is more distinctly marked than in *Eros*, and of a brown tint in the ground-colour.   Habitat, Southern districts of Asia Minor.

L. Eros v. *Myrrha*, H.-S. 508-11.—Somewhat like *Eroides*, but much larger and paler in colour ; the spots on the under side being proportionately smaller.   Expands nearly 1·50 in.   The female pale brown, with faint hind marginal bands.   Habitat, Asia Minor.

Var. *Candalus*, H.-S. 502-5.—Expands 0·75 to 1·10 in. Fringes white.   Male blue, about the same colour as *Icarus*, with a very narrow but well-defined blackish-brown hind marginal border.   Hind wings with a row of black dots.   Female dark brown, with a row of small orange spots on the hind wings and with the bases of all the wings faintly blue.   Under side as in *Eros*, but with two basal spots.   Habitat, mountains in Syria and Asia Minor.   Staudinger catalogues this as a variety of *L. Eros*, but it certainly appears to be a distinct species.

L. Icarus var. *Persica*, Bienart.—Has the spots on the under side very small or absent.   Habitat, Persia.

L. Bellargus var. *Polona*, Z. Stett. e. z. 1845 ; H.-S. 432-3.— About the size of *L. Corydon*, which it greatly resembles in general appearance, having rather broad black hind margins and a row of spots on the hind wings.   In colour, however, it resembles the typical form of *Bellargus*.   The under side also is the same as in the type.   Habitat, mountains in Asia Minor.

L. Corydon var. *Corydonius*, H.-S. 595-6 ; *Olympica*, Ld. z. b. v. 1852.—In this variety the wings are deep blue, and not pale as in the typical form, which in other respects it resembles.   Habitat, mountains in Asia Minor.

Var. *Caucasica*, Ld. Ann. Soc. Belg. xiii. 23.—This form is very similar to the last, being rather smaller, but nearly the same

colour. Habitat, Armenia. It will be seen by these varieties that
L. Bellargus and *Corydon* are really much more intimately allied
than appears at first sight.

L. Marcida, Ld. Hor. 1871.—Expands about 1 in. The
male has the wings pale shiny bluish grey, with a narrow brown
hind marginal border. Fore wings with a very faint discoidal
spot. Female uniform brown, with the exception of a slight band
of a lighter colour along the hind margins. Under side brownish
grey ; all the wings with a hind marginal row of black spots,
rounded by white rings but without any orange band ; all the spots
ocellated. Fore wings without any basal spots ; three basal spots
on the hind wings. Habitat, Persia. (Descr. from Lederer's fig.)

L. Hylas var. *Armena*, Stgr. Cat. p. 12.—A variety with the
white marginal fringe very broad, the hind wings beneath having
the spots almost absent.

L. Damon var. *Poseidon*, Ld. z. b. v. 1852, p. 37.—The wings
of the male are rather darker blue than in the type. Habitat,
valleys in Asia Minor.

Var. *Damocles*, H.-S. 214-7.—Smaller than the type. The
male is deeper blue (nearly as blue as *Eroides*) and the hind margins
are less broadly brown. Habitat, Asia Minor.

Var. *Iphegenia*, H.-S. 354.—A variety rather resembling the
last, but smaller. The spots on the under side are very small, and
the white basal streak large and distinct. Habitat, Asia Minor.

Var. *Carmon*, H.-S. 506-7.—In this variety the wings of the
male are deep blue, with a well-defined dark brown border. The
under side is pale grey. Fore wings with a discoidal spot and an
irregular central row ; central row of hind wings even ; there is a
faint trace of a hind marginal band, and there are two faintly
marked white basal streaks.

Var. *Actis*, H.-S. 496-9.—About the size of *L. Icarus*, which,
judging from the figure of Herrich-Schäffer, it much resembles in
the colour of the male, the hind margins having an exceedingly
narrow brown border. Female dark brown, slightly blue at the
base. Under side as in var. *Carmon*, but with a single basal streak.

L. Hopfferi, H.-S. 512-14.—Expands 1·40 in. Intermediate
between *L. Dolus* and *Damon*. It is rather bluer than *Dolus*, and
is hardly if at all marked with the brown basal patch that is found

in that species. Under side much as in *Dolus*, but the white streak
on the hind wings is much more distinct. Habitat, Alpine regions
in Asia Minor.

    L. Donzelii var. *Hyacinthus*, H.-S. 345-8.—Resembles the
type in size and shape, but a deep rich blue replaces the ordinary
pale colour in the wings of the male. Female with an orange
hind-marginal band on the hind wings.

    L. Biton, Brem. Lep. O. S. p. 30.—About the size of *L. Icarus*.
Wings in both sexes brown, shining, with silvery greenish blue
scales. Fringes white. Hind wings with yellow marginal lunules.
Under side brownish grey. Fore wings with a faint orange hind-
marginal band, a central row and a discoidal spot. Hind wings
with three basal spots in addition, and with a strongly marked
yellow hind-marginal band ; all the spots are ocellated. Eastern
Siberia. June (from Bremer's fig.)

    L. Argiolus var. *Hypoleuca*, Koll. Ins. Pers. p. 11.—A variety
occurring in Persia and Cyprus, in which the black spots are
absent from the under side.

    L. Semiargus var. *Antiochena*, Led. Wien. Mts. 1861, 148.—
The female has an orange band on all the wings above. The male
resembles the type in colour, but is somewhat larger and lighter.
Habitat, Mountains in Syria and Lydia.

    L. Cyllarus var. *Tristis*, Gerh. 15, 4.—Characterized merely
by the greater size of the spots beneath. Habitat, Asia Minor.

    L. Astræa, Frr. 583, 1, 2, vi. 161.—Size of *L. Melanops*.
Male deep rich blue, with a broad black hind-marginal border to
all the wings. Female with some light blue spots towards the
apex of the fore wings. Hind wings with some black spots near
the edge of the hind-marginal band. Wings duller blue than in
the male. Under side as in *Melanops*, but the spots, especially
those of the fore wing, are smaller. Habitat, Amasia.

    L. Arion var. *Cyanecula*, Ev. Bull. Mosc. 1848; H.-S. 593-4.
The same size and shape as the type ; wings somewhat lighter
above. Under side :—Hind wings have the ground colour, greenish
blue throughout the entire area between the base and the hind-
marginal row of black spots.

Species discovered since the last Edition of Staudinger's Catalogue.

L. Glaucias, Led. Hor. Ent. Ross. 1871, p. 10, pl. viii.— Larger than *Marcidas*. Expands 1·20 in. Male very pale blue, with a row of faintly brown hind-marginal spots on all the wings. Fore wings with a brown discoidal spot. Female dark brown, slightly paler along the hind margins. Under side pale grey. Fore wings with a hind-marginal band of black spots; all the wings with the usual central row and discoidal spots, two basal spots on the fore wing and two or three on the hind wing, which have also a hind-marginal orange band and some white streaks. Habitat, Persia.

L. Staudingeri, Christoph. Hor. Ent. Ross. x. 26.—" Alis supra mas cæruleis anguste fusco marginatis, posterioribus striga macularia submarginali nigra, spatio inter illam et marginem cyanescente; fœm. fuscis, copiose cæruleo-pulverulentis; ciliis albis; subtus canis, lineolis disci angustis, anticis ante medium sine ocellis, striola media e maculis ocellatis nigris, posticis fascia macularia obsoleta fusca maculaque subanali atra, cæruleo squamato, 27 mm." Habitat, Persia.

L. Ægargus, Christoph. Hor. Ent. Ross. x. 24.—" Alis supra mas pallide cæruleo-argenteis, nigro-limbatis, fœm. fuscis canescente variis, maculis disci ocellatis, mas subtilioribus, fœm. crassioribus nigris; ciliis albis; subtus ut in *L. Dardano*, 25—28 mm." Habitat, Persia.

L. Sieversii, Christoph. Hor. Ent. Ross. x. 23.—" Alis supra mas violaceo-cæruleis, anguste nigro-limbatis, posterioribus punctis tribus subanalibus nigris; fœm. fuscis, cæruleo mixtis, striga ante marginem exteriorem cærulescente obsoleta, striola disci nigra; subtus mas et fœm. ut in *L. Hyrcana*, sed macula subanali unica atra, cæruleo-squamata, cinereo-circumdata, 25—29 mm." Habitat, Persia.

L. Tengstræmi, Ersch. Lep. Turk. p. 11, n. 32 (1874).— Expands 0·75 to 1 in. Wings uniform brown above, with brownish marginal fringes. Under side brownish grey. All the wings with a hind-marginal row of light spots; there are no spots upon the area of the wing. Habitat, Turkestan. Probably to be referred to the Genus *Lœsopis.*

L. *Mirza,* Staud. Stet. Ent. Z. 1874, p. 90.—Habitat, Persia.

L. *Prosecusa,* Ersch. Lep. Turk. p. 13, n. 38 (1874).—Habitat, Turkestan.

L. *Palæstina,* Staud.—Habitat, Palestine.

L. *Scylla,* Staud.—Habitat, Asia Minor.

Lycæna Lycidas, Trappe, occurring in Switzerland, will be described and figured at the end of this work.

The following North American Lycænidæ may be compared with some of the European species.

Genus *THECLA.*

These species are very close to the European group of brown *Theclæ,* such as *T. Spini, W-album,* &c. :—

T. *Damon,* Cram. Pap. Ex. iv. t. 390.

T. *Spinetorum,* Hew. Ill. Diurn. Lep. p. 94.

T. *Favonius,* Smith Abb. Lep. Georg. i. t. 14.

T. *Acadica,* Edw. Proc. Ac. Nat. Sc. Phil. 1862, p. 55.

T. *Californica,* Edw. Proc. Ac. Nat. Sc. Phil. 1862.

T. *Sylvinus,* Boisd. Ann. Soc. Ent. Fr. 1852.

T. *Ametorum,* Boisd. l. c. p. 287.

T. *Sæpium,* Boisd. l. c. p. 287.

T. *Dumetorum,* Boisd. Ann. Soc. Ent. Fr. 1852, p. 291, is probably the American form of *Thecla Rubi;* it inhabits California.

Genus *POLYOMMATUS.*

P. *Sirius,* Edw. Trans. Am. Ent. Soc. vol. iii. p. 270.

P. *Rubidus,* Behr. Proc. Ent. Soc. Phil. vol. vi. p. 208.—Both these species seem to be allied to P. *Virgaureæ, Thetis,* &c., their wings being of a brilliant copper-colour.

P. *Xanthoides,* Boisd. Ann. Soc. Ent. Fr. p. 292, 1852, and P. *Mariposa,* Reak. Proc. Ent. Soc. Phil. vi. p. 149.—Are both large species, inhabiting California. They appear to be close to the European *Gordius* and *Alciphron*. In the same group may be placed also *P. Gorgon*, Boisd. Ann. Soc. Ent. Fr. 1852, p. 292, which is also a native of California.

P. Epixanthe, Boisd. and Sec. Lep. Am. Sept. p. 127, from Labrador and Canada.

P. Helloides, Boisd. Ann. Soc. Ent. Fr. (1852) 292.

P. Thoë, Gray (1832), Griff. An. Kingd. t. 56, f. 4.

P. Dorcas, Kirb. Faun. Bor. Am. iv. p. 299.

L. Tarquinius, Fabr. Ent. Syst. iii. 1, p. 319 (1793).

L. Porsenna, Scud. Proc. Eos. Inst. iii. p. 163, and *L. Phlæas* var. *Americana*, D'Urb. Canad. Nat. v. p. 246 (1857).—All these species, which represent the European group containing *P. Phlæas*, *P. Dorilis*, and *P. Helle*, inhabit Canada and the Northern States.

Genus *LYCÆNA.*

The Genus *Lycæna* contains a large number of North American species, all closely allied to those found in Europe and Western Asia, some being apparently identical with them. I can do no more than barely mention some of the commoner forms. The American *Lycænæ* will be found treated fully in Herrman Strecker's 'Lepidoptera of North America,' page 82, &c.

L. Comyntas, Godt. Enc. Meth. ix. p. 660, 1823.—This and *L. Amyntula*, Boisd., are very close to *L. Argiades*.

L. Pseudargiolus, Boisd. et Leconte, Lep. Am. Sept. p. 118 (1823).—This species is so very close to *L. Argiolus* that in some specimens it is really difficult to distinguish them.

L. Harmo, Stoll. Suppl. Cram. t. 39.—Somewhat resembles *L. Optilete*, Hab. Florida.

L. Exilis, Boisd.—Is the smallest North American species, and seems very close to *L. Lysimon*. Habitat, Georgia.

L. Pheres, Boisd. Ann. Soc. Ent. Fr. p. 297.—Found in California, is close to the European *L. Pheretes*.

L. Rustica, Edw. Proc. Ent. Soc. Phil. iv. 203.—Is considered by Strecker to be identical with the European *L. Orbitulus*. It inhabits high mountains in Colorado and California. The variety *Aquilo* occurs in Labrador.

L. Scudderi, Edw. Proc. Acad. Nat. Sci. Phil. p. 164.—Is a species much resembling *L. Argus* or *Ægon.* It occurs in Canada and the Northern States.

Obs.—Under the head of the Genus *Lycæna* I have not thought it necessary to make more than the present note concerning the Genera *Lampides* and *Scolitandides,* given by Hübner in his Verz. bek. Schmett., and recently revived in Butler's 'Catalogue of Fabrician Butterflies,' as I consider these divisions much too artificial to be really useful, and I do not think that any characters have been discovered which render it desirable to break up the genus *Lycæna* into smaller groups. The Genus *Lampides* would include the following European species :—*L. Bætica, Telicanus, Balcanica, Fischeri* and *Argiades* those which have tailed hind wings ; *Scolitantides* the following :—*L. Orion, Baton, Zephyrus, Pylaon* and *Bavius, i.e.,* those with brown and white spotted fringes and very large clearly defined spots beneath.

Fam. 4.—**ERYCINIDÆ**, Swainson, Phil. Mag. 1827.
Lemoniidæ sub-fam. Nemeobiinæ, Kirby, Syn. Cat.
Diurn. Lep. p. 284; Bates, Journ. Linn. Soc. 1867.

Larva.—Onisciform as in *Lycænidæ.*
Pupa.—Short and blunt as in *Lycænidæ,* attached by a girth in an upright position, as well as by the tail.
Imago.—With the anterior legs rudimentary in the male, but perfect in the female. Palpi short.

This family contains a very great number of Exotic species, the greater proportion being found in tropical America. Many are decorated with beautiful and brilliant colours, though nearly all are of small size, as in the last family.

It will be seen that the *Erycinidæ* do not differ in any essential character from the *Lycænidæ,* excepting in the rudimentary condition of the anterior legs. In all other respects this family resembles the preceding. There is but one European species.

Genus 1.—**NEMEOBIUS**, Steph. Ill. Brit. Ent. Haust. i. p. 28 (1827).

MELITÆA et ARGYNNIS, Auctorum.

CHARACTERS.—Eyes hairy. Antennæ slender, with a flattened triangular club. Head small. Thorax stout, and larger than the head. Abdomen tolerably long, and usually not entirely concealed by the wings when these are closed. The apices of the anterior wings very sharp; the hind wings indented along the hind margin. General character of the colour and pattern of the wings closely resembling that of the Genus *Melitæa.*

Imagines frequenting woods in the spring and autumn; larvæ feeding on *Primulæ* and *Rumices.*

1.—**N. Lucina**, Linn. Syst. Nat. x. 480; Faun. Suec. 280; Esp. 16, 2; Hüb. 21, 22; Frr. B. 43, 1.

Expands from 1·10 to 1·25 inch. Fore wings dark brown, with three irregular transverse bands of bright orange-brown spots, that parallel to the hind margin having the greatest number and enclosing black dots. Hind wings dark brown, with a hind-marginal row of orange-brown spots, enclosing a row of black ones; on the area of the wing are one or two small orange-brown spots. Under side:—Fore wings brown, with three rows of orange spots as above, but the colour, both of the ground and of the spots, is lighter. Hind wings reddish brown, with two rows of white spots, one row near the base and the other crossing the centre of the wing; along the hind margin is a row of black dots, one on each cell, except at the anal angle. Fringes of all the wings white, chequered with dark brown. Pl. XXXIII., 1.

TIMES OF APPEARANCE.—May and June, and again in August.

HABITAT.—The greater part of Western and Central Europe, the South of Sweden, North Italy and the North of the Balcans. It is exclusively a European species, frequenting open places in woods.

LARVA.—Reddish brown, with a dorsal row of black dots; the sides are brownish white with dark lines. The intersegmental divisions are dark and well-defined; there are tubercles on each segment emitting small bristly hairs. In shape the larva is somewhat more elongated than those of the Genus *Lycœna*, otherwise it resembles them, and the head is retractile as in that genus.

It appears in June and September, feeding upon the leaves of *Primula veris* and other species of primroses.

The Pupa is short and rounded, resembling that of a *Lycœna;* it is yellowish white, spotted with black, and is attached by a girth round the middle. Pl. XXXIII., 1.

Three genera of *Erycinidœ* occur in North America, but none of them are close to *Nemeobius*.

Fam. 5.—**LIBYTHEIDÆ**, Westwood, Gen. Diurn. Lep. 412 (1851).

Lemoniidæ sub-fam. Libythœinæ, Bates' Journ. Ent. ii. p. 176 (1864).

LARVA.—Cylindrical, not spiny, but with a few fine hairs. Head not retractile, but smooth and round.

PUPA.—Elongated, slightly angular, and suspended by the tail alone.

IMAGO.—With the palpi elongated to an extent very unusual amongst butterflies, forming a beak-like projection four times as long as the head. The anterior legs rudimentary in the male, but perfect in the female, a character which has induced some Entomologists to associate this family with the last in spite of its different larva, pupation, &c.

This family, like the last, is represented in Europe by one genus and species; the remaining species are not numerous, and with the exception of one or two, which occur in North America, are confined to the tropical regions of the Old World.

Genus 1.—**LIBYTHEA**, Fab. Ill. Mag. vi. p. 284 (1807); Lat.
Enc. Meth. ix. p. 10 (1819); West. Gen. Diurn.
Lep. p. 412.

CHARACTERS.—Antennæ of moderate length, gradually becoming
thick towards the tips, but without any very distinct club. The
wings are large in proportion to the body, and usually of a sombre
colour, and with lighter brown markings. In the only European
species they are angulated and with dentate margins.

1.—L. **Celtis**, Esp. 87, 2, 3 (1783); Hüb. 447-9; O. I. 2, 192;
Godt. ii. 6, 5.

Expands 1·50 to 1·75 in. Fore wings with a strongly marked
angular projection (as in some of the *Vanessæ*) dark brown, in some
specimens nearly black, with a bright fulvous patch filling the discoidal cell, excepting where it is deeply notched with a dark brown
triangle, having its base towards the costa; at the termination of
the discoidal cell this orange patch extends downwards and outwards, becoming expanded into an ovoid blotch below, which is a
smaller round one; near the costa of the wing is a round spot
much lighter in colour than the rest, and below and external to it
one of almost quadrate form and orange in colour. Hind wings
rather rectangular in outline, with shallow indentations along the
hind margin; the ground colour is dark brown, and near the centre
is a fulvous patch sometimes divided into two parts. Under side:—
Fore wings much the same as above, but slightly paler. Hind
wings mottled with brown and grey, with a faint narrow white line
or streak in the centre. Pl. XXXII., 2.

TIME OF APPEARANCE.—March, and again from January to the
end of July.

HABITAT.—The South of Europe and Asia Minor, its presence
being determined by the distribution of the nettle-tree, *Celtis
Australis*, which is the food-plant of the larva. This plant is not
wild on this side of the Alps.

LARVA.—When full grown green, with a dark dorsal stripe,
the spiracles black, the lateral stripe purplish. Pl. XXXII., 2.

PUPA green, with lighter dorsal and lateral stripes.
The larva appears in April and July on *Celtis Australis*.

Three species of *Libythea* occur in North America, and are thus
given by Edwards :—

1.—*L. Carinenta*, Cram. ii. pl. 108.—Habitat, New Mexico,
Arizona.

2.—*L. Motya*, Boisd. Lec. pl. 64.—Habitat, Southern States.

3.—*L. Bachmannii*, Kirtland Sellman's Journal, xiii. p. 336.—
Habitat, Middle and Western States.

Fam. 6.—**APATURIDÆ**, Boisd.

CHARACTERS.—Larva smooth, limaciform, the head with non-
retractile horn-like appendages.   In the European species green
is their prevailing colour.

Pupa thick in the region of the anterior abdominal segments,
generally rather pointed towards the extremities, but not exhibiting
angular projections or metallic spots. Suspended by the caudal end.

Imago with the eyes not hairy. Antennæ long and with thick
well-formed clubs. Wings large and very powerful, especially in
the males. Hind margins of anterior wings, concave. Posterior
wings dentated or tailed.   Discoidal cells of the hind wings open.

This family, which is separated from the one that follows, on
account of the peculiar slug-shaped larvæ, contains only three
European species.

Genus 1.—**CHARAXES**, Ochs. Schmett. Eur. iv. p. 18, 1816.
NYMPHALIS, *Auctorum*.

CHARACTERS.—Larva with four short cephalic horns.
Pupa rather rounded, and not tapering much towards the head.
Imago, with the hind margin of the anterior wings strongly
concave. Hind wings each with two long tails.

x

The butterflies of this genus are all rather large, some attaining a considerable size ; the prevailing colour of the wings is rich brown and black in most of the species, though other colours are found, such as blue or white. Their flight is very powerful. There are about sixty known species, all found in the Old World and chiefly in the African and Indian Regions. The only European species inhabits those districts of South Europe and North Africa which are in immediate proximity to the Mediterranean.

1.—**C. Jasius**, Linn. Syst. Nat. xii. in erratis ultimæ, pag. ; Esp. 99, 1, 2, 104, 2, 8 ; O. I. 1, 51.
  Jason, Hbst. T. 64, 3-5.
  Rhea, Hüb. 111-4, 580-1.
  Unedonis, Hüb. Text. p. 19.

Expands 2·60 to 3·25 in. All the wings deep brown. Fore wings with an orange hind-marginal band, narrowest near the apex and divided by dark brown bars following the course of the nervures, becoming broadest near the anal angle. Internal to this band and parallel to it is a row of about six orange spots, the spots being distinctly marked near the costa and becoming very undefined as they approach the centre of the wing. Hind wings much indented and with two tails ; a hind-marginal border extends from the costa to the anal angle; it is orange-yellow at its commencement, and as it approaches the anal angle it becomes strongly tinged with green ; it also becomes narrower at this part of the wing; above this and extending inwards from the anal angle is a row of four blue spots ; external to the band is a narrow border of black which is continued into the tail. Under side :—The central area of the wings has the ground colour grey, splendidly marked with square spots and streaks of a rich maroon-red bordered with white ; a broad irregular band of white runs along the centre of the wings throughout their whole length, that is, from the costa of the fore wing to the inner margin of the hind wing; external to this on the fore wing is a series of orange lunules, and on the hind wing a rich brown band; next in order on all the wings comes an irregular band of grey and black, and external to this an orange border, that on the hind wings being again narrowly bordered with black; on the hind wings near the anal angle is a row of violet-blue spots, the space around them being tinged with green; the head, thorax and abdomen are stout, and the antennæ long, tapering into well-marked clubs. Pl. XXXIV., 1.

TIMES OF APPEARANCE.—May and June, and August and September.

HABITAT.—Those countries bordering on the Mediterranean where the food-plant of the larva (*Arbutus Unedo*) grows wild. In Europe, the South of France (as at Heyères, Nice, &c.), Spain and Portugal, Italy, Greece, Turkey, and Dalmatia: beyond Europe, the North of Africa, bordering on the Mediterranean. The African specimens are often very much larger than the European. This insect is very rapid in its flight, and difficult to take.

LARVA.—Green, head lighter, with four short horn-like appendages tipped with red. The seventh and ninth segments have each a lateral yellow ring centred with bluish green, and extending from the second segment to the tail is a yellow lateral line. Legs green.

PUPA rather stout and rounded at the head, green in colour; wing-cases slightly tinged with blue. (From Hubner's figs.).

The larva feeds on the strawberry-tree, *Arbutus unedo*. It is found from May to August. Pl. XXXVI., 1.

Genus 2.—**APATURA**, Fab. Ill. Mag. vi. p. 280 (1807);
Westwood, Gen. Diurn. Lep. (1850).
DOXOCAMPA, Hüb. Verz. bek. Schmett. p. 49 (1816).

CHARACTERS.—Larva with two cephalic horns.

Pupa tapering rather strongly towards the head.

Imago with the hind margin of the fore wings slightly concave. Hind wings without tails, but with a dentated margin. In the two European species the wings are dark with light bands, and are shot with violet in the males.

The Genus *Apatura* contains over forty known species; they are all dark coloured and have generally bands of white or light colour running across the wings, and spots near the apices of the

fore wings arranged much as in the European species. Most of
the species are found in the tropical regions of Asia and South
America. Some of those from the latter are very beautiful, such
as for instance *A. Lucasii*, Doubld., and *A. Laurentia*, Godt., from
the Brazils, which have their nearly black wings banded with
brilliant green, changing according to the direction of the light to
a fine blue or purple.

1.—A. Iris, Linn. Syst. Nat. x. 476; Hüb. 117-8; Esp. 71-4.

Expands 2·25 to 3·25 in. Fore wings dark brown, with ten
white spots arranged as follows:—One oval-shaped spot rather
larger than the rest in the centre of the wing, and below this a
crescentiform spot with its convex side turned outwards, and then
below this and nearly touching the inner margin a small round
one; external to these and near the hind margin are two round
spots placed one above the other; above and internal to these last
are three costal spots placed one above the other, and external to
them two small ones close to the apex. Hind wings with a broad
white band running across their centre from the costa to the inner
margin, broadest at its costal end, and becoming gradually narrow
towards the inner margin; on its inferior edge is a spur-like projection just where the band crosses the discoidal cell; near to the
anal angle is an orange ring, the centre of which is black or
bluish; the anal angle itself is marked with orange. The wings in
the male are short, with rich violet, excepting along the hind
margins, which are brown, sometimes slightly tinged with fulvous.
Under side of all the wings light brown, with white bands and
spots arranged as on the upper surface. The fore wings are
strongly tinged with reddish orange towards the costa, and towards
their centre is a black spot with a bluish pupil and surrounded by
an orange ring; the hind wings are tinged with reddish brown
along the edges of the white band, and the anal angle is a round
black spot with a bluish pupil. The clubs of the antennæ are
black without orange tips. Pl. XXXIV., 2.

Times of Appearance.—June and July.

Habitat.—South Western and Central Europe (including

Britain, Holland, Belgium, &c., but not found in Scandanavia), in the East very sparingly, though it apparently occurs in Dalmatia and some parts of Asia Minor.

LARVA.—Green, slightly bluish towards the ventral surface, studded all over with minute yellow spots, each segment having a yellow line running upwards and backwards in an oblique direction. The head, which is flattened in front, has a pair of horn-like appendages, which are not retractile, and in colour are dark green in front and lighter posteriorly.

PUPA.—Green, darkest on the wing-cases and on the ventral surface; dorsally it is slightly tinged with yellow. Pl. XXXVI., 2.

The Larva feeds on the leaves of the Sallow and on several kinds of Poplar (such as *Populus alba* and *P. tremula*). It is full fed by the middle of June at the latest.

The Imago is remarkable for its habit of frequenting oak woods, where it keeps itself at a great distance from the ground ; it may, however, sometimes be attracted by sugar, in dry weather by moisture or by decaying animal matter.

VARIETY.

Iole, Schiff. S. V. p. 172 ; Hüb. 622-3 ; Esp. 46, 1. ; Frr. 385. —*Beroë*, Fab. Ent. Syst. iii. p. 111.
The central band of white spots on the fore wings, and the white fascia of the hind wings are entirely obsolete.

2.—A. Ilia, Schiff. S. V. p. 172 ; Hüb. 115-6.—*Iris*, Esp. 11, 2 ♀, 37, 1 ♂.

Expands 2·0 to 2·50 in. The typical form of this species is very similar to *Iris* at the first glance, but on examination will be found to present many important differences ; the chief points of distinction are—1. The presence in both sexes of an orange ring on the upper surface of the fore wings near the anal angle. 2. The white fascia of the hind wings has not, as in *Iris*, a spur-like projection, but is nearly straight on its outer edge. 3. The under side will be seen to differ in the distinctness of the markings, and the hind wings are delicately tinted with bluish pink or grey, especially along the hind margins. 4. The hind margins are much more indented in the female of this species than in that of *Iris*. Lastly, the clubs of the antennæ are tipped with yellow or orange. Pl. XXXV., 1.

TIMES OF APPEARANCE.—June and July.

HABITAT.—Part of Central Europe (including Germany, Switzerland and France), also a part of Northern Asia and Armenia.

LARVA.—Very similar to that of *Iris*, but somewhat smaller and of a yellower colour, except on the ventral surface. The cephalic horns are bordered with yellow.

PUPA.—Somewhat rounder than that of *Iris*, and less pointed anteriorly.

The larva is found in May, and feeds on Sallow and on several kinds of Poplar, chiefly *Populus alba*. Pl. XXXVI., 3 (from Hübner).

VARIETIES.

a. **Clytie**, Schiff. S. V. p. 321; Hüb. 113-4; Frr. B. 31.—*Iris Rubescens*, Esq. 71, 2, 3.—*Astasia*, Hüb. 812-3. This is the usual form of the species in some parts of Europe, the typical form only occurring occasionally; it is smaller than the type and has the white fascia and spots replaced by yellow or orange; sometimes the ground colour is entirely orange as in the female specimen, figured (Pl. XXXV., 2). This specimen was taken on the shores of Lago Maggiore, where this extremely light form seems to be constant in the autumn brood.

HABITAT much more extended than that of the type; Germany, Switzerland, France, North Italy, Finland, and Eastern and South-Eastern Europe. It is double-brooded in the South. The imago occurring in June and the end of August.

b. **Metis**, Frr. B. 67, 1, ii. p. 61; H.-S. 539-41. Rather like *Clytie*, but the fore wings have their hind margins more deeply concave, and the hind wings more indented. The ocellated spot of the hind wing is altogether wanting.

HABITAT.--South-East Russia. (Sarepta). Pl. XXXV., 3.

c. **Bunea**, H.-S. p. 45.   Like *Melis*, but with the fasciæ and all the spots white, except the marginal ones, which are orange ; probably a variation of the last.

HABITAT.—South Russia. (Sarepta). Pl. XXXV., 4.

Of the North American species of *Apaturidæ*, the three under-mentioned species are most nearly allied to those of Europe :—

A. Idyja, Hüb. Samml. Ex. Schmett. (1816).

A. Alicia, Edw. Butt. N. Amer. i. (1868).

A. Celtis, Boisd. Lec. Lep. Amer. Sept. p. 210 (1833).

The latter two species inhabitat New Orleans and Virginia respectively. These three species remind one somewhat of *A. Clytie* in colouring, but the wings in all of them are more pointed in shape, and the hind wings generally have a row of ocellated spots. The larvæ are much more elongated and more Satyriform than those of the European species.

Fam. 7.—**NYMPHALIDÆ**, Swainson, Phil. Mag. Ser. ii. vol. i. p. 187 (1847) ; West. Gen. Diurn. Lep. p. 143 (1852).

CHARACTERS.—Larva generally cylindrical and armed with long spines or with tubercular elevations armed with bristles.

Pupa usually with angular projection and often decorated with metallic spots, always suspended by the tail.

Imagines.—Middle-sized or large butterflies usually brilliantly coloured, and with the wings nearly always more or less emarginate; the pattern of the wings in some genera greatly resembles that prevalent in the last family, as in *Limenitis*, *Neptis*, &c.; in others there is a tendency towards a design composed of black spots on a tawny ground, as in *Argynnis* and *Melitæa*, and in some of the *Vanessæ*, other *Vanessæ* are strikingly brilliant, and more varied in their coloration than any other European Butterflies. The discoidal cell of the hind wings open. The fore legs rudimentary or atrophied in both sexes.

We saw among the *Papilionidæ* a prototype of this family in the Genus *Thais*, all the species of which possess many nymphaloid characters both in their larval and imaginal states.

This family is of wide distribution, some species, such as *Vanessa Cardui*, being absolutely cosmopolitan.

The genera *Argynnis* and *Melitæa* are very numerously represented in Europe; they, in fact, are almost confined to the colder and temperate region of the earth. Among the circumpolar butterflies there are more species of *Argynnis* than of any other genus.

There are a great number of gorgeous species belonging to tropical genera of *Nymphalidæ*, but the brilliancy of their colouring is very nearly equalled by the common European *Vanessæ*, such as *Atalanta, Io, Urticæ*, &c., while the metallic decoration of *Argynnis Lathonia* may be said to excel in beauty that of the majority of allied tropical species.

Genus 1.—**LIMENITIS**, Fab. Ill. Mag. vi. p. 281 (1807);
West. Gen. Diurn. Lep. p. 274 (1850).
NYMPHALIS, Boisd. Ind. Meth. p. 14 (1809).

CHARACTERS.—Larva cylindrical, with tubercular excrescences covered with bristles or with branching spines.

Pupa with a few angular projections, the wing-cases large.

Imago.—Large or middle-sized butterflies of a dark brown or black colour, with white spots and bands. The hind margins are without angular projections, but that of the fore wing being slightly concave, whilst in the hind wing it is somewhat dentated but not markedly so. Antennæ long, and with tapering clubs. Only three species occur in Europe, the Oriental Region being the chief habitat of the genus. The Nearctic Region also contains many fine and large species.

1.—**L. Populi**, Linn. Syst. Nat. x. 476 ; Esp. 12, 1 ; Hüb. 108-10 ;
Frr. B. 37.

Expands from 2.50 to 3 in. All the wings dark brown. Fore wings with a central row of large white spots, an oblong white discoidal spot and a row of three or four near the apex. Hind wings with a central white band, narrow in the male, but broad and distinct in the female. All the wings have the hind margin tinged with green, and with a row of orange lunules, complete on the hind wings, but on the fore wings interrupted towards the anal angle. Under side orange, with the spots and stripes greenish; hind wings with a row of blackish spots parallel to the hind margin, and with the inner margins greenish grey.

Times of Appearance.—June and July.

Habitat.—Central Europe (excepting Great Britain and Holland), also the Southern parts of Scandanavia and Finland. Pl. XXXVII., 1.

Larva.—Green, mixed with reddish brown and yellowish white, with two rows of fleshy elevations covered with bristles.

Pupa.—Dirty white; thorax with a dark brown dorsal patch; wing-cases dark brown.

The Larva feeds on *Populus alba* and *tremula*, in May. It appears first in the autumn, and then hybernates between dead leaves.

In its habits this species very much resembles *Apatura Iris*, frequenting the higher branches of tall trees, and occasionally descending on moist ground or attracted by carrion, &c. It is said to be much more common in some seasons than in others.

VARIETY.

a. **Tremulæ**, Esp. 114, 3, 4. This is a form of the male, in which nearly all the white markings are absent, sometimes even the apical spots scarcely appearing at all. This aberrant form may be found at the same time and in the same localities as the type. Pl. XXXVII., 2.

2.—**L. Camilla**, Schiff. S. V. p. 182; Hüb. 106-7; Godt. i. 6, 3. Lucilla, Esp. 38, 2; Bkh. i. 25.

Expands from 1·50 to 2·25 in. The ground colour of all the wings is bluish black, and all have a row of blue-black or purple dots running parallel to the hind margin. Fore wings with a white spot near the apex, and with a distinct white discoidal spot;

Y

between these a row of four or five spots which have no tendency
to form a continuous band. Hind wings with a central row of
large white spots, in most specimens coalescing to form a broad
white band. Under side brown, with a bright burnt sienna or
light red tinge on the costa and at the apices of the fore wings.
The hind wings have two rows of roundish spots of this colour
running parallel to the hind margin, and between them is a row
of black dots; the base is broadly lilac-blue. All the white
markings of the upper side are visible beneath. The head, thorax,
and abdomen are black above and bluish white beneath. Antennæ
black, except at the extreme tips, which are reddish brown. Pl.
XXXVII., 3.

TIME OF APPEARANCE.—June to August.

HABITAT.—Woods and bushy places in Southern and Central
Europe, not further north than 51°, Western Asia and Persia.

LARVA.—Green, with green tubercles tipped with red. The
head, legs and ventral surface are red; the head is covered with
white dots. It feeds on *Lonicera* &c., in May.

OBS.—This species and the next were confounded by some of
the older writers, which has led to *Camilla* being erroneously
reputed British.

3.—L. **Sibylla**, Linn. Syst. Nat. xii. 781 ; Esp. 14, 2 ; Hüb.
　　103-5.—*Camilla*, Linn. loc. cit. n. 187 ; Esp. 14, 3.

Expands from 2·0 to 2·25 in., being much less variable in size
than the last. All the wings brownish black ; the margins slightly
dentate, and the fringes black and white. Fore wings with a very
indistinct light brown discoidal spot; the remaining spots have
much the same arrangement as in *Camilla*, but show a greater tendency to form a band ; there is no marginal row of purple dots, but
there are some faint black ones. The hind wings much resemble
those of *Camilla*, but the white band is concave on its outer edge for
about two-thirds of its length, whilst in *Camilla* it is straight.
Under side brown, with a yellowish or fulvous tinge instead of the
red colour found in *Camilla*. The fore wings have a greenish grey
discoidal spot. The hind wings have a double row of black spots

running parallel to the hind margin; the base of the wings is greenish, with a pearly lustre and with some black spots which are not found in the last. Head, thorax and abdomen brownish black above, greenish white beneath. Antennæ black above, with the extreme tips light brown, beneath they are fulvous. Pl. XXXVIII., 1.

TIME OF APPEARANCE.—June and July.

HABITAT.—Woods in Central Europe, and in Spain, Britain, and South Russia.

LARVA.—Head pinkish brown, darker laterally. Body almost cylindrical, but with the intersegmental spaces strongly marked; the dorsal surface is rough and tuberculated, with small branched spines; the colour on the back is dark green, the sides being paler and the tubercles yellow. The spiracles are white, and below them is a narrow white lateral stripe. The spines are pink at the tip and brown at the base; their bristles are black. The ventral surface is apple-green, the claspers being paler and the legs brownish green. (The above description is condensed from that given in Newman's 'British Butterflies').

PUPA brown, with the wing-cases green; the head and ventral surface are decorated with metallic spots.

Time of Appearance:—The larva is hatched about the beginning of August, hybernating and appearing again in April, becoming full fed and pupating at the beginning of June. It feeds on *Lonicera*.

OBS.—Aberrations of this and of the preceding species occur, in which the white markings are either wholly or partially absent from the upper surface, thus corresponding to the var. *Iole* of *Apatura Iris*. Such a variety of *L. Camilla* is figured by Millière (Ic. 1859, pl. 4, 1), and named by him *Pythonissa*. The black form of *L. Sybilla* is given by Hübner (Btr. ii. 3, 1, b.); also in Newman's 'British Butterflies.' Specimens of this may be seen in almost every good collection of British Lepidoptera.

Genus 2.—**NEPTIS**, Fab. Ill. Mag. vi. p. 282 (1807); Westwood, Gen. Diurn. Lep. p. 270 (1850).

CHARACTERS.—Larva cylindrical with spiny excrescences. Pupa with a few angular projections.

Imago of medium size ; discoidal cell of fore wings open (in *Limenitis* it is closed) ; hind margin of fore wings not concave, but rather rounded. Hind wings slightly dentate along the hind margin. Wings longer in shape than those of *Limenitis*, and though in colour and the general character of their markings they greatly resemble those of that genus, they differ in having white basal spots or streaks on the fore wings.

Most of the species are found in the tropical and sub-tropical regions of the Old World, principally in India. Two occur in Europe, and a few in the Amur.

1.—N. Lucilla, S. V. p. 173 ; Fab. Mant. 55 ; Hüb. 101-2 ; Frr. B. 13 ; O. I. 1, 38.
 Camilla, Esp. 59, 1. (Nec Linn.).
 Rivularis, Scop. Ent. Carn. p. 165 (1763).
 Sappho, Kirby Cat. Diurn. Lep. p. 239.

Expands from 1·75 to 2 in. All the wings brownish black, much the same as in *L. Sybilla*. Fringes black and white. Fore wings with a row of white spots arranged much as in *Sybilla*, but they are rather larger and more banded ; there is an indistinct basal streak sometimes absent, and outside this one or two white spots. Hind wings with one broad white band running straight across the centre and divided into seven spots by the nervures. Under side marked as above, but the ground colour is reddish brown, and there are some indistinct whitish spots along the hind margin. Pl. XXXVIII., 2.

Time of Appearance.—June and July.

Habitat. — Damp woods in Eastern and South-Eastern Europe (but not in the Balcans) ; it occurs in Piedmont and in the Swiss Canton Tessin. It is also found in Western Asia and the Altai. Local, not common.

Larva.—According to Freyer reddish brown, yellowish laterally ; four of the segments have each two thick spines. It feeds on *Spiræa Salicifolia* in June.

2.—**N. Aceris,** Lepechin, i. 203, t. 17, 5, 6 (1768-70); Esp. 81,
3, 4; O. I., 1, 136; B. Ic. 18, 2.
SAPPHO, Pall. Reis (1771).
PLAUTILLA, Hüb. 99, 100.

Expands from 1·25 to 2·25 in., being more variable in size
than *N. Lucilla.* All the wings black as in *Lucilla,* from which it
differs in the markings; the fore wings have the white basal streak
very distinct, the central row of spots more evenly and regularly
placed than in the last, and near the hind margin is a row of small
white spots. Hind wings pearly grey at the bases, with a broad
central white band not divided into spots by the nervures; beneath
this is a narrower second band divided into five or six oblong white
spots. Under side :—Markings of fore wings as above, but the
ground colour is light reddish brown; the inner margins pearly
grey. Hind wings reddish brown, with two white bands as above,
but between them a narrow white line and another external to
them, both running parallel to the lower band; there is a streak
of white on the costa near the base and another below it and
parallel to it. Pl. XXXVIII., 3.

TIMES OF APPEARANCE.—June, July and August.

HABITAT. — Moist woods and shaded river-banks in South-
Eastern Germany, Hungary, North-Eastern Turkey, the South-
East of Russia, Central Asia, the Amur, and Japan. Scarce and
local in Europe, from the Western parts of which it is quite
absent.

LARVA.—Reddish brown, with a paler lateral line; the 3rd, 4th,
and 11th segments have two spiny tubercles. Feeds on *Orobus
vernus* in the spring, and again in July.

Genus 3.—**VANESSA,** Fab. Ill. Mag. vi. 281 (1807); Lat. Enc.
Meth. ix. p. 10 (1819).
POLYGONIA, EUGONIA, ET INACHIS, Hüb. Verz. bek.
Schmett. pp. 36, 37 (1816).
ARASCHNIA, GRAPTA, VANESSA, ET PYRAMEIS, *Auctorum.*

CHARACTERS.—Larva cylindrical, covered with long spines;
often gregarious, at least through some of their stages.

Pupa with angular projections; head double pointed, usually decorated with metallic spots.

Imago middle-sized or large butterflies, with the hind margins of the wings generally very dentate or with prominent angular projections; the upper surface of the wings is generally very brilliantly and variously coloured, whilst their under surface is usually streaked with sombre brown, though some of the species, especially those referred by authors to the genera *Pyrameis* and *Araschnia*, have under sides coloured more or less brilliantly. The antennæ are long and terminated by well-formed clubs, somewhat pyriform. The palpi are of moderate length and moreor less hairy. Eyes hairy.

Most of the species are strong flyers, and are found in gardens, woods, fields, &c. The majority of the European species are also common; indeed the genus is represented in every part of the world. Out of the twelve species described as European, seven occur in Britain.

I have followed Staudinger in referring all the European species to the old Fabrician genus *Vanessa*; the divisions commonly received by Entomological writers seem to be founded mainly on the shape and markings of the wings. They are as follows:—

ARASCHNIA, Hüb. Verz. bek. Schmett. 37 (1816); Doubl. Gen. D. L. p. 187 (1848). — Expanse of wings averaging 1·25 in. Brown, spotted with black, or black with white bands. Under side with the nervures white; inner margin of the fore wings straight. Two slight projections on hind margin of fore wings and one on the hind wings. One European species, *Levana*, L. Two others closely allied from Japan and the Amur.

GRAPTA, Doubl. Gen. D. L. p. 195 (1848).—Expanse of wings reaching 2 in. Brown, spotted with black. Under side with a distinct letter-like mark on hind wings. Inner margin of fore wings more or less concave; hind margin deeply dentated; fore wings with two and hind wings with one very prominent projection. Two European species, *Egea*, V., and *C. Album*, L.

VANESSA.—Doubl. Gen. D. L. p. 198 (1848).—Expanse of wings reaching 3 in. Brown, spotted with black, or variously black, brown, blue, and yellow. Under side dark brown, sometimes with a white mark on hind wings. Inner margin of fore wings straight ; hind margins dentated ; fore wings in most species with two marked projections, and hind wings with one. Six European species :—*Polychloros*, L. ; *Xanthomelas*, S. V. ; *Vau-Album*, W. V. ; *Urticæ*, L. ; *Io*, L. ; and *Antiopa*, L.

PYRAMEIS, Hüb. Verz. bek. Schmett. p. 32 (1816); Doubl. Gen. D. L. p. 202 (1849).—Expanse of wings reaching 2·75 in. Black or dark brown, banded or spotted with red or orange. Under side coloured as above, sometimes with ocelli. Fore wings black towards the apex with conspicuous white spots. Inner margins straight ; hind margins very little angulated. Hind wings rounded and scalloped. Three European species, *Atalanta*, L. ; *Callirhoë*, F. ; and *Cardui*, L.

1.—V. **Levana**, Linn. Syst. Nat. x. 480 ; Esp. 15, 2 ; Hüb. 97, 98, 728-9 ; O. I. 1, 132.

Expands from 1·0 to 1·25 in. (the smallest species). The wings are fulvous ; fore wings with a slight projection at the anal angle, yellow on the costa ; they have several rows of small black spots towards the base and larger ones on the central area, and at the apices near which there are one or two white spots. Hind wings dentated along their hind margins, fulvous, spotted with black, and with a hind-marginal row of very small blue lunules. Fringes black and white. Under side ground colour purplish brown, the nervures white. The fore wings have a small white circle in the discoidal cell near the base, and outside this four white lines ; near the apex are some yellowish white and brown markings ; towards the anal angle are some light brown markings, and above these two pure white spots surrounded by light blue. The hind wings have an oblong white spot near the base touching the costa ; across their centre extends, from the costa to the inner margin, a brownish white band, and a line of the same colour runs parallel to the hind margin throughout its entire length, except opposite

the angular projection of the wing, where it is obscured by a large
light blue blotch.   The pattern of the under side, resulting from a
combination of the white lines of the nervures and the delicately
defined markings, is exceedingly intricate, and has earned for the
species its common French name of "Carte geographique."   A
description is very difficult, as also is its representation in a figure.
Mr. Horace Knight has done his best to represent it in the
present work, and I think I may say has succeeded pretty well.
Pl. XXXVIII., 4.

   TIMES OF APPEARANCE.—April and May.

   In this species we have a very well-marked example of seasonal dimorphism ; the typical form *Levana*, L., described above, is
only found as the first brood, but the second brood differs considerably, and will be thus described :—

   Var. a. *Prorsa*, Linn. Syst. Nat. x. 480; Esp. 15-3, 59, 4;
Hüb. 94-6 ; Frr. B. 55.—About the same size and the same shape
as *Levana*, but all the wings are brownish black.   Fore wings with
a white central band interrupted much in the same way as that of
Limenitis Sybilla.   External to this is a row of small white spots
beginning near the apex ; near the anal angle is a red line, which
is continued along the margin of the hind wings.   Hind wings
with a well-defined white band across their centre, and besides the
red line mentioned above, two others, one on either side of it, less
distinct.   Marginal fringes black and white.   Under side as in
Levana, but the ground colour is darker and the yellow spots are
replaced by white ones; the central white band is distinct, and the
light blue blotches are absent.   Pl. XXXIX., 2.

   TIME OF APPEARANCE.—July to September.

   Var. b. *Porima*, O. I. 1, 134.—An intermediate form, which
has the black ground colour of *Prorsa*, but with the white markings
and bands yellow and much narrower.   It is supposed by some
entomologists (Geyer amongst others) to be a form of the
second brood modified by cold weather during its earlier stages.
Pl. XXXIX., 1.

   TIMES OF APPEARANCE.— June and September.

HABITAT.—Moist woods, throughout the greater part of Southern and Central Europe (absent from Britain and Denmark); it also occurs in Armenia, Southern Siberia and the Amur.

LARVA.—Black or dark grey, sometimes with brown stripes; spiny, the spines being black or brownish yellow. Gregarious on *Urtica dioica*.

OBS.—This species is to my mind much nearer to the genus *Pyrameis* of authors than to any of the other *Vanessæ*.

2.--V. **Egea**, Cramer, Pap. Exot. T. 78, C.D. (1778); Herbst.
160, 5-6.
TRIANGULUM, Fab. E. S. 125 (1793); O. I. 1, 123.
V-ALBUM, Esp. 52-1; Bkh. i. 17.

Expands from 1·75 to 2 in. The wings are pale fulvous, with black spots; brown along the hind margins. Fore wings with two black spots touching the costa, two smaller ones in the discoidal cell, and three in the centre of the wings. Hind wings with two or three black spots towards the base. All the wings have, immediately internal to the hind-marginal brown band, a row of indistinct light crescents. The hind margins of the wings are rather deeply incised, the fore wings showing two and the hind wings one well-marked projection. The inner margin of the fore wings is concave near the anal angle. Under side of all the wings light brown, with lighter and darker streaks; in the discoidal cell of the hind wings is a white mark composed of two lines, forming an acute angle like the letter V. Pl. XXXIX., 3.

TIMES OF APPEARANCE.—April to June, and again in September, the species being probably double-brooded.

HABITAT.—Gardens and sunny places in the South of Europe (excepting Spain, Portugal, Hungary and South Russia). It likewise occurs on the shores of the Caspian, in Persia, and in Asia Minor.

LARVA blue and spiny, second segment striped transversely with black and yellow; said to be solitary on *Parietaria officinalis* in July.

Z

3. — **V. C-Album**, Linn. Faun. Suec. p. 279, n. 159 (1761) ;
Syst. Nat. x. 1, 2, p. 778 (1767) ; Esp. Schmett.
i. 1, t. 13, f. 3 (1777) ; Hüb. Eur. Schmett. i. 92,
93 ; H.-S. i. 159, 160 (1844).

Expands from 1·80 to 2·30 in. Has all the characters of
the last species intensified. The hind margins of the wings are
deeply incised, presenting a jagged outline with strongly marked
projections. The inner margin of the fore wings is deeply concave,
much more so than in *Egea*. All the wings are deep fulvous, dark
brown along the hind margins, varied with light brown lunules ;
the black spots are larger and more intense than in *V. Egea*, and
there is one more on the hind wing below the discoidal cell.
The pattern of the under side consists of various shades of brown
varied with dark green or greenish yellow. In the centre of the
wings, immediately below the discoidal cell, is a very distinct
white C-like mark. This species exhibits seasonal dimorphism,
the vernal form having the under side light brown, almost as pale
as that of *V. Egea*, whilst the æstival form is richly variegated in
the under side of the male and very dark brown in the female.
The C-like mark is subject to great variation, sometimes being
shaped like a G, and at others like an O. XXXIX. fig. 4. (The
under sides are those of males of the spring and summer broods,
showing the difference in colouring).

Time of Appearance.—Throughout the fine season from April
to September.

Habitat.—The whole of Europe and the greater part of
Northern Asia (excepting both European and Asiatic Polar Regions),
also Asia Minor, Persia, and Armenia. In Great Britain it is a
very local species, being chiefly found in the Western and Midland
districts of England ; rare in Ireland, and altogether absent from
Scotland. Pl. XXXIX., 4.

Larva.—Greyish brown, a whitish dorsal stripe reaching from
the 7th to the 13th segment. Head with two horn-like projections.
The colour of the head is black, and also that of the second segment, which is covered with minute bristly warts ; the rest of the
body is covered with branching spines, brown and white in colour.
The spiracles are black, surrounded by white and red-brown.

Pupa angulated, brown, decorated with metallic spots on the back of the thorax. The Larva is found from June to August in this country—"on elm, currant, sloe, hop and nettle," Stainton.

4.—V. Polychloros, Linn. Syst. Nat. x. 477 ; Faun. Suec. 278 ; Esp. 13, 1 ; Hüb. 81, 82.

Expands from 2·0 to 2·75 in. Hind margin of all the wings dentated ; two well-marked angular projections on the fore wings and one on the hind wings ; inner margin of fore wings straight. All the wings deep fulvous, with a dark border. The fore wings have the costa yellowish, with three large black spots ; in the centre of the wings are two smaller round ones, and below these, near to the inner margin, two others. The hind wings are darker brown at the base and clothed with hairs ; on the costa is a large triangular black spot, and along the hind margin a row of blue lunules. Under side streaked, various shades of brown darker at the bases, and having a darker line running across the centre of the wings ; the hind margins have a row of faint blue lunules. Pl. XXXIX., 5.

TIME OF APPEARANCE.—Throughout the year in the South ; in Britain, and the more northern parts of its area of distribution, it occurs in the spring after hybernation, its first appearance being however in the summer and autumn from July to September.

HABITAT.—Gardens, fields and road-sides throughout Southern and Central Europe, Asia Minor, Armenia, Syria and Siberia.

LARVA.—Brown, with yellowish lateral and dorsal stripes. The spines are brownish yellow. It is found from May to July on elm and sallows ; rarely in this country on cherry ; on the Continent, commonly, sometimes causing great damage.

VARIETIES.

a. Testudo, Esp. 73, 1.—*Pyrrhomelæna*, Hüb. 845-6. An aberrant form in which the spots on the fore wings are confluent. This is a peculiarity that not unfrequently occurs in this and other allied " tortoiseshells."

b. **Pyromelas,** Freyer, 139, 2, p. 75. Nothing else but a small form of the species ; one meets with a similar peculiarity in the form *Ioides* of *V. Io* ("Vix nomen conservandum" Staudinger).

5.—V. **Xanthomelas,** Wien. Verz. p. 125; Esp. 63, 4; Hüb. 85, 86 ; O. I. 1, 117.

Expands from 2·0 to 2·75 in. To be distinguished from *Polychloros* by the somewhat sharper angular projections of the hind margins, the redder ground colour of the wings; and by the presence of a whitish spot on the apical side of the exterior costal black spot. The blue lunules on the hind wings are rather more definite than in *Polychloros*. The under side greatly resembles that of the last species, but the markings are more defined, and there is on the centre of the hind wing a faint light spot. The legs are lighter than in the last species. Pl. XL., 1.

TIME OF APPEARANCE.—July to September.

HABITAT.—Woody places in the neighbourhood of willows, on which the larva feeds; it is confined to the Eastern parts of Europe, including Germany, the North-East of Switzerland, Hungary, and Central Russia to 60°; it also occurs in Northern India.

LARVA.—Black, with white dots, yellowish white dorsal and lateral stripes, the spines black. Gregarious on various kinds of *Salix* in June.

6.—V. **Vau-Album,** Wien. Verz. p. 176 (1776) ; Hüb. 83, 84 ;
　　　　Fab. Mant. 50 (1787); Fr. 133 ; O. I. 1, 112 ; B. Ic. 24, 1.
　　L-ALBUM, Esp. 62, 3, a. b. (1780); Hbst. T. 62, 3, 6.

Expands from 2·0 to 2·50 in. The hind margins of all the wings are much more strongly dentated than in the last two species, but the inner margin of the fore wings is straight as in those species, and not concave as in *V. C-album*. All the wings deep fulvous, with black spots similar to those of the last two species ; there are no blue lunules either on the fore or the hind wings. The fore wings are black towards the apex, with a conspicuous white spot. The hind wings have, near the costa, a large white spot between two black ones. Under side more varied than in the last, being dark brown towards the bases, and bluish grey varied with brown and green from the centre of the wings nearly to the hind margins, which are darker. On the hind wings, immediately below the discoidal cell, is a whitish letter-like mark, not so distinct as in *V. Egea* or *C-Album*. Pl. XL., 2.

TIMES OF APPEARANCE.—June and July.

HABITAT. — Woods, &c., in the East of Europe, namely, North-Eastern and East-Central Germany, Hungary, &c.; Central Russia to 60°, and Northern Asia.

LARVA brownish red, with a dark dorsal line and yellowish white lateral stripes; the spines are yellowish with darker tips. Gregarious on poplar and sallow in June.

7.—V. Urticæ, Linn. Syst. Nat. x. 477; Esp. 13, 2; Hüb. 87-9: O. I. 1, 120.

Expands from 2·0 to 2·30 in. Hind margins of all the wings dentate. The fore wings have only one angular projection, that near the anal angle being absent in this species. All the wings are reddish orange, with rather narrow dark border inclosing a row of bright blue lunules. The fore wings have three black spots on the costa, two small ones in the centre and one larger one on the inner margin, not two as in the preceding species; the space between the costal spot is yellow; external to the third one is a white spot; there is also a yellow patch external to the inner-marginal spot. Hind wings blackish at the base, on the costa a large black spot with a yellow patch placed externally to it. Under side: The markings have much the character of those of the preceding species, but are much lighter; the figure very accurately represents them. Pl. XL., 3.

TIME OF APPEARANCE.—Throughout the year in the South of Europe; all through the fine season in Britain and the more Northern regions.

Habitat.—Gardens, fields, waste places, in fact almost any kind of country throughout the whole of Europe and North Asia; it is, however, absent from North America and from North Africa; in Britain it is about the most widely distributed, and one of the commonest butterflies, occurring everywhere.

Larva yellowish grey, with a black dorsal line and with a broad brown lateral stripe, beneath which is a yellow line. The head and spines are black. It feeds on the common nettle (*Urtica dioica*) in June and July.

Obs.—*V. Urticæ* is subject to a great many variations in the arrangement of spots and in the breadth of the dark marginal band; indeed it would be possible to fill a dozen plates with figures of aberrations of this kind. I have had many beautiful and interesting specimens kindly brought to my notice by Entomologists, and I only regret that want of space prevents their being figured in the present work.

VARIETIES.

a. **Turcica**, Staudinger, Cat. p. 16 (1871). Inhabits the Balcan provinces and Asia Minor, and is intermediate between the typical form and the next; it is redder in colour than the type, has the two small central spots nearly absent, and the yellow patch on the inner margin is wanting. The eyes are red instead of brown. Pl. XL., 4.

b. **Ichnusa**, Bon. descr. t. 3, 2; Dup. i. 23, 4; B. Ic. 24, 2; Hüb. 840; Rbr. Ann. Soc. Fr. (1832), pl. 7, 3. This important variety or sub-species differs from the type in the shape of the wings as well as in the markings; the wings are much less angulated than in true *Urticæ*, the central and inner-marginal spots are either entirely absent or very nearly so. The under side is more strongly marked. Pl. XL., 5.

Habitat.—Corsica and Sardinia.

Obs.—Varieties of *Urticæ* sometimes occur in Britain and elsewhere, in which the dorsal and median spots are in abeyance on the fore wings, but these have not the same outline as the true *Ichnusa* which is confined to the above islands.

c. **Polaris**, Staudinger, Cat. p. 16, 1871.—*Urticæ*, Stett. Ent. Zeit. 1861, 345. This is a constant Polar form of the insect in which the wings are very dusky, the inner-marginal or dorsal spot being joined to the costal spot.

8.—**V. Io**, Linn. Syst. Nat. x. 472 ; Faun. Suec. 274 ; Esp. 5, 2 ; Hüb. 77, 78 ; O. I. 1, 107.

Expands from 1·75 to 2·75 in. All the wings are dull red, with the hind margins brown. The fore wings have two black costal marks, external to which is a large eye-like spot composed of various colours, principally blue, and below this two small spots of light blue. The hind wings have a large blue and black eye surrounded by light brown. The under side is very dark brown, varied with nearly black lines. Pl. XLI., 1.

TIME OF APPEARANCE.—April to October ; the imagines hybernating.

HABITAT.—Gardens, woods, &c., throughout Europe, with the exception of the Polar Regions, Andalusia, and Sicily ; it is also found throughout temperate North Asia and Asia Minor ; common in England, and well known as the " Peacock " ; scarce in Scotland.

LARVA.—Black, with minute white spots and covered with black spines. Feeds on the common nettle, *Urtica dioica*, from June to August.

VARIETIES.

a. **Ioides**, O. I. 1, 109. A small form of the insect not confined to any particular locality.

b. **Sardoa**, Staud. Cat. p. 16, 1871. A variety found in the Island of Sardinia, and differing from the type in being larger and more deeply coloured.

9.—**V. Antiopa**, Linn. Syst. Nat. x. 476; Faun. Succ. 277; Esp. 12, 2; Hüb. 79, 80; O. I. 1, 110.

Expands from 2·30 to 3·30 in. Wings with strongly marked angular projections; reddish brown, with a purplish tinge. Hind margins broadly white or light yellow; internal to this light border is a narrower deep black one enclosing a row of conspicuous blue or lilac spots. The fore wings have two whitish costal spots, and at the basal end of the costa a row of small white dots. Under side dark brown varied with black, and with a broad whitish hind-marginal border. Pl. XLI., 2.

TIMES OF APPEARANCE.—From the beginning of August to the middle of October, and again from March to May, after hybernation in the colder regions of its distribution; in the warmer countries it is on the wing nearly throughout the year.

HABITAT.—Woods, gardens, river-banks, &c., throughout the greater part of the Northern Hemisphere, occurring throughout Europe with the exception of Andalusia and the Russian Steppes, Northern Asia, Asia Minor, North Africa, and North and Central America; in some parts of Europe, as in Britain, it is rare or very capricious in its appearance; in other countries, as in Switzerland for example, it is as common as *V. Io* or *Atalanta*; I have seen it in the streets of Paris.

LARVA.—Black, with white dots; from the fifth to the eleventh segments is a row of dorsal light red spots. The spines are black or dark brown; it feeds on *Salix alba*, sometimes on nettle or on birch, in the spring and early summer.

ABERRATION.

a. **Hygiæa**, Hdrch. Cat. p. 7.—*Antiopa*, Hüb. 933; Fr. 145.— An aberration occurring occasionally in the same localities as the typical form. The white or light yellow hind-marginal band is very broad, and the blue spots are either completely or partially absent. This form, which is very rare, has occurred in Britain, but seems to be commoner in North America than in Europe.

10.—V. Atalanta, Linn. Syst. Nat. x. 478; Esp. 14, 1; Hüb.
75, 76; O. I. 1, 104; Frr. 181.

Expands from 2·50 to 2·80 in. Fore wings very little angulated.
All the wings black; fore wings with one large and five small
white spots towards the apex, and crossed in the centre by a
brilliant scarlet-red band. Hind wings with a broad hind-marginal
band of brilliant red enclosing four black spots; at the anal angle
is a blue spot surmounted by a black crescent. Marginal fringes
black and white. Under side: The hind wings are variously
mottled with different tints of grey, brown, purple, and blue, the
red bands appearing of a lighter colour than above; the entire
pattern of the under side is very intricate and difficult to describe.
Pl. XLI., 3.

Times of Appearance.—June to October, and after hybernation, in the spring.

Habitat.—Woods and gardens throughout Europe (with the
exception of the Polar Regions), Asia Minor and North Africa; it
also occurs in North America. In India and China it is replaced
by the next species which is closely allied, and in New Zealand
and the Sandwich Islands there are also species very much resembling it with black red-banded wings. (*V. Cordelia*, Doubl., and
V. Gonerilla, Fabr.).

Larva.—Greenish grey sprinkled with black, and with a pale
yellow lateral line. The legs are shining black; the claspers reddish
brown; spines black or reddish brown; very variable in colouration.
Feeds on the common nettle (*Urtica dioica*) in June and July.

Pupa angulated, yellowish grey varied with brown or purple,
and decorated on the thorax with metallic spots, as is usual in this
genus.

11.—V. Callirhoë, Fab. Syst. Ent. 473 (1775); Mill. Icon. Pl.
88, i. p. 354.
Atalanta Indica, Hbst. T. 180, 1, 2.
Vulcanica, Godt. Encyc. Meth. ix. 320.

Expands from 2·25 to 2·50 in. Close to *Atalanta*, but rather
smaller; and the red band of the fore wings is larger, more sinuous,
and of a deeper red; it is also marked on its internal edge with

2 A

three black indentations of different sizes. The white spots at the apex are smaller than in *Atalanta*. The markings of the under side differ somewhat from those of *Atalanta*, as will be seen in the figure, the most notable difference being the almost entire obliteration of the light yellow patch on the costa of the hind wings. Pl. XLII., 1.

TIME OF APPEARANCE.—August.

HABITAT.—China and Northern India; the Canaries, chiefly the Isle of Teneriffe, whence it has been imported and has become acclimatised in the South of Portugal and Andalusia; so that it is now always considered a European butterfly, being included in Staudinger's Catalogue and in that of Berce.

LARVA.—Undescribed. Milliere says it feeds on nettles, like that of *Atalanta*.

OBS.—Never having seen a European example of this species, I am obliged to make use of Milliere's representation of a Portuguese example received from Dr. Staudinger.

12.—V. **Cardui**, Linn. Syst. Nat. x. 475; Esp. 10, 3; Hüb. 75, 76; O. I. 1, 102.

Expands from 2 to 2·80 in. All the wings are pale reddish orange, sometimes with pinkish tinge, spotted and mottled with black; bases blackish. The fore wings are black at the apex, with four or five white spots. The hind wings have two rows of black spots running parallel to the hind margin. Under side: The fore wings are pink, the apex grey, tinged with yellow, and with large white spots. The hind wings are yellowish grey, marbled with different colours, and with several white spots, a triangular one in the centre being the most conspicuous. Along the hind margin is a blue line, interrupted with yellowish white; and internal to this a row of dark blue eyes in pale rings, outlined with dark brown. Pl. XLII., 2.

TIMES OF APPEARANCE. —March to June, after hybernation; and from July to the end of autumn.

HABITAT.—A truly cosmopolitan species, being found in every part of the world, excepting perhaps the Polar regions. It occurs on high mountains, in valleys, plains, woods, fields, &c. As a British insect it is extremely capricious in its appearance, sometimes not being seen for some years and then suddenly appearing in great numbers more or less throughout the country. It appears to be very fond of clover-fields and of ivy-bloom in the autumn.

LARVA.—"The colour of the head is dull black; the dorsal surface of the body is black; the spines paler, with black tips and branches; the hairs are white; the skinfold separating the dorsal and ventral surface is yellow; the ventral surface, legs, and claspers are pitchy red; the spiracles above the skinfold are pale in the middle, then surrounded with black, then again with paler; In many individuals the dorsal surface is irrorated with yellowish white dots, which are more conspicuously collected in a double series along the back, interrupted by a narrow medio-dorsal stripe intensely black; in these examples the bulbous base of each spine is pitchy red."—Newman, Brit. Butt., page 65.

PUPA.—Yellowish grey, striped with brown, and with metallic gold-coloured decorations.

The LARVA feeds on various species of *Carduus*, generally the common field-thistle.

Genus 4.—**MELITÆA**, Fab. Ill. Mag. vi. p. 284 (1807); Doubl. Gen. D. L. p. 177 (1848).

ARGYNNIS, Lat.

Larvæ covered with spine-bearing tubercles; they are gregarious under a common web until after hybernation; they feed on various low plants, such as *Scabiosa*, *Plantago*, and *Veronica*.

Pupæ short and thick, not angulated as in *Vanessa*, blunt anteriorly; they are without metallic markings, but generally spotted with black or yellowish.

Imagines usually below the medium size. The wings are subdenticulate, not angled; the discoidal cell of the fore wings generally half closed, that of the hind wings open; in colour the wings are fulvous, generally brightly so, with black square spots arranged in transverse rows. The hind wings have on their under side a central band of pale yellow, sometimes broken up into detached spots; there are always similarly-coloured spots clustered near the base of the wing, and a row placed along the hind margin.

The antennæ are long and slender, with pyriform clubs—rounded at the apex in living specimens.

The eyes are smooth. The palpi are generally longer than the head and are hairy, especially on the middle articulation.

This genus is confined to the Palæarctic and Nearctic regions; in the latter its range is restricted to the South Western States.

The genus *Melitæa* is full of difficulties for the zoologist, as the exact limits of species and varieties have not yet been satisfactorily determined; the local forms of some of the species are very numerous and perplexing, and in many instances are no doubt specifically distinct.

Staudinger gives sixteen species as being found in Europe; these for the sake of convenience may be divided into three groups, of which our three British species, *M. Aurinia*, *Cinxia*, and *Athalia* may be taken as representatives respectively.

1st Group.   Central light band of under side hind wings not spotted:—*M. Cynthia, Iduna, Maturna, Aurinia, Bertica.*

2nd Group.   Central light band of under side hind wings with a row of black spots:—*M. Cinxia, Arduinna, Phœbe, Trivia, Didyma.*

3rd Group.   Central light band broken up into distinct blotches, without black spots, but more or less varied with white or very pale yellow, and almost coalescing with the basal spots, which are large:—*M. Dictynna, Deione, Athalia, Aurelia, Parthenie, Asteria.*

1.—**M. Cynthia**, Hüb. 569, 570 (fœm.), 608-9 (mas.); O. I. 1. 21; Dup. 1, 21, 25; Frr. 247.

Expands from 1·40 to 1·50 in.   The male has all the wings white, with a bluish gloss; the hind margins are broadly dark brown, with rows of large reddish fulvous spots, forming a double band on the fore wings, and a single one enclosing a row of black dots on the hind wings.   The bases are blackish; the fore wings have two reddish discoidal patches, and external to these a short black band.   The hind wings have an irregularly-shaped basal blotch of reddish brown, enclosing two white spots. The female has all the wings dull fulvous; the bases are blackish. Fore wings with two dusky bands and two spots in the discoidal cell, brighter fulvous than the rest of the wing. Hind wings like those of the male, but dull fulvous instead of white. Under side: all the markings have a cloudy appearance; the fore wings are dull white in the male, fulvous in the female, with bright fulvous markings. The hind wings are bright fulvous, with a central white band, and externally to this a reddish fulvous one enclosing a row of black spots; hind-marginal and basal spots white. Pl. XLII., 3.

TIMES OF APPEARANCE.—July and August.

HABITAT.—Pastures in the higher Alps of Switzerland and North Italy.

LARVA.—Black, with greenish yellow rings and yellowish lateral stripes; the spines are black, and the head brownish. It feeds on several species of plantain in June.

2.—**M. Iduna**, Dalm. Pap. Suec. p. 75, n. 2 (1816); Frr. 611, 1, 2; Wallengren, Scandinaviæ Rhopalocera, p. 70.
     MATURNA, Hüb. 600-1, 807-8.

Expands from 1·16 to 1·30 in. Wings dull ivory-white, hind margins narrowly dark brown; nervures rather strongly black; bases dark brown. Fore wings with a broad fulvous band placed parallel to the hind margin; internal to this is a narrow waved dusky band; in the discoidal cell are two bright fulvous spots bordered with black, and another immediately below them at the base. Hind wings with a bright fulvous band placed similarly to that of the fore wings, and not spotted as in *Cynthia;* near the base is a large fulvous patch enclosing a white spot. Under side: Very similar to that of *M. Cynthia*, but more distinctly marked, and with the markings much more distinct; the fulvous band of the hind wings is without black spots; head, thorax, and abdomen black above, fulvous beneath; antennæ black and white, clubs black, with the apices fulvous; palpi with black and fulvous hairs. The female is somewhat larger than the male, and has the wings less marked with white. Pl. XLII., 4.

TIME OF APPEARANCE.—July.

HABITAT. — Lapland (at Quickjock, *Wallengren*) ; a rare and local species—supposed by some to be a variety of the following, but apparently perfectly distinct. Besides Lapland, Staudinger gives the Altai as a habitat for this species.

LARVA.—Unknown.

3.—M. **Maturna**, Linn. Syst. Nat. x. 480; Faun. Suec. 280 ; Hbst. 275, 1-4; Frr. 307 ; H. S. 132.
CYNTHIA, Esp. 37, 2 ; Hüb. 1, 2.

Expands from 1·45 to 1·75 in. All the wings bright reddish fulvous, blackish at the base. The hind margins are black, and have immediately internal to them a row of black crescentic spots. The fore wings have several rows of coalescing spots of a light yellow colour, surrounded by black. The hind wings have spots of a similar character, and also some near the base. Under side of all the wings bright reddish fulvous, with a hind-marginal row of crescentic yellow spots; fore wings with a patch of whitish yellow extending across the centre of the wings from the costa; in the discoidal cell are two yellow spots, the external one being surrounded with black. Hind wings with a very distinct and broad light yellow band running across their centre, divided throughout its whole length by a narrow black line, but not spotted; midway between this and the base are three or four irregularly-shaped light yellow spots, and at the base is a similarly-coloured smaller one. Head, thorax, and abdomen black above, bright fulvous beneath ; antennæ tipped with fulvous. The female is larger than the male, and somewhat brighter on account of the black markings not being so distinct. Pl. XLIII., 1.

TIME OF APPEARANCE.—June.

HABITAT.—Open places in woods throughout Central Europe (excepting Great Britain and Denmark), the South of Sweden, Russia (excepting the Polar portions), and Central Asia to the Altai.

LARVA.—Black; head and spines black, with pale yellow spots arranged in three rows. On plantain and scabious in May.

Uralensis, Ld. in Coll.; Stgr. Cat. p. 17. "Albo-nigro-rufoque variegata, transitus ad Idunam."—Stand.

HABITAT.—The Ural. I have not seen a specimen.

4.—**M. Aurinia**, Rott. Naturf. vi. p. 5 (1775).
> ARTEMIS, S. V. (1776); Hüb. 4-6, 653; Lewin, Ins. T. 15,
>   1-4; Frr. B. 7; H. S. 364-5, ab.
> MATURNA, Esp. 16, 2, 41, 3.
> LYE, Bgst. Nom. T. 82, 4, 5.

Expands from 1·30 to 2·16 in. Wings reddish fulvous, with lightish yellow spots surrounded by black. The hind margins of all the wings with a row of brownish lunules, external to which is a very narrow black border. Under side glossy (whence the common English name of the insect, "Greasy fritillary"), much duller and less distinct than in *Maturna*, the yellow spots on the fore wings being scarcely visible. The hind wings have a central light band and light basal spots; along the hind margin is a row of light lunules, and between these and the central band a row of light spots enclosing black dots. Pl. XLIII., 2.

TIMES OF APPEARANCE.—May to August.

HABITAT.—Marshy meadows in Central and Northern Europe (excepting the Polar regions), and in Armenia.

LARVA.—Black, with a lateral row of white dots; the head as well as the spine black; legs reddish.

PUPA.—Light yellow, spotted with black; the wing-cases white. It feeds on plantain, scabious, &c., from April to September.

VARIETIES.

a. **Merope**, Prun. Lep. Ped. p. 73, O. iv. 104; Dup. 1, 21, 1; B. Ic. 22, t. 7; Frr. 13, 1; H. S. 134-5. Expands from 1·20 to 1·40 in. Much duller than the typical form, being clouded with yellowish or greenish brown, and showing but few fulvous spots; the wings have thus a dark appearance. Pl. XLIII., 3.

TIME OF APPEARANCE.—July.

HABITAT.—The higher Alps of Switzerland and North Italy; perhaps also the Altai.

b. **Provincialis,** Boisd. Gen. p. 20.—*Maturna, Var.* Esp. 97, 4, Suppl. 1, p. 23. Larger than the type, the females measuring more than 2·75 in. Paler and more uniform in colour, the bands being brown instead of yellow. The hind wings have along their hind margin a row of white lunules. The under side has the markings much more distinct than in the type, especially the light central band of the hind wings. The var. *Provincialis* replaces true *Aurinia* in the South of France, Piedmont, Dalmatia, and Bithynia. Pl. XLIII., 5.

    N.B.—In the lettering at the foot of pl. 43, the numbers belonging to the figures of var. *Provincialis* and *Desfontainii* have unfortunately been transposed. Figure 5 correctly represents *Provincialis*, and fig. 4 *Desfontainii.*

c. **Desfontainii,** Godt. Enc. Méth. p. 278; Ramb. Cat. And. Pl. I. 1, 2.—*Beckeri,* Ld. 2, b. V. 1852, p. 39.—*Desfontainesii,* H. S. 1, 2. Somewhat larger than the type, but not so large as the last. The wings are very rich dark brownish fulvous, the yellow bands being more distinct than in *Provincialis.* The hind wings have brown, and not white, hind-marginal lunules. The markings of the under side are very distinct, somewhat as in the last. This form of the species is found in Spain and Portugal, and has been often confounded with the next species, from which, however, it is perfectly distinct. Pl. XLIII., 4, *vide supra.*

d. **Orientalis,** H. S. 265-6. — *Artemis* var. Frr. 571, 1. — *Sareptana,* Stgr. Somewhat more varied in colour than the type, and paler on the under side.

    HABITAT.—South Russia (*Sarepta*); also Asia Minor.

5.—**M. Bœtica,** Rambr. Cat. S. And. p. 11, pl. I. 3, 4.
        DESFONTAINESII, H. S. 569-70; Kirby, Man. Europ. Butt.
        p. 24.
        DESFONTAINESI, B. Ic. 23, 1, 2.

    Expands from 1·80 to 2·25 in. All the wings bright fulvous; fore wings with two narrow bands of rather bright yellow, edged with black as in the allied species; hind margin dark brown, with a row of yellow crescentic spots; internal to this a row of indistinct yellow dots or spots. Hind wings as in *Aurinia*, but the yellow band more decided and brighter; the hind margin has a row of yellow spots. Under side not so glossy as in *Aurinia*. Fore wings somewhat resembling that species in markings, but much lighter, the light markings being nearly white; the black spots are very distinct. Hind wings very light, with the markings indistinct and diffused, and quite unlike any of the allied species. Head, body, antennæ, palpi, &c., as in the last species. Pl. XLIII., 6.

TIME OF APPEARANCE.—June.

HABITAT.—Andalusia, very local.

LARVA.—Not described.

6.—**M. Cinxia**, Linn. Syst. Nat. x. 480; Hufn. B. Mag. ii. p. 66 (1766); Frr. B. 103.

CINXIA MAJOR, Esp. 25, 2.

DELIA, Bkh. i. 50, ii. 193, fig. 9, 10; Hbst. 276, 5-8; Wallengr. Rhopal. Scand. 73.

PILOSELLÆ, Esp. 47, 3.

Expands from 1·50 to 2·0 in. All the wings fulvous, with a network of black spots and lines; the hind margins black; fringes spotted black and white. Under side: Fore wings pale fulvous, with a few small black spots; apex pale yellow. Hind wings with three pale yellow bands, and two of a bright fulvous colour. All the pale yellow bands with a row of black spots. Pl. XLIV., 1.

TIMES OF APPEARANCE.—May and June, and again in August.

HABITAT.—Europe generally (excepting the Polar regions and the southern part of the Spanish Peninsula); also Asia Minor, Armenia, and Siberia, frequenting meadows. In Britain it is extremely local, being found only in the South of England, near the sea-shore. ("Glanville Fritillary.")

LARVA.—Black, dotted with white in transverse rows; the spines are black; the head and legs reddish brown.

PUPA.—Orange-brown, dotted with black.

The Larva feeds on *Hieracium*, *Pilosella*, *Veronica*, and *Plantago*, in May and June, and from August to September.

2 B

7.—M. Arduinna, Esp. 87, 4 (1783), ii. p. 169; Hbst. 276,
9, 10; Frr. 277, 1; H. S. 319-21.
URALENSIS, L. V. Lep. Ross. p. 77, T. x. 1, 2.

Expands from 1·50 to 2·0 in. Wings brighter fulvous than in
Cinxia. Black markings more in the form of spots than of a
reticulation, and less numerous; otherwise having very little the
character of those of that species. Under side: Fore wings with
a distinct central row of black spots; the apical patch of light
yellow extending for some distance along the hind margin Hind
wings greatly resembling those of *Cinxia,* but the basal light band
is more broken up; the central band broader; the hind-marginal
band without black spots, the absence of which at once distinguishes
this species from the last. Pl. XLIV., 2.

TIMES OF APPEARANCE.—May and June.

HABITAT.—Very limited. The South-Western Ural, Armenia,
Siberia, Northern Turkey, and Croatia.

8.—M. Phœbe, Knoch. Btr. iii. 124 (1783); Fab. Mant. 58;
Hüb. 13, 14, O. I. 1, 39; Frr. 325; H. S. 263-4.
CORYTHALLIA, Esp. 61, 4, 5, 72, 2.
ATHALIÆ VAR. Esp. 61, 6.

Expands from 1·50 to 2·0 in. The fore wings fulvous, much
duller than in the preceding species, with several rows of black
spots arranged in lines. Hind wings black, with four rows of
fulvous spots, and a group of fulvous patches near the base.
Under side: Fore wings much as in the last, but paler fulvous;
and the apical light markings are generally yellower. Hind wings
fulvous, with yellowish white bands; the central band is more
irregular than in the preceding species, and has a row of black
spots running at first along its centre from the costa and then
touching its basal edge. On its hind-marginal side the pale band
extends beyond the usual edging of black spots, which thus appear
on the light band as another row. The fulvous band between this
and the hind-marginal light band is very narrow, and the hind-
marginal band itself is not spotted. Pl. XLIV., 3.

TIMES OF APPEARANCE.—May to August.

HABITAT. — Southern and Central Europe (except North Germany and the British Isles), Western Asia, the Altai, the Amur, and North Africa. Local, frequenting meadows near woods.

LARVA.—Black, spotted with white, and with a lateral band brown or fulvous. Spines black, except those on the fulvous band, which are fulvous. The ventral surface is reddish grey (Guenée). It feeds on *Centaurea jacea* from May to September.

VARIETIES.

a. **Ætheria**, Ev. Lep. Ross. p. 73, pl. ix., 5, 6. Generally somewhat larger than the type, though not so in the case of the specimen figured in the present work. Much paler in colour, and with fewer spots. It occurs in South-East Russia, and Siberia. Pl. XLIV., 4.

b. **Caucasica**, Stg. Hor. 1870, p. 29, T. 1, 2. A variety figured by Staudinger from Armenia and Greece. Larger and darker than the type.

c. **Occitanica**, Stg. Cat. p. 18. Mentioned in Staudinger's Catalogue, with this note, "Forma magis variegata," and with the habitat, *Italy*.

d. **Ætherie**, Hüb. 875-8. As may be seen by Hübner's figure, a more unicolorous and deeper coloured form of the species— probably originating in North Africa, but now found on the sea-coast of Andalusia, as well as on the African side of the Mediterranean.

9.—**M. Trivia**, Schiff. S. V. P. 179 ; O. I. 1, 36 ; Frr. B. 91 ;
  Hüb. 11, 12.
  IPHIGENIA, Esp. 77, 1, 2 ; Bkh. i. 61.

Expands from 1·35 to 1·60 in. All the wings deep reddish fulvous, bases blackish. Fore wings with a double hind-marginal row of black crescents, a central row of black spots, and some less distinct black markings near the base. Hind wings similarly marked, there being three rows of black spots, besides some basal marks, and a black hind-marginal border. Under side marked as in *Cinxia*; but the colour is much deeper, and all the markings are more distinct; the light apical patch of the fore wing extends almost along the whole length of the hind margin; the black spots are very strongly marked. Hind wings deep reddish fulvous, with three straw-coloured bands spotted with black. Pl. XLIV., 5.

TIMES OF APPEARANCE.—June to August.

HABITAT. — South-Eastern Europe, and Siberia; Western Asia, and the Amur.

LARVA.—Grey, with bluish grey spots mixed with brown ones. Head red, with black dots; legs white, with black spots; spines reddish, tipped with white (Guenée). Feeds on *Verbascum Thapsus* and other kinds of mullein in June.

VARIETIES.

a. **Fascelis**, Esp. ii. p. 171; O. iv. 107; Hüb. 871-2.—Larger than the typical form, the female having the wings dusky. A variety of the spring brood found in South Russia (Sarepta).

b. **Nana**, Staud. Cat. p. 18; *Phœbe*, Esp. 88, 5, 6 (1783); *Fascelis*, Hüb. 873-4.—Smaller than the type, only having about two-thirds of its expanse; it is also much paler, having the ground colour of the wings pale fulvous. It inhabits the South of Turkey and other places in the South-East of Europe, occurring as the second brood.

10.—**M. Didyma**, O. I. 1, 30; Frr. B. 85, 104, 1, 2; H. S. 560-2; Esp. 61, 1.

Expands from 1·35 to 1·60 in. All the wings deep reddish fulvous. Hind margins black. The male has on the fore wings a row of crescentic black spots running parallel to the hind margin, a somewhat irregular central row; there is a black discoidal mark, almost ring-shaped, and some basal spots. Hind wings with a hind-marginal row of black spots, and some black basal patches and spots. The female is slightly duller in colour; and the fore wings have a faint row of spots between the hind-marginal and central rows, and traces of a similar row on the hind wings. Under side almost exactly as in *Trivia*, but not so brilliant in colour or so strongly marked; the apical light patch of the fore wings does not extend along the hind margin as in that species.

Obs.—The female figured in this work is taken from a Swiss specimen, and is rather duller than usual; it in fact approaches somewhat to the var. *Alpina.* The species is very variable, especially as regards the female, there being a tendency, especially in the southern forms, towards a dark coloration in that sex, as will be seen in the descriptions of varieties. Pl. XLIV., 6.*

TIMES OF APPEARANCE.—June and August.

HABITAT.—Germany, Switzerland, France, and Hungary. In the more southern and eastern portions of its habitat the species varies greatly, as will be seen by the long list of varieties.

LARVA.—Bluish grey, with white dots and black bands; lateral stripes yellow; spines white and fulvous. Feeds on plantain, &c., from April to June.

VARIETIES.

a. **Alpina**, Staud. Cat. p. 8; *Fascelis*, var. H. S. 267-8.—The common alpine form of the insect in which the female has the fore wings clouded with a dusky greenish colour, the hind wings being of the normal colour.

HABITAT.—Switzerland, in the higher alpine regions.

b. **Meridionalis**, Staud. Cat. Hor. 1870, 60.—The male deeper fulvous; the female clouded with grey, especially on the fore wings. This is a common Southern form of the species.

HABITAT.—Sicily, Turkey, Greece, Asia Minor, and Syria.

c. **Neera**, F. de W. Bull. In. 1840, p. 31; *Didyma* var. H. S. 133.—Male deeper fulvous, and with fewer spots than the type. (Pl. XLIV., 7). Female pale fulvous or greenish.

HABITAT.—South Russia (Sarepta).

d. **Græca**, Staud. Hor. 1870, 60, T. 1, 3.—Male deeper fulvous, with a broad black border; female generally suffused with greenish grey.

HABITAT.—Greece.

e. **Caucasica**, Staud. Cat.; *Trivia* var. H. S. 588-90.—Larger than the type, both sexes deeper fulvous; the female with white spots on the costa of the fore wings near the apex.

HABITAT.—Armenia and the Caucasus.

* Wrongly written 5 at foot of Plate XLIV.

f. **Occidentalis**, Staud. Cat. Hor. 1870, 60; *Cinxia*, Hüb. 869-70; *Trivia*, var. H. S. 326-7. — Differs from the type only in being of a lighter fulvous in both sexes.

Habitat.—South-Western Europe, Dalmatia, Armenia, and North Africa.

g. **Persea**, Koll. Ins. Pers. p. 11; *Dalmatina*, Staud. Cat. Hor. 1870, 60. — Something like the last, but smaller, both sexes being pale fulvous and having very few spots, as in the Russian var. *Neera.*

Habitat.—Dalmatia and Persia.

11.—M. **Dictynna**, Esp. i. 1, 48, 2 a, b (1799); Godt. i. 4, 3; Frr. 319.

Corythalia, Hüb. 15, 16.

Expands from 1·25 to 1·50 in. Male : Fore wings dark blackish brown, with three rows of fulvous spots, and several basal spots of the same colour. Hind wings black, with two rows of fulvous spots. Female : Fore wings fulvous, with black bands ; hind wings black, with three rows of light fulvous spots, sometimes nearly white. Under side : Fore wings fulvous, with a few black spots, and some yellowish ones at the apex. Hind wings straw-colour, with a central row of light spots, some of them white, and almost as bright as those met with in the small species of *Argynnis;* on either side of this row is a bright brown band, that nearest the hind margin being the broadest. Marginal fringes brown and white. Pl. XLV., 1.

Times of Appearance. — June to the middle or end of August.

Habitat. — Meadows and open places in woods, in Central Europe (with the exception of Great Britain), Southern Scandinavia, Dalmatia, Turkey, and South Russia.

Larva.—Dark brown, with bluish grey dots, and with a black dorsal line. The spines are reddish brown, with black tips. Head black, with greyish blue spots. It feeds on *Veronica* in May.

12.—M. Deione, H.-G. 937-40; Dup. i. p. 276, pl. 44, 1, 2.

Expands from 1·50 to 1·75 in. Wings orange-fulvous, the nervures distinctly marked; bases of all the wings black; hind margins narrowly black; marginal fringes black and pale yellow. Fore wings with three narrow black bands running across from the costa to the inner margin; near the base are also several narrow black lines. Hind wings marked in a similar manner to the fore wings, having three narrow wavy black bands and some basal lines. Under side : Fore wings fulvous, with some semilunar spots near the apex; pale yellow, narrowly edged with black; three or four black spots in the centre of the wings near the costa, and one large one near the inner margin ; in the discoidal cell are several black lines. Hind wings pale fulvous, with two broad straw-coloured bands, the central one having a row of very faint black spots ; along its centre there are four or five large straw-coloured spots, narrowly edged with black near the base. Pl. XLV., 2.

TIME OF APPEARANCE.—June.

HABITAT.—The South of France and Andalusia.

LARVA.—Undescribed.

13.—M. Athalia, Rott. Naturf. vi. p. 5 (1775); Esp. 47, 1, a, b, I. p. 377 ; O. I. 1, 44 ; Frr. B. 49.

MATURNA, Hüb. 17, 18.

LEUCIPPE, Schn. Syst. B. p. 209, a, b ; Bkh. ii. 197.

Expands from 1·50 to 1·75 in. All the wings deep brownish fulvous, with black markings arranged as in the last species, but much broader, the whole insect having a darker appearance. Under side : Fore wings pale fulvous, a row of black spots running across the wing near its centre, several black lines and spots internal to this ; near the apex and along the greater part of the costa is a patch of straw-coloured spots. Hind wings with the light bands very broad, the central one often nearly white, without spots. Pl. XLV., 3.

TIMES OF APPEARANCE.—May to the end of August.

HABITAT. — Woods and meadows near woods throughout Europe ; Asia Minor, in the mountainous parts of Armenia, and Siberia. Local in Britain. ("Heath Fritillary.")

LARVA. — Black, dotted with white; head and prolegs black; spines rust-coloured (Dup.)　Feeds on *Plantago*, &c., in May and September.

The following aberrations have been named:—

a. **Corythalia**, Hüb. Btr. ii. 2, T. 3. — The fore wings more broadly fulvous than in the type.

b. **Navarina**, Selys-Longchamps, Catalogue des lépidoptères de la Belgique, Liege, 1837, p. 19.—Almost entirely black above, with hind-marginal fulvous band.

14.—M. Aurelia, Nick. Syn. p. 12; Frr. 641, 2, 3.
PARTHENIE, Hbst. T. 283, 1-4, x. p. 238.
ATHALIA MINOR, Esp. 89, 2 (1).
ATHALIA, Hüb. 19, 20.

Expands from 1·25 to 1·30 in.　Very close to *M. Athalia*, but rather smaller; the wings are lighter in colour both as to the ground colour and the markings.　The under side is much the same as in *Athalia*, but the white spots are smaller and less distinct.　Pl. XLV., 4.

TIMES OF APPEARANCE.—June and July.

HABITAT. — Northern and Eastern Germany, Switzerland, Lapland, Russia, and Armenia.

LARVA. — Black, dotted with white, and with yellow lateral spots; spines black.　Feeds on *Veronica* in June.

VARIETY.

Britomartis, Assman. Bregl. z. 1847; *Veronicæ*, Dorfmeister z. b. 1863, p. 136. — According to the original description this variety differs from the type in having the exterior margin of the fore wings fulvous.　I confess to my inability to see sufficient difference in specimens which have been shown to me as *Britomartis* to constitute a variation; nevertheless this form is supposed to exist, and inhabits Eastern and South Germany and the Valais in June and July.

15.—M. Parthenie, Bkh. ii. 194; Godt. ii. 9, 7, 8; Frr. 295, 1; H. S. 136-7 ; O. I. 1, 48.

Expands from 1·25 to 1·30 in. Smaller than *Athalia*. The upper surface much resembles that of *Deione*, the black markings being narrow, but the ground colour is deeper fulvous. Under side much more uniform in colour than in *Athalia ;* hind wings with the light spots small ; the central row narrow and uniformly yellow, without inclining to white ; the fulvous portions lighter and proportionately broader than in *Athalia.* Pl. XLV., 5.

TIMES OF APPEARANCE.—June and August.

HABITAT. — South-West Germany, Switzerland, France, and Piedmont; Central Spain, and Andalusia. It often frequents mountainous districts. I have taken it on the St. Gothard, Furca, and other Swiss Passes at moderate elevations.

LARVA.—Black, dotted with white and lateral yellow spots ; it feeds on *Plantago.*

VARIETY.

Varia, Meyer-Dür. Tagf. p. 133; *Athalia* var. H. S. 370-4. —This is the Alpine variety of the species, being only found at considerable elevations in the higher Alpine Passes. It is smaller than the type ; the female is darker, being clouded with greenish black. Both sexes have the central row of spots on the under side of the hind wings white instead of yellow.

16. M. Asteria, Frr. B. 36, 1, i. p. 115.
ASTERIE, H. S. 3, 4, 568.

Expands from 1·0 to 1·20 in. The smallest species. Rather variable in colour, sometimes, especially in the male, being dark brown, with a few fulvous bands and spots ; at others, especially in the female, almost as light as *Athalia* or *Aurelia*. Under side much as in the allied species, but duller in colour ; the external light band of the hind wings has not the hind-marginal black line found in the other species. This line is nearly obliterated in some specimens of *Aurelia*—a fact worthy of note, as some entomologists consider *Asteria* a local form of that species. Pl. XLV., 6.

TIMES OF APPEARANCE.—June and July.

HABITAT. — Mountain pastures at a considerable elevation in the Alps of Eastern Switzerland and the Tyrol. Scarce and local. LARVA.—Unknown.

Genus 5. — **ARGYNNIS**, Fab. Ill. Mag. vi. p. 283 (1807); Latr. Enc. Méth. ix. p. 10 (1819); Doubl. Hew. Gen. D. L. p. 171 (1848).—*Brenthis* et *Argynnis*, Hüb. Verz. bek. Schmett. 30-32 (1816).

Larva cylindrical, spiny; the prothoracic segment with two spines longer than the rest. Generally of a brown colour, with lighter or darker longitudinal lines. They are solitary, and in the greater number of species feed on plants belonging to the order *Violaceæ*.

Pupa angular, with a distinct dorsal depression. The pupæ of the smaller species are less angular than those of the larger, thus somewhat resembling those of the last genus. The pupæ of the larger species, such as *Argynnis Paphia*, &c., are decorated with metallic spots even brighter than those of *Vanessa*.

Imago large or of medium size. The eyes are smooth. The antennæ are long and slender, with short oval clubs. The palpi are longer than the head, three-jointed, the middle articulation being more or less swollen. The wings on their upper surface are bright fulvous in the male; in the female generally duller, and sometimes suffused with dark olive-green; this is seen in the var. *Napæa* of *A. Pales*, and in *A. Paphia* var. *Valezina;* whilst in *Argynnis Sagana*, a Siberian species, it is the normal condition. The wings are generally fulvous with a pattern of black spots. The spots on the area of the wings are narrow or quadrate; parallel to the hind margin is a row of rounded spots, whilst along the hind margin itself is always a row of crescentic markings. In some of the larger species the males have the nervures of the fore wings broadly black. These markings are very similar in the different species, and it is chiefly to the under side of the hind wings that we must look for distinctive markings. They generally consist of yellow, metallic silvery, or pearly spots, arranged on a ground of purple, brown, or green; sometimes the metallic markings are disposed in stripes, as in *A. Paphia;* whilst in some of the smaller species there are few or no metallic spots, as in *A. Aphirape, Hecate, Daphne,* &c.; or they are more pearly than silvery, as in *A. Euphrosyne,* &c. In outline, the wings in the smaller species are simple, the fore wings being slightly pointed and with the hind margins convex; whilst the hind wings are more or less rounded. In some of the larger species the fore wings have the hind margins concave, and the hind wings are slightly dentated. Discoidal cell of fore wings closed.

The genus is widely distributed, being represented almost throughout the extra-tropical portions of the Northern Hemisphere from the North Africa, the Himalayas, South China, California, &c., to the Polar Regions as far north as explorers have penetrated. Out of the Palæarctic and Nearctic Regions, however, it is not found—the genera *Atella, Cethosia, Dione,* &c., replacing it in the tropics.

The European species at first sight have the appearance of being easily divided into two groups, and it has been at various times proposed to place the smaller species in a separate genus; the difference between the larger and the smaller species is, however, more apparent than real, the genus as it stands in its Fabrician sense being well defined, and not capable of any advantage by being divided. In the 'Genera of Diurnal Lepidoptera' the genus is arranged in two sections thus:—

Sect. I.—Second joint of palpi mostly much swollen; sub-costal nervule thrown off before the end of the cell. Containing the larger species, together with *A. Polaris,* which resembles those of the next section.

Sect. II.—Second joint of palpi not remarkably swollen; second subcostal nervule thrown off beyond the end of the cell. Containing the *Euphrosyne* group, *Frigga, Pales,* and its allies, excepting *A. Polaris.*

In the larger species, *Niobe, Adippe, Aglaia, Paphia, Pandora,* &c., the thighs are smooth inferiorly, whilst in the remaining section they are densely hirsute.

For obvious reasons the description of the species of this genus will refer chiefly to the under surface of the wings.

1.—A. **Aphirape**, Hüb. 23-25; O. I. 1, 52; Godt. ii. 9, 3, 4;
   Frr. B. 1, 2, 6.
   EUNOMIA, Esp. 110, 5.
   TOMYRIS, Hbst. (Lasp.) x. p. 102, T. 270, 6, 7.

Expands from 1·40 to 1·50 in. Wings above bright fulvous,
spotted with black; rather dark towards the base. The black
spots are somewhat narrow. Under side: Fore wings as above,
but fainter, and with hind-marginal straw-coloured spots. Hind
wings brownish or yellowish fulvous, without silvery markings, but
with straw-coloured spots near the base and across the centre of
the wing, like those seen in the last genus; hind margin with a row
of almost triangular yellow spots, and internal to these a row of
round ones with black outlines. Pl. XLVI., 1.

TIME OF APPEARANCE.—June.

HABITAT.—Bushy places in North-Eastern and West-Central
Germany, Belgium (the provinces of Liege and Luxembourg),
Bavaria, Poland, Central Russia, and Armenia. Local.

LARVA.—Grey, with lighter spines, a pale dorsal line, and
white lateral streak. Feeds on *Polygonum bistorta* in May.

VARIETY.

Ossianus, Hbst. x. p. 98, T. 270; Frr. 355, 1, 2; Dup. I.
20, 5, 6; H. S. 322-3. — Smaller than *Aphirape*. Expands from
1·16 to 1·25 in. Duller above, and with darker markings; bases
dusky. Under side: Hind wings dull red, with white or silvery
markings, not yellow as in the type. Pl. XLVI., 2.

HABITAT. — Lapland; North Russia, Eastern Siberia, and the
Amur.

OBS. — This variety, and also *Triclaris* from the American
Polar regions, seem at first sight perfectly distinct from *Aphirape;*
compare, however, their relations with those, for instance, of
Argynnis Adippe to *Cleodoxa, Niobe* to *Eris,* &c.

2.—A. **Selenis**, Ev. Bull. Mos. 1837, 1, 10; Frr. 277, 2, 3;
   H. S. 154-5.

Expands from 1·25 to 1·50 in. Fulvous with black spots,
which are rather large and distinct; bases blackish; marginal

fringes black and white. Under side: Fore wings with some straw-coloured and purple markings at the apex. Hind wings straw-coloured, with a light fulvous band near the base, some purple markings near the hind margin, and across the centre of the wing a pearly streak with some purple dots. Pl. XLVI., 3.

TIMES OF APPEARANCE.—May and June.

HABITAT.—The East of Russia (Central and Southern Ural), Southern Siberia, and the Amur.'

Life-history unknown.

3.—A. Selene, Schiff. S. V. p. 321; Hüb. 26, 27; O. I. 1, 55.
EUPHROSYNE, Bgstr. Nom. 42, 1, 2; Esp. 30, 1.
THALIA Esp. 97, 2 a b.

Expands from 1·50 to 1·75 in. Bright fulvous, spotted with black; bases blackish. Under side: Hind wings straw-coloured, marked with dark purplish brown; a row of black spots parallel to the hind margin. There are pearly or silvery spots arranged thus: two or three at the base; a central row of several, always more than one; a hind-marginal row of triangular pearly spots, sometimes a streak between these and the central row. Pl. XLVI., 4.

TIMES OF APPEARANCE.—June and August.

HABITAT.—The whole of Europe, excepting Andalusia, the Mediterranean Islands, and Greece; also Bithynia, Armenia, the Altai, and the Amur. Generally distributed in Britain; commonest in the South of England, frequenting heaths and open places in woods.

LARVA. — Black, with paler spines, sometimes with whitish dorsal and lateral stripes; anterior legs red. It feeds on *Viola canina* in June and September.

VARIETY.

Hela, Stgr. Stett. Ent. Zeit. 1861, 347.—A small and darker form of the species, having the superior surface blackish at the base, and with large spots. It is a boreal variety, occurring in Lapland, Eastern Siberia, and the Amur (conf. *A. Euphrosyne* var. Fingal).

4.—A. Euphrosyne, Linn. Syst. Nat. x. 482; Faun. Suec. 282; Esp. 18, 3, 72, 3; Hüb. 28-30; O. I. 1, 58; Frr. B. 139.

Expands from 1·75 to 1·80 in. Undistinguishable with certainty from *Selene* above, but on the under side the markings are sufficiently characteristic. The brown patches are light reddish, instead of dark as in *Selene*, and the silvery spots are fewer; there is only one near the base, and only one in the central row.

TIMES OF APPEARANCE.—May and June, and again in August.

HABITAT. — The whole of Europe, except Portugal, Sicily, Sardinia, and Corsica; also Armenia, the Altai, and the Amur. Common in Britain, frequenting open places in woods, &c., but, like the last, rarely double-brooded here.

LARVA. — Black, with bluish white lateral stripes, two white dorsal lines, and short yellowish or black spines. Feeds on *Viola canina* in June and September.

VARIETY.

Fingal, Herbst. Naturs. Schmett. x. t. 270, f. 1-3; *Dia Lapponica*, Esp. Schmett. 1, 2, t. 108, f. 5. — Darker and more strongly marked than the type. A boreal form found in Lapland. Pl. XLVI., 5.

5.—A. Pales, Schiff. S. V. p. 177; Hüb. 34, 35; Trsk. x. 1, 11; Frr. B. 115, 1; O. I. 1, 63; Frr. 187, 1. ARSILACHE, Esp. 56, 4 (1780), ii. p. 35.

Expands from 1·16 to 1·40 in. Bright fulvous, spotted with black; bases dark. Marginal fringes plain. Under side: Fore wings with very indistinct black spots; hind wings with a marginal row of pearly or silvery spots; the area of the wing variegated with yellow, purple, and reddish brown, and with silvery spots mostly with a triangular outline. Pl. XLVII., 1.

TIMES OF APPEARANCE.—July and August.

HABITAT. — Elevated Alpine meadows in Switzerland, the Pyrenees; Northern Europe, generally on moors and plains, being found in such localities in North Russia, Scandinavia, &c.; also in the Altai and North-Eastern Siberia.

LARVA.—Grey, with a pale yellow dorsal streak; spines flesh-coloured, and set on black elevations. On violets, *V. montana* and *canina*, generally in July.

VARIETIES.

a. **Napæa**, Hüb. 757-8; *Pales*, 1, 9, 964; *? Isis*, Hüb. 563.—The male is tinged with sulphur-yellow beneath; the female greenish or dusky above. Variable, all degrees of colouring intermediate between this and the type being found. Pl. XLVII., 2.

HABITAT.—Same as type.

b. **Lapponica**, Stgr. Stett. Ent. Zeit. 1861, 347.—Smaller than the type; much darker above, and more intensely marked beneath. Appears to be close to the next form. Pl. XLVII., 3.

HABITAT.—Lapland, &c.

c. **Arsilache**, Esp. 56, 5, ii. p. 35; Hüb. 36, 37; Trsk. x. 1, 12; Frr. B. 115, 2, 121, 2; *Napæa*, Düp. i. 48, 5, 6.—Larger than the type, and generally with the hind wings more rounded; colour brighter fulvous, and the spots larger; on the under side the fore wings differ from those of the type in having large and well-defined black spots; especially plain is an X-shaped marking situated near the inner margin. The hind wings have the markings more variegated, and the silvery spots brighter. Pl. XLVII., 4.

HABITAT. — Germany; valleys in Switzerland, Scandinavia, North and Central Russia, and Siberia; always at lower elevations than *Pales*.

d. **Caucasica**, Stgr. Hor. 1870, p. 61; *Arsilache*, H. S. 259-62. — A mountain form, inhabiting the Southern Caucasus and Armenia; the male being brighter fulvous than the type above, and beneath paler.

e. **Græca**, Stgr. Hor. 1870, p. 62. — Another local mountain form found in Northern Greece. It is paler than the type beneath, and has the marginal fringes varied with black and white.

OBS.—*Arsilache* was formerly considered distinct from *Pales*, but, as will be seen by the above varieties, there are many connecting links between the two forms. Elevation and latitude seem to be very influential in producing variations, *Pales* or something like it being probably the original Palæarctic type of the species.

6.—A. **Chariclea**, Schn. N. Mag. v. p. 588 (1794); Hbst. 272,
5, 6; Hüb. 769-70; O. iv. 114.

ARCTICA, Zett. Ins. Lapp. p. 899.

BOISDUVALII, Dup. i. 27, pl. 20, 4; Boisd. Ic. 20, 5, 6.

Expands from 1·50 to 1·70 in. Wings more angular in outline
than in the preceding species. Marginal fringes black and white.
All the wings above are dull fulvous, with black spots, and black at
the base; the central row of spots has a tendency to form a band.
Under side: Fore wings marked with reddish brown along the hind
margin; hind wings reddish brown, darkest at the base, near
which are some very small silvery spots; across the centre of the
wing is a yellow band, marked with two triangular silvery spots,
and bordered with a black zigzag line; external to this is a
whitish streak, and parallel to it a row of black spots; lastly, there
is a hind-marginal row of silvery spots enclosed by a row of angular-
shaped markings of a reddish brown colour. There is no white
discoidal spot. Pl. XLVII., 5.

TIME OF APPEARANCE.—July.

HABITAT. — Russian Lapland (Wallengren Rhop. Scand.
p. 97), frequenting mountains; also in Labrador and Greenland. It was taken during the Polar Expedition of 1876, between
lat. 79° to 81° 52′ N., at Hayes' Sound, Port Foulke, Walrus
Island, Discovery Bay, &c. (See Mr. M'Lachlan's report in the
'Journal of the Linnean Society,' already referred to at p. 48 of
this present work.

LARVA, &c.—Unknown.

OBS. — This species is very variable, some dark specimens
approaching somewhat to the variety *Lapponica* of the preceding
species, others being hard to distinguish from *A. Polaris* or *Freija;*
in fact all the circumpolar species of Fritillaries are very closely
allied, and seem to run into one another.

VARIETY.

Duponchel's variety *Boisduvalii* appears to differ chiefly in
being less varied on the under side of the hind wings, and in
the zigzag border of the light band being indistinct or absent.

7.—A. **Polaris**, Boisd. Ind. p. 15; Dup. i. 20, 1-3; Boisd. Ic. 20, 1, 2; H. G. 1016-9; Frr. 439, 1, 2; Wallengr. Rhop. Scand. p. 91.

Expands from 1·50 to 1·70 in. Marginal fringes not variegated. Wings dull fulvous, black at the base, with the usual black spots, the central row being banded. Under side : Hind wings dull rusty brown ; near the base are some white patches, and there is a very distinct white discoidal spot in typical specimens ; external to this is a broad yellowish grey central band, with white zigzag borders ; along the hind margin is a row of conspicuous spots white or grey in colour, and having a lozenge or T-shape ; internal to this is a light streak, upon which is a row of black spots. Pl. XLVII., 6. This is the most conspicuous of the Polar Fritillaries, the milky-white markings of the under side being very striking in dark specimens, especially the marginal row of spots, which frequently have a distinct T-shape. It is a variable species, closely allied to its immediate congeners ; but attention to the above-mentioned points, and particularly to the presence of a *white* discoidal spot, will assist the diagnosis.

TIME OF APPEARANCE.—July.

HABITAT. — North Lapland, ? Iceland, Northern Siberia, Labrador, and Greenland. This species has probably been taken further north than any other butterfly, being reported as having been taken by Capt. Feilden at lat. 81° 52′ N.

LARVA, &c.—Unknown.

In connection with the subject of Arctic butterflies, I quote the following passage from Mr. M'Lachlan's * report, already several times referred to (Journ. Linn. Soc., vol. xiv., p. 98, 1878):—
" Capt. Feilden, in answer to questions, gave me some valuable and interesting information on the habits of Lepidoptera in those latitudes. He informed me that, during the short period when

---

* This gentleman has, since these pages were written, kindly called my attention to an allusion in Markham's ' Polar Reconnaissance,' p. 351 (1881), concerning the occurrence in Novaya Zemlya of *Argynnis Improba*, Butler. There are two specimens of this European butterfly in the British Museum (those originally alluded to by Markham). It is my intention to give a figure of it in a supplementary plate.

there is practically no night, butterflies are continuously on the wing, supposing the sun's face not to be obscured by clouds or passing snow-showers. Furthermore, he told me that about one month in each year is the longest period in which it is possible for these insects to appear in the perfect state, and that about six weeks is the limit of time allowed to plant-feeding larvæ; during all the rest of the year the land being under snow and ice. This latter fact is suggestive as showing the conditions under which the species maintain their existence. We have, however, much yet to learn respecting their life-history. The intense cold is not of great importance. We know already that larvæ may be frozen till they are as brittle as rotten twigs, and still suffer in no way. The principal point may be put as follows :-- Is there sufficient time in each year for a larva to hatch from the egg, feed up, and change to a chrysalis? The continuous day no doubt acts beneficially in this respect on the larvæ of butterflies, such as *Colias* and *Argynnis,* which probably feed only in the daytime; but it must act in a contrary manner on those of Noctuæ, &c., which practically feed only at night. Upon reviewing all these conditions, I am disposed to think that more than one year is necessary in most of the species for the undergoing of all their transformations. This indeed is already suspected in certain species that inhabit the boreal and alpine portions of Europe."

8.--A. **Freija,** Thub. Diss. Ent. 2, p. 34, T, fig. 14 (1791); Hüb. 55, 56, 771-2; Esp. 109, 1; O. I. 1, 78; Dup. i. 19, 6, 7; Frr. 295, 3.

    Papilio Dia lapponica, Esp. 97, 3.

Expands from 1·30 to 1·75 in. Wings dull fulvous, black at the base, and with the usual black spots rather larger than in the last species. Under side : Hind wings reddish brown; there are generally some yellowish spots near the base, and a discoidal spot of yellowish white, with a black spot; on the central light band are two or three white triangular spots, the band itself being bordered with a more or less distinct zigzag dark line; hind-marginal white spots triangular; the row of spots internal to these not black, but very nearly the same as the ground colour. Pl. XLVII., 7.

Times of Appearance.—June and July.

Habitat. — Lapland; mountains in Central Norway; North Russia, down to 60° N. lat.; North-Eastern Siberia; Labrador and Greenland. The commonest and most widely distributed of the Arctic species.

Larva, &c.—Unknown.

9.—**A. Dia**, Linn. Syst. Nat. xii. 785; Esp. 16, 4; 61, 2: Hüb. 31-3; O. I. 1, 61; Frr. 211.

Expands from 1·25 to 1·50 in. All the wings bright fulvous, with the usual black spots large and distinct; marginal fringes plain; base black. Under side: Fore wings with the black spots very distinct. Hind wings purple, darkest at the base; central light band yellow, with three triangular silvery spots; parallel to the hind margin is a row of conspicuous black spots, generally with light centres, and external to these a hind-marginal row of silvery spots. Pl. XLVIII., 1.

Time of Appearance.—May to September.

Habitat.—Woods and bushy places throughout Central and South-Eastern Europe, Northern and Central Italy, Asia Minor, Armenia, and the Altai. One frequently meets with accounts of the existence of British specimens, but I never remember to have seen any record from an actual captor.

Larva. — Grey, with darker lateral lines, and a black dorsal line; spines greyish white; legs black. Feeds on *Viola canina, tricolor*, &c., in July and September.

10.—**A. Amathusia**, Esp. 88, 1, 2 (1783), ii. p. 170; Hüb. Zett. p. 11; O. I. i. 75; Frr. B. i. 1.
Titania, Hüb. 47, 48.
Dia Major, Esp. 93, 2, 3.
Diana, Hüb. 51-4.

Expands from 1·25 to 1·75 in. Wings bright fulvous, black at the base; hind margins with a black border, and with the fringes variegated; black spots arranged very much as in *A. Dia*. Under side: Fore wings reddish purple at the apex, mixed with yellow; hind wings marbled with purple, red, and yellow; central light band enclosing three triangular spots not so silvery as in *A. Dia;* the light band is bounded externally by a zigzag black line; the hind-marginal spots are triangular and yellow, surmounted by black angular marks; parallel to the hind margin is a row of black spots with faintly light centres. Pl. XLVIII., 2.

TIME OF APPEARANCE.—June to September.

HABITAT. — Mountain woods at a moderate elevation in Switzerland and Germany; plains in Russia and the East of Europe (it does not inhabit the Polar regions); the Altai.

LARVA. — Nearly black, with black dorsal and lateral stripes; spines yellow. Feeds on *Polygonum bistorta* in May.

11.—A. **Frigga**, Thub. Diss. Ent. 2, p. 33 (1791); Hüb. 49, 50; O. I. 1, 74; Dup. i. 19, 3, 5; Boisd. Ic. 19, 6, 7.

Expands from 1·50 to 1·70 in. Wings pale fulvous, deeply black at the base, the black spots large, the central rows forming bands, and on the hind wings merging into the black basal shading, especially in the female; hind margins pale fulvous; fringes not spotted. Under side: Basal portion of wing deep red, with a white spot; central light band yellow, broken up into a chain of more or less rounded spots; with two white spots, one near the costa and another near the centre. Marginal portion of wing light yellow, tinged with purple, with a faint row of spots running parallel to the hind margin; there are faint traces of yellow hind-marginal spots. Pl. XLVIII., 3.

TIME OF APPEARANCE.—July.

HABITAT.—Lapland; North Russia, to 60° N. lat. (Staudinger); North-Eastern Siberia; Labrador. A rare and local species, though circumpolar. In Swedish Lapland it occurs on the banks of the River Tornea, and at Quickjock in June and July (Wallengr.).

LARVA, &c.—Unknown.

12.—A. **Thore**, Hüb. 571-3; O. iv. 111; Dup. i. 19, 1, 2; B. Ic. 20, 3, 4; Frr. B. 104, 3; Frr. 294, 4.

Expands from 1·40 to 1·60 in. Wings deeply fulvous, very dark at the base, hind generally almost completely clouded over with dark brown, with a central fulvous streak; black spots large; hind margins black, fringes not spotted. Under side : Fore wings with the spots large and distinct; hind wings brown, light at the base, with a central band composed of yellow spots; external to this are some pale blue spots without perceptible metallic tinge; the hind-marginal spots are of the same colour; there is a row of very faintly indicated brown spots running parallel to the hind margins. Pl. XLVIII., 4.

Time of Appearance.—May to July.

Habitat. — Alpine meadows at a considerable elevation in Switzerland, Germany, and Piedmont; and is scarce and local.

Larva.—Unknown.

VARIETY.

Borealis, Staud. Cat. 9; Stett. Ent. Zeit, 1861, 351; Boisd. Gen. 18. — Very much lighter than the Central European form. Inhabits Lapland, the Altai, and the Amur.

13.—A. **Daphne,** Schiff. S. V. p. 177; Hüb. 45, 46; O. I. 1, 72; Godt. ii. 8, 1, 2.

Expands from 1·60 to 2·0 in. Bright fulvous; hind margins black, fringes not variegated; wings with the usual black spots, central row very irregular, two rows running parallel to the hind margin. Under side : Hind wings greenish yellow at the base, central band yellow; external to this the wing marbled with reddish purple, and has a row of black spots with light centres; hind margin light yellow. Pl. XLVIII., 5.

Time of Appearance.—May to August.

Habitat. — Elevated woods in Southern and East Germany, France, Spain, Switzerland, Italy, the South-West of Europe, Asia Minor, Armenia, the Altai, and the Amur.

Larva. — Blackish brown, with yellow lateral lines, and a double yellow dorsal stripe; the spines are yellow, tipped with black. Feeds on *Rubus Idæa* and *Violaceæ* (?) in May.

14. — **A. Ino,** Esp. 76, 1, a, b (1782), ii. p. 125; O. I. 1, 69; Godt.
ii. 3, 3, 4; Frr. 409.
DICTYNNA, Hüb. 40, 41.

Expands from 1·25 to 1·75 in. Very much resembles the last species, but is smaller, the ground colour of the wings is darker fulvous, the black spots are closer together, and the hind margins are more broadly black; the marginal fringes are narrowly streaked with black. The wings are much darker at the base than in *Daphne*. Under side: The markings of the hind wings greatly resemble those of *Daphne*, but are more defined on account of the darker colour of the purple marbling, which is very dark violet, the light-centred spots showing out upon it very distinctly. Pl. XLVIII., 6.

TIMES OF APPEARANCE.—June and July.

HABITAT. — Meadows, &c., throughout Europe, but not in Britain, Holland, the Peninsula, or Greece. It occurs also in Asia Minor, the Altai, and the Amur.

LARVA. — Yellowish, with brown lateral stripes, dorsal stripe brown, spines whitish yellow. Feeds on *Urtica urens* and *Spiræa* in May.

15.—**A. Hecate,** Esp. 76, 3, a, b (1782), ii. p. 127; Kn. Btr. iii.
128, T. vi. 5, 6 (1783); Hub. 42-4; O. I. 1, 67; Godt.
ii. 9, 5, 6: Frr. B. 121, 3; H. S. 138-9.

Expands from 1·40 to 1·60 in. Somewhat similar to *Ino* above, but darker, especially the female, which has often a very dusky appearance, and frequently a slight purple iridescence. Under side: Hind wings brownish yellow, with straw-coloured basal spots, and a central band of the same colour; hind-marginal spots absent, but there are two submarginal rows of black spots on a pale ground.

TIMES OF APPEARANCE.—June and July.

HABITAT. — Elevated meadows in South-Eastern Europe (except Greece), the South of France, North Italy, Asia Minor, and the Altai. Pl. XLIX., 1.

LARVA.—Unknown.

Caucasica, Staudinger, Cat. p. 21. — Larger than the type, the male lighter.

HABITAT.—Armenia and Turkey.

16.*—A. Lathonia (rect. *Latonia*), Linn. Syst. Nat. x. 481; Faun. Suec. 282; Esp. 18, 2; Hüb. 59, 60; O. I. 1, 80; Frr. 13, 25.

Expands from 1·80 to 2.15 in. Hind margin of fore wings distinctly concave; hind wings dentate, less rounded in outline than in the other species. All the wings bright fulvous, with distinctly rounded black spots, two rows running parallel to the hind margins; the base is blackish. Under side: Fore wings yellowish, with black spots, and with several silvery spots at the apex; hind wings fulvous, inclining to yellow, with a row of large oval silvery spots placed parallel to the margins; internal to these is a row of reddish brown spots, each enclosing a central silvery dot; there are large basal and central silvery spots, more or less oval in shape, and very brilliant. Pl. XLIX., 2.

TIME OF APPEARANCE.—May to September.

HABITAT.—The whole Palæarctic territory, excepting the Polar and Higher Alpine Regions—that is, all Temperate Europe and North Asia, North Africa, Persia, &c. It is rare in Britain, occurring principally in the South of England. It inhabits woods, meadows, roadsides, gardens, &c., and is generally common.

LARVA.—Greyish brown, with a white dorsal line. It has sixty spines, four on the first and last segments and six on each of the others, those of the first two are shorter than the rest, the central ones being the longest. Feeds solitarily on *Viola tricolor* and *Onobrychis* in May and August.

PUPA.—Grey anteriorly, green posteriorly, with gold spots.

---

* The statement on p. 195 that the genus *Argynnis* only occurs in the Nearctic and Palæarctic Regions requires some modification, since one if not two species occur in Chili which are closely allied to *A. Lathonia*; another species also occurs in Buenos Ayres. Some zoologists, however, consider the temperate portion of South America to form part of a region which includes the Palæarctic and Nearctic Regions of Wallace, &c.

Valdensis, Esp. 115, 4; *Lathonia* var. Frr. 671; H. S. 152. — An aberrant form, in which the silver spots are confluent. It has occurred in Britain.

17.—A. Eliza, Godt. Enc. Méth. p. 817; Dup. 1, 18, 3, 4, p. 111. CYRENE, Bon. Descr. p. 175, T. 1, 1; Frr. B. 69, 1; B. Ic. 21, 1-3; Trsk. x. 1, 17.

Expands from 2·0 to 2.25 in. Wings rather light fulvous; hind margins black; parallel to them on all the wings is a row of black spots; the fore wings have a row of spots internal to these, running obliquely from the apex towards the centre of the wing, a few more scattered about the basal portion, and a row of narrow spots along the costal margin. Hind wings with a few spots thinly scattered over the area of the wing. Under side: Fore wings greenish at the apices, with some indistinct silvery spots; hind wings yellowish green throughout the basal half, the rest being light yellow or straw-colour; the silver spots are disposed very nearly in three rows; along the hind margin is a row of distinct semilunar spots, another row across the centre, between this and the base another row of smaller ones, and a few more scattered about the base; between the marginal and central row are some round dark brown spots with minute silver centres, like those seen in *Adippe* and *Niobe*.

TIMES OF APPEARANCE.—June and July.

HABITAT.—Mountains in Corsica and Sardinia.

LARVA, &c.—Unknown. Pl. XLIX., 3.

OBS.—This interesting insular form seems to be a perfectly good species, and quite distinct from *Aglaia*. It is in fact much more nearly allied to *Adippe* or *Niobe* than to *Aglaia*, as may be seen by the presence of the round, silver-punctured spots between the marginal and central rows.

18.—A. Alexandra, Mén. Cat. Rais. p. 246; H. S. 417-8; Ev. Lep. Ross. ii. 4.

Expands from 2·10 to 2·30 in. Fringes light brown. Bright fulvous above, with black spots, almost exactly resembling *Aglaia* or *Adippe.* Under side: Fore wings reddish, with black spots; yellowish bronze towards the apices, which are without silver spots. Hind wings rich greenish bronze in the basal portion, brown towards the hind margin; marginal row of silver spots wanting, but those in the central row are very brilliant; they are rather irregular in shape and arrangement; near the base are also five or six silvery spots of different sizes, two at least of them being large and brilliant. Pl. XLIX., 4.

TIME OF APPEARANCE.—July.

HABITAT. — Mountain woods in the Caucasus, Armenia, and Persia (North-West). It seems to be a rare species, at least in collections. Some entomologists consider it a variety of *Aglaia;* all the specimens, however, that I have seen look perfectly distinct.

LARVA, &c.—Unknown.

19.—A. Aglaia, Linn. Syst. Nat. x. 481, xii. 786; Esp. 17, 3, 60, 2; Hüb. 65, 66; O. i. 1, 91; Godt. i. 3; sec. 3; Frr. 241, 205; H. S. 140-1.

Expands from 2·25 to 2·50 in. Fringes black and whitish. Nervures of fore wings with black lines in the male. All the wings bright fulvous in the male, duller in the female. Hind margins black; bases dusky, darkest in the female; the markings consist of the usual black spots and black marginal lunules; the square spots near the centre of the wings are joined so as to form a narrow wavy band. Under side: Fore wings light reddish brown, spotted with black; apices with green and silver marginal spots. Hind wings bronze-green, with a band of straw-colour tinged with green running parallel to the hind margin; silver spots large and distinct, arranged in three rows: first a row of semilunar spots along the hind margin, then a central row, internal to this a row of three spots, and then three more at the base. There are never any brown, silver-centred spots between the marginal and central rows, or any costal silver spot between the central and internal rows. Pl. L., 1.

TIME OF APPEARANCE.—May to September.

HABITAT. — The Palæarctic territory, excepting North Africa, the Canaries, Asia Minor, Syria, and Persia. Frequents woods

2 E

and heathy places, ascending to a great elevation in the mountains. As a British insect local, but common where it is found.

LARVA.—Blackish brown, with two pale yellow dorsal lines and reddish brown lateral spots; spines blackish. Feeds on *Viola canina* in May and June.

VARIETY.

Charlotta, Haw. Lep. Brit. i. 32 (1803); Sow. Brit. Misc. i. t. 11.—*Caroletta*, Jerm. Vade Mecum p. 107 (1827).—An aberrant form in which the silver spots on the under side are larger than in the type, there being especially three large basal silvery blotches; occasionally the spots of the central row coalesce to form bands. Described by the old English authors as a distinct species, and I believe found nowhere else but in Britain. Staudinger does not notice it in his Catalogue.

20.—A. Niobe, Linn. Syst. Nat. x. 481; xii. 786; Esp. 18, 4; Hbst. x. p. 56; O. i. 1, 83; Godt. ii. 7, 3, 4.

Expands from 1·75 to 2·0 in. The wings above very much resemble those of the last species; but the female has the ground colour duller fulvous and the markings darker, especially towards the base. Under side: Fore wings duller in colour than in the last species, with the black spots strongly marked and with greenish yellow markings upon the costa, and extending from the apex for a short distance along the hind margin; but there is usually an absence of any silvery blotch. Hind wings yellowish, with a slight tinge of green mixed with brown. The pale spots are only slightly silvery in the typical form, and between the central and hind-marginal rows there is a row of reddish brown spots with silvery centres; near the bases, between the two rows of spots, there is always a dot or spot, either silvery light yellow or black. Pl. L., 1.

TIMES OF APPEARANCE.—July and August.

HABITAT. — Throughout Europe, and Northern and Central Asia. It is very doubtful whether it has ever occurred in Britain, but it has lately been reputed British and introduced into some lists of British Lepidoptera; the probability is that specimens of the var. *Cleodoxa* of *Adippe* have been confounded with it. *A. Niobe* is found in woods, and ascends to a much greater elevation in the mountains than *A. Adippe*.

LARVA.—According to Guenée, brown, with a black dorsal stripe bordered with white; lateral stripe black; between the lateral and dorsal stripes are some white spots. Feeds on *Viola tricolor* and *odorata* in May. Pl. LIII., 5.

VARIETIES.

a. **Eris**, Meig. i. p. 64, T. 14, 5, 6.—*Cleodoxa*, Esp. 94.—*Niobe*, Linn. Faun. Suec. 281; Hüb. 61, 62; Frr. 199, 337; Hbst. T. 26, 3, 10; H. S. 142-3. — This is the most common form of the species. It resembles the type above, but beneath it has the hind wings brown mixed with greenish grey, the light spots being light yellow, with a slightly greenish tinge, and without any trace of silver; the inner margin is greenish. Sometimes the reddish spots outside the central row have very small silvery centres, and occasionally a slight silvery tinge is seen on the light markings, especially on those along the hind margins; there are in fact intermediate forms between *Niobe* and *Eris*. This variety is very like the analogous variety of *Adippe*, *Cleodoxa*, but on comparing them the differences will be at once seen. Pl. L., 3.

b. **Pelopia**, Bkh. i. 36; Hbst. 269, 3, 4. — An aberration thus named and figured by Borkhausen sometimes occurs in which the wings are more or less suffused with black by the extension of the black markings. All the larger species of *Argynnis* are subject occasionally to this melanic variation, as well as the smaller ones— such as *Euphrosyne* and *Selene*. It may also be mentioned here that it is not a very uncommon thing to find on the wings of species of this genus white patches, owing to the absence of pigment in the scales. Sometimes the marks are placed irregularly, and sometimes are perfectly symmetrical ; I look upon them as the result of accidental causes, and in fact as analogous to the pathological condition sometimes found in the human skin and known as leucoderma.*

* Mr. A. E. Hudd, of Clifton, has recently shown me a specimen of *Melitæa Didyma*, taken in Switzerland, in which the ground colour of the wings is entirely white. The specimen is not a worn one, and has all the appearance of having emerged in its present condition.

21.—A. **Adippe,** Linn. Syst. Nat. xii. 786; Esp. 18, 1, 43, 2, 74, 1, 2, 4; Hüb. 63, 64; O. i. 1, 88.

Expands from 1·90 to 2·20 in. The male closely resembles that of the last, but the black lines along the nervures of the fore wings are thicker. The female is brighter fulvous, and the dark markings are not so intense as in *Niobe*, neither is there the black shading towards the base of the wings found in that species. Under side : Fore wings brighter fulvous than in *Niobe*, with distinct silvery markings near the apex. Hind wings light yellow, with a fulvous tinge; spots well defined and silvery; between the hind-marginal and central rows is a series of reddish spots with silvery centres; the inner margin is greenish, with a silvery gloss. Pl. L., 1.

TIME OF APPEARANCE.—July.

HABITAT. — Heaths and woods throughout Europe, with the exception of the Polar Regions. Also in Asia Minor, Armenia, the Altai, and the Amur. It is found in the mountains, but not at such great elevations as the last. A somewhat local British species, but commoner than *Aglaia*.

LARVA. — Reddish brown, with a whitish dorsal stripe, and a black spot on either side of this on every segment; the spines light brown. Feeds on *Viola tricolor* and *odorata* in May and June.

VARIETIES.

a. **Cleodoxa,** O. iv. p. 118.—In this variety the silver is absent from the light spots beneath, excepting that it is sometimes slightly apparent near the apex of the fore wings and along the hind margin of the hind wings; the series of red spots between the hind-marginal and central rows also frequently have silvery centres. The ground colour is pale brown, and the light spots are straw-colour; not sharply defined, as in *Niobe*, var. *Eris;* nor is there any tinge of green. Pl. LI., 2.

HABITAT. — The more southern parts of Europe, especially Greece and Sicily. I have taken it commonly on the Lago Maggiore. It sometimes occurs as an aberration in the more northern parts of Europe, and has occasionally been taken in England.

b. **Chlorodippe,** H. S. vi. p. 5.—The fore wings are tinged with green at the apices on the under side, and the ground colour of the hind wings is replaced by a bright green as deep as that seen in *A. Adippe.* The arrangement of the silver spots and the silver-centred brown ones is, however, exactly the same as in the type; it is therefore impossible to confound it with *Aglaia*, to which in colour it bears a striking resemblance. Pl. LI., 3.

HABITAT.—Andalusia and Central Spain.

c. **Cleodippe,** Staudinger, Cat. p. 21,—*Adippe,* Hüb. 859-60.— A form of the last, occurring in Central Spain, in which the silver spots are nearly or altogether absent.

22.—**A. Laodice,** Pallas, Reis. i. 470 (1771); Esp. 93, 1, 102, 4; O. i. 1, 95; Dup. i., 18, 1, 2; Boisd. Ic. 21, 4-6; Frr. 331, 2.

CETHOSIA, Hüb. 67, 68; Hbst. 263, 1, 2.

Expands from 2·25 to 2·50 in. Hind margins of fore wings slightly concave, especially in the female. Hind margins of hind wings rather more dentate than in the preceding species. All the wings fulvous in both sexes, brightest in the male, spotted with black as in the other species, but the hind-marginal spots are more distinctly separated and less crescentic in form. Under side : Fore wings fulvous, spotted with black, tinged with yellowish green at the apex and along the hind margins. Hind wings with all the silver spots absent, except the central row ; the basal portion of the wing as far as this is yellow, marked with one or two fine red lines; the silvery spots are small and indistinct in the male, but more strongly defined in the female ; immediately external to these is a dark brownish purple band, shading off into lighter purplish brown, upon which are two parallel rows of dark spots ; the hind margin is rather broadly light yellow. Pl. LI., 4.

TIMES OF APPEARANCE.—June and July.

HABITAT. — Woods in North-Eastern Germany, Central and South-Eastern Russia to 60°, Poland, Armenia, Siberia, the Amur, and Japan.

LARVA.—Unknown.

23.—A. Paphia, Linn. Syst. Nat. x. 481; Esp. 17, 1, 2; Hüb. 69, 70; O. i. 1, 96; Frr. B. 25.

Expands from 2·25 to 2·75 in. The male has all the wings bright fulvous, spotted with black; bases blackish; fore wings with thick black lines running along the nervures; hind wings with the spots large and distinct. The female is much duller fulvous, the fore wings have their hind margins distinctly concave, and the nervures are not marked with black lines. Under side: Fore wings fulvous, spotted with black, greenish at the apex. Hind wings shining green, with two short silver streaks near the base, and another extending right across the centre of the wing; the hind margin has a silver streak throughout its entire length. The silvery markings are much more distinct in the female than in the male. Pl. LII., 1.

TIMES OF APPEARANCE.—July and August.

HABITAT. — Woods throughout Europe (excepting the Polar Regions), Western Asia, Siberia, and the Amur. It is a common British species (Silver-washed Fritillary), being found in almost every large wood in the South of England, frequently settling on the flowers of the bramble.

LARVA. — Dark brown, with two narrow light yellow dorsal stripes, and lateral reddish yellow lines; spines dark brown, the two nearest the head being longest. Feeds on *Viola canina* from the end of May to the middle of June.

The PUPA is very beautifully decorated with shining metallic golden green spots.

VARIETIES.

a. **Valezina,** Esp. 107, 1, 2; Frr. 331, 1.—*Paphia*, Hüb. 767-8. —A variety of the female in which the ground colour of the wings is entirely dark greenish, with black spots, the fore wings having some light yellow or white spots near the apex. Under side : The fore wings as in the type, but the ground colour is reddish and the apex deeper green. Pl. LII., 2.

HABITAT.—This dimorphic condition of the female is apparently most frequently met with in the South. I have taken *Valezina* in Switzerland, and have seen it from Germany and Italy. In England it is very local, being almost entirely confined to the New Forest district.

b. **Anargyra**, Staud. Cat. p. 22.—A variety in which the silvery streaks are absent from the under side of the hind wings.

HABITAT.—The South of Europe, and Western Asia.

24.—A. **Pandora**, Schiff. S. V. p. 176 (1776); Esp. 58, 1, 2, 73, 3; Hüb. 71, 72, 606-7; O. i. 1, 99; Frr. 715.
CYNARA, Fab. Gen. Ins. p. 266 (1777); Hbst. Nat. Schmett. ix. t. 261.
MAIA, Cram. Pap. Ex. i. t. 25 B. C. (? nomen vetustius).

Expands from 2·25 to 2·90 in. The colour of the wings in both sexes is greenish fulvous. The arrangement of the black spots is much the same as in *Paphia*, but those on the hind wings are somewhat smaller. Both sexes have the hind margins of the fore wings slightly concave. Under side: Fore wings in fresh specimens rosy red, with deep black spots; the apex light yellow, with green spots; costa light green. Hind wings deep green, with basal central and hind-marginal silver streaks; in the male less distinct than in *Paphia*, but in the female, which is always larger and more brightly coloured than the male, the markings are generally very clear and distinct, as in the figure (Pl. LII., 3); and between the central and hind-marginal stripes there is a row of faintly-defined reddish spots with minute silvery centres.

TIMES OF APPEARANCE.—June and July.

HABITAT.—Woods and shrubby places in the South of Europe; South-Eastern Germany, very rarely in the Valais; Western Asia; Algeria; and probably the Canaries. It is entirely absent from Northern and North-Central Europe.

LARVA.—Brownish purple, with black dorsal transverse marks. Feeds on *Viola tricolor*.

— — —

OTHER SPECIES AND VARIETIES OF NYMPHALIDÆ DESCRIBED IN STAUDINGER'S CATALOGUE:—

Genus *LIMENITIS*.

L. Schrenkii, Mén. Schrk. Reis. p. 31, T. iii. 2.—Expands from 3·0 to 3·10 in. A large and handsome species, probably rightly belonging to the genus *Adolias*. Wings black; fore wings with a central row of five white spots; between these and the inner margin is a dash of fulvous, and near the inner margin itself a patch of light blue. Hing wings with a central white patch bordered with blue. Under side : Fore wings black, variously spotted with white, blue, violet, light green, and orange. Hind wings pearly white, with a narrow orange band bordered with black, limiting the basal portion ; hind margin with a similarly-coloured band throughout its entire length. Habitat : The Amur.

L. Sydyi, Ld. z. b., V. 1853, p. 7, T. 1, 3.—Expands 1·75 in. All the wings dark brown. Fore wings with a band of white spots running outwards from the centre of the costa; in the centre of the inner margin is a large white spot, and near the apex three small ones. All the white markings have a violet gloss. Under side : very like that of *L. Sibylla,* but duller ; central band of hind wings narrower and more oblique ; there is only a single row of black dots ; the hind margin is rather broadly greenish white, with a broad brown line. Habitat : The Altai.

Var. *Latefasciata,* Mén. Schrk. p. 29.—A variety of the last, in which the white spots coalesce to form broad white bands, with a violet gloss ; the under side is brighter in colour than in the type. Habitat : The Amur and Japan.

L. Helmanni, Ld. z. b., V. 1863, p. 6, T. 1, 4.—Expands from 1·75 to 2·0 in. Blackish brown ; fore wings with a central row of six white spots ; there are three or four near the apex ; there is a white basal streak, and at its outer extremity a triangular white spot, with the apex outwards. Hind wings with a central band made up of six white spots. Under side : Very like that of *L. Sibylla,* but the fore wings have a white basal streak and a triangular spot, and the spots of the ventral row are smaller. Hind wings as in *L. Sibylla ;* but the central band is narrower and bluer, and the submarginal rows of black spots are wanting. Habitat : The Amur, Japan, and the Altai.

L. Amphyssa, Mén. Schrk. p. 30, T. iii. 1. — Expands from 1·75 to 2·0 in. Wings dark brown. Fore wings with three white spots near the apex, then six larger ones arranged in a crescentic figure extending from the centre of the costa to the centre of the inner margin, and two more white spots near the basal end of the costa. Hind wings with a white central band like that in *L. Sibylla.* Parallel to all the hind margins is a row of light spots. Under side: Fulvous, with markings similar to those above; hind wings with some bluish green basal spots. Time of Appearance, July. Habitat: The Amur.

Genus *NEPTIS.*

N. Nycteis, Mén. Schrk. p. 28, T. ii. 11.—*Apatura Nycteis*, Kirby, Cat. Diurn. Lep. p. 260.—Expands 2·50 in. Hind margin of fore wings concave. All the wings dark brown, with a slight purple gloss; fore wings with a central row of white spots arranged in a crescentic form, three white spots at the apex, and a whitish basal streak. Hind wings with a central white band, crossed by the black nervures, and parallel to the hind margin a row of distinct white spots. Under side: Fore wings brown, the basal portion with a slight violet gloss, a triangular white basal patch enclosing a black spot, a central row of white spots as above, and some silvery white spots at the apex and along the hind margin. Hind wings olive-green, with three bands of silvery white, a row of four white spots between the two outer bands, and another row along the hind margin. Habitat: The Amur.

N. Lucilla, var. *Ludmilla*, H. S., 546, vi. p. 6.—Rather smaller than the type. It appears to be intermediate between *Lucilla* and *Aceris*, the fore wings having a faint basal streak and hind-marginal spots. The hind wings also show traces of a second light band. On the under side these extra markings are quite distinct, and the ground colour is lighter brown than in the type. Habitat: Armenia and the Altai.

N. Philyra, Mén. Schrk. Reis. p. 25, T. ii. 8.— Expands from 1·75 to 2·0 in. Somewhat resembles *Aceris*, but larger; the white basal streak of the fore wings is long and narrow, and is continuous with the white spot, which appears separate from it in *Aceris;* there is also a white spot nearly in the centre of the costa which is not found in that species. Habitat: The Amur and Japan.

N. Thisbe, Mén. Schrk. Reis. p. 26, T. ii. 9. — Expands from 2·50 to 2·75 in. Wings dark brown. Fore wings with a yellow streak extending from the base to the centre of the wing, and with three large yellow spots, one near the apex, another in the centre

2 F

nearly touching the hind margin, and a third almost in the centre of the inner margin. Hind wings with a straight central yellow band, and with a light brown streak parallel to the hind margins. Under side: Reddish brown, with some light blue spots, and with the markings of the upper surface somewhat wider. Hind margins yellowish brown.

N. Raddei, Brem. Lep. O. S. p. 18, T. i. 9. – Expands 2·50 in. Hind margin of fore wings concave. All the wings smoke-coloured, broadly veined with black, and without any of the white markings usual in *Neptis*. It belongs in all probability to another genus. Habitat: Eastern Siberia.

Genus *VANESSA.*

V. Burejana, Brem. Lep. Ost. Sib. p. 15, T. i. 8. — Resembles *Levana* and *Prorsa* in coloration and pattern, but the apices of the fore wings are rounded. The hind wings have their hind margins rounded, and without angular projections. Habitat: The Amur.

V. Progne, Cr. P. Ex. T. 5, E. F.—This species, which is close to *V. C-Album*, is included by Staudinger on account of its occurrence in Kamschatka. It is, however, rather Nearctic than Palæarctic, being a common North American species.

V. Charonia, Drury, 1, T. 15, 1, 2; Brem. Lep. Ost. Sib. p. 17. —About the size of *V. Polychloros*. The wing strongly angled, deep black, with a bright blue fascia running parallel to the hind margins. The fore wings have a large light blue spot on the costa near the centre, and another smaller one at the apex. Under side: Mottled with various shades of brown. Hind wings with a light yellow triangular mark in the centre. Habitat: The Amur, &c.

V. Œnone, Linn. Syst. Nat. xii. 770; *Junonia Œnone*, Kirby, Cat. Diurn. Lep. &c.—Expands from 1·50 to 2·0 in. Wings light brownish yellow. Fore wings angulated, black at the apex and along the costa; female with a black ocellated spot near the anal angle; inner margins black. Hind wings deep black at the base, with a large bright blue spot; hind margins dull brown. Habitat: Syria.

V. Huntera, Fab. Syst. p. 499; *Virginiensis*, Drury; *Pyrameis Virginiensis*, Kirby, Cat. Diurn. Lep. 186. — About the size of *V. Cardui*, which species it greatly resembles in shape and colour.

The anterior wings, however, are more rounded at the apices, the ground colour of the wings is richer, and there are fewer black markings. The fore wings on the under side are deeper pink than in *Cardui*, the hind wings are veined with white, and have the ocellated spots very large, especially two near the hind margin. This beautiful species is common in North America, and has become naturalised in the Canaries (conf. *V. Callirhoë*). Stray specimens are sometimes taken in Europe, being probably imported by American vessels. It has even been taken in England, and is sometimes referred to as British and called "the Scarce Painted Lady."

Genus *THALEROPIS*, Staud. Cat. p. 17.

Dr. Staudinger has founded this genus to contain a single species, which was considered a *Vanessa* by Eversmann and an *Apatura* by Herrich Schaffer. I have not seen the original definition of the genus, which appears to be very closely allied to *Apatura*.

T. Ionia, Ev. Lep. Ross. p. 111, pl. xiii. 1, 2; *Ammonia*, H. S. 542-5. — Expands from 1·50 to 1·75 in. Hind margins of fore wings deeply concave, apex prominent, and marked with white spots on a black ground. Hind wings somewhat dentate and prominent at the anal angle, as in *Apatura*. All the wings are fulvous, sometimes with an addition of white and yellow; the bases of the wings are black, and the general pattern of the wings is composed of black spots. The hind wings have an ocellated spot at the anal angle. Under side: Fore wings fulvous, marked with black; an eye-like spot near the anal angle. Hind wings bluish grey, with a dark central band, and a small ocellus at the anal angle. Habitat: The North-East of Asia Minor, principally Amasia and Tokat; it seems to be a rare species, and does not occur in Europe.

Genus *MELITÆA*.

M. Aurinia var. *Sibirica*, Staud. Cat. 1861, p. 7; 1871, p. 17; var. *Desfontainesii*, Ev. Lep. Ross. p. 92. — A very pale-coloured variety, inhabiting Eastern and South-Eastern Siberia.

M. Arduinna var. *Rhodopensis*, Frr. 193, 1, iii. p. 3; H. S. 5, 6. —Larger than the type, and deeper fulvous. Habitat: Asia Minor and Armenia.

M. Casta, Kolv. Ins. Pers. p. 11.—A species but little known, and much resembling *M. Trivia*, of which it is probably a local form. Habitat: Persia.

M. Didyma var. *Didymoides*, Ev. Bull. Mosc. 1847, iii. 67, T. i. 3, 4; H. S. 597-8.— *Sibirica*, Staud. Cat. 1861.—The male has only a few black spots, the female dark greenish or yellow. A variety much resembling the South Russian *Neera*. Habitat: Eastern Siberia and the Amur.

Ab. *Latogenia* (♀), H. S. 599-600: Brem. Lep. Ost. Sib. p. 14.—An aberration in which the female has the wings very dark brown, almost black, with white spots. Habitat: The Amur.

M. Dictynna var. *Erycina*, Led. Z. b. v. 1853, p. 7. — The upper surface of the wings very dark, with yellowish white spots. Habitat: The Altai and the Amur.

M. Protomedia, Mén. Schrk. Reis. p. 23, T. ii. 6, 7.—A species described by Ménetries from Eastern Siberia and the Amur; very close to *Dictynna*.

M. Arcesia, Brem. Lep. Ost. Sib. p. 15, T. i. 7. — Described by Bremer from Eastern Siberia; hardly distinct from *M. Athalia*.

M. Athalia var. *Orientalis*, Mén. Schrk. Reis. p. 23, T. ii. 5.— Under side with the yellow spots replaced by white. Habitat: North-East Siberia.

Genus *ARGYNNIS.*

A. Aphirape var. *Triclaris*, Hüb. Ztr. Exot. vol. ii.—Somewhat resembles *Ossianus*, figured on Pl. XLVI., but is much lighter above, like *Selene* or *Euphrosyne*. The under side is also brighter, the hind wings light red, with conspicuous white markings. Habitat: Labrador.

A. Oscarus, Ev. Bull. Mosc. 1844, iii. 588; H. S. 603-4. — Expands from 1·70 to 1·80 in. Somewhat resembles *A. Amathusia* above. Under side: Hind wings reddish brown, with a light yellow central band, some basal spots of the same colour; between these and the central band is a small round yellow spot with a black centre. Hind margin with a row of pearly lunules; internal to these, near the costa, is a light purple streak, and then a row of black spots with whitish centres (described from fig. in H. S.). Habitat: Eastern Siberia and the Amur.

A. Amphilochus, Mén. Schrk. Reis. p. 21, T. ii. 1. — Expands 2 in. Wings smoky black, with a slight tinge of fulvous. Fore wings with a row of round black spots parallel to the hind margin, surrounded by fulvous rings; a similar row on the hind wings. Under side: Fore wings light reddish brown, spotted with black; marginal rows with white rings; hind wings deeper reddish brown, a large white spot at the base, and another at the centre of the wing; between these is a round white spot with a black centre. The central white spot forms part of a band composed of light yellow spots; parallel to the hind margin is a row of large white spots with round black centres, and outside these a marginal row of white triangular marks. Habitat: The Amur and North-East Siberia.

A. Angarensis, Ev. Bull. Mosc. 1870, i. p. 112. — Size and shape of *A. Thore.* Pattern of the wings above like that of *A. Selenis,* except that the black spots are somewhat larger. The under side resembles that of *A. Thore,* except that the hind wings have a row of pearly white lunules bordered internally by black triangles. Habitat: Eastern Siberia (Irkoutsk).

A. Eugenia, Ev. Bull. Mosc. 1847, ii. 68; Ev. Lep. Ross. T. ii. 1, 2; H. S. 605. — Expands from 1·75 to 1·90 in. Wings rather light fulvous, blackish at the base, and deeply black at the hind margins. Hind wings dark brown, marked with green along the nervures: in the centre of the wing are three rather large silvery spots; between the two nearest the costa are two yellowish white ones; near the base are four small silvery spots, and parallel to the hind margin a row of seven or eight; between these and the medial silver spots is a row of yellow ones with green centres (described from fig. in H. S.). Habitat: Eastern Siberia.

A. Niobe var. *Gigantea,* Staud. Cat. p. 21. — *Niobe,* Ld. Hor. 1869, 82. — Very much larger than the typical form. The male is brighter fulvous, and the female is greenish. Habitat: The North of Persia.

A. Anadyomene, Feld. Wien. Mts. 1862, p. 25. — *Ella,* Brem. Lep. Ost. Sib. p. 94, T. viii. 1. — *Ruslana,* Motsch. Bull. Mosc. 1866, iii. — Expands from 2·25 to 2·50 in. The wings of the male are fulvous, spotted with black, the bases are not dark, and the

spots of all the wings are distinct and do not coalesce; the
nervures are not lined with black. The hind margin of the fore
wings is deeply concave. Under side as in *A. Paphia*, but lighter,
and the silvery markings less distinct. Female expands from 2·25
to 2·70 in. The wings are somewhat duller fulvous than in the male,
and near the apex of the fore wing, on the costa, is a white spot; in all
other particulars the sexes are similar as regards the wings.
Habitat: The Amur and Japan.

A. Sagana, Dbld. Gen. T. 24, 1 (1850); Feld. Wien. Mts.
1862, p. 24; Brem. Lep. Ost. Sib. p. 10. — *Paulina*, Nord. Bull.
Mosc. 1851, iv. p. 440. — Male expands from 2·25 to 2·50 in.
Greatly resembles *Paphia* above, but is somewhat larger and lighter;
hind wings with only one black band between the base and the central
row of spots. The bases are not dusky. Under side almost as in
Laodice, but the fore wings are tipped with purplish brown. The
basal half of the hind wings is light brown, without the yellow
tinge seen in *Laodice*; the outer half is very much as in that
species, only it has faint silvery markings as in *Paphia*. Female
expands from 2·25 to 2·75 in. Dull greenish grey; fore wings
spotted with black, with two large central white spots, and a short
band of similar ones on the costa; there is a white spot near the
apex, and several along the hind margin. Hind wings with a
central white band, outside which is a double row of black spots,
and outside these a row of white ones. Under side: Fore wings
green, black, and white. Hind wings green, with silvery markings.
Habitat: The Amur, Japan, and Eastern Siberia.

NORTH AMERICAN NYMPHALIDÆ ALLIED TO THOSE OF THE PALÆARCTIC REGION.

Genus *LIMENITIS*.

Limenitis Ursula, Fab., and *L. Proserpina* are two dark species
marked with shining bluish green along the hind margins. The
under sides are spotted with light red. These species seem to be
the nearest North American allies to *Limenitis Populi*, and inhabit
the Northern States.

L. Artemis, Dru., is a species something like the above, but has a broad white band on all the wings; inhabits Canada.

L. Weidemeyerii, Edwards, a black species with white bands to the wings, and measuring 2·50 in., is the nearest allied species to the European *L. Sibylla;* it is found near Pike Peak, in Colorado.

Genus *VANESSA.*

There are a number of North American *Vanessæ* closely allied to the European *V. Egra and C-Album;* they are—

V. Fabricii, Edw. Trans. Ent. Soc. Am. 1870. Habitat: California. — *V. Interrogationis,* Fabr. Ent. Syst. Suppl. p. 424 (1798). Habitat: The United States. Larger than any European forms; hind wings with a long tail, and marked with a metallic silvery C.

V. Comma, Harris, Ins. Mass. p. 241 (1852). Habitat: States of New York and Massachusetts. — *V. Faunus,* Edw. Proc. Ent. Soc. Phil. i. p. 183. Habitat: Canada and the Northern States. These two species are very close to the European *V. C-Album.*

V. Satyrus, Silenus, and *Zephyrus,* of Edwards, are also nearly allied.

V. Dryas, Edwards, found in the Southern States, is something like *Egea,* but has the wings less angled and the hind wings darker, whilst beneath they are marked with a metallic silvery C.

(*V. G-Argenteum,* Doubl., from Mexico, is like these, only much larger.)

V. Progne, Cramer, Pap. Ex. 1, t. 5, E. F. — Is a common North American species, darker brown, and less spotted than *C-Album.* Under side black, with a white C-like mark.

V. J-Album, Boisd. Lec. Lep. Am. Sept. p. 185, t. 50 (1833). —Resembles the *V. Vau-Album* of Eastern Europe and North Asia, but is slightly larger and more brightly coloured; it is probably a form of that species. Habitat: Canada, New Brunswick, and the Northern States.

V. Milberti, Godt. Enc. Méth. ix. p. 307 (1819). — A small species somewhat resembling *V. Urticæ,* which it seems to replace in North America; it is common in Canada, Newfoundland, and the Northern States.

V. Californica, Boisd. Ann. Soc. Ent. Fr. 1852, p. 366.—This species is very close to *V. Xanthomelas* or *Polychloros,* but there are very few black spots on the fore wings and only one on the hind wings near the costa. The hind margins are entirely without blue lunules above. Habitat: California.

V. Huntera, Fab. Syst. Ent. p. 499, n. 240.—This species has been alluded to above as having occurred in Europe; it replaces *V. Cardui* in the Nearctic Region.

The North American species of the genus *Melitæa* are not very like those of Europe, with the exception of a few Californian species such as *M. Hoffmannii,* Bhr.; *M. Palla,* Boisd.; *M. Minuta,* Edwards; *M. Arachne,* Edwards.

Among the smaller species of *Argynnis,* the Arctic species *A. Polaris, Charielea,* and *Freija* occur in North America, in Greenland, and Labrador. *A. Freija* has a special variety, *Tarquinius,* Curt. Other North American species are *A. Epithore,* Edw., found in the Rocky Mountains. — *A. Montinus,* Scudd. Habitat: The White Mountains. — *A. Myrina,* Cram., allied to *Selene,* inhabiting the Northern States. — *A. Aphirape,* var. *Triclaris,* has been already alluded to. The following belong to the group of larger species:—

Argynnis Cybele, Fab. Ent. Syst. iii. 445. Common in New York, Pennsylvania, and Virginia. — *Argynnis Aphrodite,* Fab. Ent. Syst. iii. 443. Habitat: New York and Canada. Both these species resemble *A. Adippe,* but are much larger; they have been erroneously reputed British by some of the older writers.

A. Atlantis, Edwards, Proc. Acad. Nat. Sci. Phil. 1862. — Habitat: Mountains in the Northern States and Canada.

A. Hesperis, Edwards, Proc. Ent. Soc. Phil. ii. 1864.—Habitat: Colorado. To this group belong also *A. Zerene,* Boisd. (California), and *A. Behrensii,* Edwards (California).

Allied to *A. Aglaia* are several Californian species, especially *A. Edwardsii,* Reakirt, Proc. Soc. Ent. Phil. 1867.—*A. Nevardensis,* Edw., Trans. Am. Ent. Soc. iii. p. 14. — Both these species are very close to *Argynnis Aglaia* of Europe and Asia.

Family 8. — **DANAIDÆ**, Doubl. Gen. Diurn. Lep. 1847;
Fcld. Wien. Ent. Mosc. vi. p. 74.

CHARACTERS. — Larva cylindrical, smaller towards the head,
furnished on one or more of the anterior segments with a pair of
long slender flexible non-retractile tentacula; a similar shorter
pair on the twelfth segment.

Pupa suspended; rather compressed longitudinally in the
abdominal region.

Imago.—Fore wings elongated, the hind margin generally
concave, and never angulated. The discoidal cell closed; the subcostal nervure five-branched. Hind wings with their margins
entire and rounded; discoidal cell closed; abdominal fold ample.
Eyes prominent. Palpi divergent, not very prominent, triarticulate.
Antennæ gradually thickening into a club. Thorax spotted with
white or yellow on the pectoral surface. Anterior legs atrophied,
the two posterior pairs stout and strong. Abdomen slender, not
reaching as far as the anal angle of the hind wing.

Genus 1.—**DANAIS**, Latr. Ency. Méth. ix. p. 10 (1819); Boisd.
Lec. Lep. Am. Sept. p. 1833; Doubl. Gen. Diurn.
Lep. p. 89.
DANAUS, Latr. Gen. Crust. Ins. iv. p. 20.

Larva generally whitish or grey, marked with green, yellow,
or purple, and with black transverse lines. All the known larvæ
feed on *Asclepiadæ*.

Pupa.—Generally green, with gold decorations.

Imago.—Antennæ about half the length of the body, and with
distinct though gradually formed clubs. Expanse of wings generally
considerable; prevailing colours, brown, black, and white. Nervures
black and strong; costal and subcostal nervures of fore wings
widely separated. Hind wings with a patch of closely placed scales
on the first median nervule in the male.

This is a large genus, widely spread over the warmer regions
of the earth. Only one species inhabits Europe, and but three are
found in North America. They are much given to moving about

2 G

in migratory hordes, like some of the *Pieridæ*, flying with a slow but steady progress, and often soaring high into the air with expanded wings. It is well known that the species of this family are distasteful to birds, either on account of the toughness of their tissues or else of the presence of some acrid secretion. On this account their colour and markings are often strangely mimicked by other Lepidoptera, especially among the *Papilionidæ* and *Pieridæ*, and notably in the genus *Diadema* among the *Nymphalidæ*.

The species of this genus do not inhabit forest lands, but prefer open spaces, plains, and meadows. *Danais Archippus* is said to be commonly seen in the streets of large cities in the United States.

1.—D. **Chrysippus**, Linn. Syst. Nat. x. 471; Hüb. 678-9; O. iv. 120; Boisd. Ic. 18, 3; Dup. i. 17, 1, 2.

Expands from 2·50 to 3·25 in. All the wings fulvous, with a dull reddish tinge. Fore wings black on the costa, the apex black, the black colour covering about one-third of the area of the wing and extending along the hind margin. On the black portion of the wing near the apex are three or four large white spots nearly confluent; some small white spots at the apex, and some more in the centre of the hind margin. Hind wings with three black spots placed on the outer edge of the discoidal cell; hind margin with a continuous black border, spotted with white. Under side: Fore wings the same colour as above, but the apex, external to the large white spots, is yellowish, with the exception of the black border spotted with white, which extends the entire length of the hind margin. Hind wings yellowish brown, the three discoidal black spots being bordered with white; around the wing, from the centre of the costa to the anal angle, is a black border enclosing a row of white spots having a very regular arrangement; there are generally some white basal spots, and the thorax is black, spotted with white or yellow.

TIME OF APPEARANCE.—July.

HABITAT. — In Europe, Central and Eastern Greece: it was found in South Italy, near Naples, at the commencement of the century, but is said to have been destroyed by the unusual

DANAIS.

severity of the winter of 1808. It is common in Persia, in Syria, and other places on the Asiatic border of the Mediterranean. It is spread over the whole of South Asia; in fact it extends to every zoological region of the Old World, and is replaced by very closely allied species in the Nearctic and Neotropical Regions.

LARVA. — Grey, with transverse black streaks, and lateral stripes of green and yellow. It is furnished with three pairs of elastic processes, one pair on the third segment being the longest, and the other two on the sixth and twelfth segments being of about equal length. Feeds on *Asclepiadæ*.

PUPA.—Flattened in the abdominal segments, bright green, with metallic golden markings (all stages figured on Pl. LIV.).

OTHER SPECIES OF DANAIDÆ MENTIONED BY STAUDINGER.

Danais Chrysippus, var. *Alcippus*, Fab. Ent. Syst. p. 50.— *Chrysippus*, Cr. Pap. Ex. 118, E, F; Boisd. Ic. 18, 4 .— A variety occurring in Asia Minor and Syria, in which the hind wings are white, excepting along the hind margins.

Danais Dorippus, Klug. Symb. Phys. Ic. v. T. 48.—About the size of *Chrysippus*. All the wings are a light fulvous colour, with a narrow black border, on which are a few white spots; the fore wings are not tipped with black, and are without white spots. Habitat: Syria.

NORTH AMERICAN SPECIES OF DANAIDÆ.

Only three species occur in North America. One of these, *Danais Archippus* (Cramer, iii. pl. 206), is a large insect, measuring between three and four inches, with the wings brighter in colour and more strongly marked with black than *D. Chrysippus;* all the wings have a border of black, marked with white and yellow spots. Its range extends from Canada to Brazil, and from the Atlantic to the Pacific. This species occasionally migrates or gets imported into the Old World, and has several times been taken

in England. It is thus within the limits of possibility for it to
become naturalised in Europe in the same manner as the Palæarctic *Pieris Rapæ* has become in the Nearctic Region.

D. *Berenice*, Cram. iii. pl. 205. — A species allied to the last,
and found in the Southern States, Mexico and Colorado.

D. *Strigosa*, Ent. Mo. Mag. i. p. 32, 1864.—Habitat: Texas.

Family 9. — **SATYRIDÆ**, Swainson, Cab. Cycl. pp. 86, 93
(1840); Westw. Gen. Diurn. Lep. p. 352 1851.
SATYRINÆ, Bates, Journ. Ent. ii. p. 176.

CHARACTERS. — Larva without spines or projections, generally
pubescent, pisciform or at least tapering towards the anal extremity,
which is usually forked; head generally rounded, occasionally bifid.
Colour generally green or brown, with lateral lines placed above
and below the row of stigmata. The food plants consist of grasses
of various species.

Pupa suspended by the anal extremity or lying face on the
ground; scarcely or not at all angulated, and without metallic
spots; in shape rather cylindrical.

Imago.—Head small; eyes naked or hairy; antennæ generally
rather short, slender, the club distinct, but usually gradually
formed; palpi compressed and elongated, hairy anteriorly;
abdomen small. Fore legs rudimentary in both sexes; in the
male brush-shaped, and not articulated; in the female they are
longer, and have articulated tarsi. Wings proportionately large,
but generally not fitted for a sustained flight; in some species,
however, the flight is rapid. Fore wings often with the nervures
enlarged at the base; the discoidal cell generally long, and always
closed. Inner margin distinctly grooved, to receive the abdomen.
Colour of the wings usually sombre, generally brown, sometimes
black or black and white; fore wings nearly always with an
ocellated spot near the apex; all the wings more or less marked
with submarginal rows of ocellated spots. In the European species
the hind margins of the fore wings are entire, and the apices
generally rounded; the hind wings are either entire or only
moderately dentate.

This family is very large, and of wide distribution: Europe is especially rich in *Satyridæ;* very nearly a third part of the whole number of European butterflies belong to this group.

The eight genera represented in Europe are characteristic of the Palæarctic region; but *Erebia* has some representatives in the Nearctic and Neotropical regions, also in South Africa and the Himalayas; the greater number of species, however, occur in the Alpine ranges of Europe and North Asia.

Œneis has also a few Nearctic representatives, otherwise it is confined to the Arctic and Alpine regions of Europe and Siberia.

Melanargia is exclusively Palæarctic, the majority of the species being found in the Mediterranean region, one only occurring north of the Alps.

Most of the other genera have a few representatives in the temperate regions of the New World, and in the Himalayas. *Cœnonympha* is represented in the Australian region.

Genus 1.—**MELANARGIA**, Meig. Eur. Schmett. i. p. 97 (1829). ARGE, Hüb. Verz. Bek. Schmett. p. 60; Boisd. Gen. Ind. Meth. p. 25 (1840); Westw. Gen. Diurn. Lep. p. 283 (Nom. spec.).

Larva pubescent, with longitudinal stripes; the head rounded; body long and fusiform.

Pupa short and rounded, not suspended by the anal extremity, but resting on the ground.

Imago.—Antennæ long, gradually enlarging into a fusiform club. Palpi slender, separated, the anterior joint pointed and not hairy. Eyes smooth. Wings rounded and very slightly dentate on the hind margins; the costal nervure of the fore wings very slightly dilated at its base, as much above as beneath; colour of the wings white or yellowish, with black bands and marks, and often with ocellated spots which sometimes have blue centres; there is always at least one ocellated spot on the under side of the fore wings, near the apex.

Eight species are found in Europe, but only one occurs in Britain, *M. Galatea* (the marbled white); the majority are swift flyers, and somewhat difficult to capture; they mostly frequent dry grassy places, hill-sides, &c., and are generally very local in their habitats.

1.—M. Galatea, Linn. Syst. Nat. x. 474, xii. 772; Esp. 7, 3, 25, 1; Hüb. 183-5; O. i. 1, 242.

Expands from 1·75 to 2·25 in. Wings pale yellowish white, with black markings; hind margins black, with a row of semilunar white spots; marginal fringes black and white. Fore wings black at the base, and with a large black spot having an irregular rectangular outline, beginning near the costa and extending to the centre of the wing. Hind wings black at the base, and with a central irregular black band; along the hind-marginal black band, near the anal angle, are three spots, most clearly marked in the female, and having blue ocelli. Under side: Fore wings as above, but there is a black spot with a white ocellus near the apex. Hind wings without the basal black markings seen above; the central band grey, outlined with black; the hind-marginal band is grey, with a row of well-marked ocellated spots. Pl. LV., 1.

Time of Appearance.—June to August.

Habitat. — Dry chalky or limestone localities throughout Central and Southern Europe, except the Spanish Peninsula; in Britain it is local, but, where it occurs, abundant; it also occurs in Armenia. The flight of this species is, unlike that of some of its congeners, feeble and short; the butterfly is therefore easy to capture. It is very liable to be infested by an *Acarus*, of a bright red colour, of the genus *Scirus*, which fastens itself firmly to the thorax and abdomen of the insect, sometimes in such numbers as to considerably retard its movements. I have frequently found mites, probably of the same species, on other *Satyridæ*, such as *Satyrus Semele, Epinephele Hyperanthus*, and the *Erebiæ*.

Larva. — Head round and reddish brown; body green, with darker dorsal and lateral stripes; beneath the lateral streak, along the course of the spiracles, is a faintly marked reddish line.

Pupa. — Brown, marked with whitish on the head and wing-cases. The larva feeds on grasses in April and May. Pl. LXXVI., 1.

a. **Leucomelas** (♀), Esp. 81, 1, 2; Hüb. 517-8; O. i. 1, 246; Dup. i. 45, 3, 4; Boisd. Ic. 25, 3, 4.—In this aberration the markings are entirely absent from the under side of the hind wings, which is quite white, the markings of the upper surface showing faintly through. Most commonly found in the South of Europe. I have never seen a British specimen. Pl. LV., 2.

b. **Galene** (ab.), O. i. 2, 236. — An aberration in which the ocellated spots are absent.

c. **Procida** (var.), Hbst. 183, 5, 6; Hüb. 658, 9; O. i. 1, 246; Dup. i. 45, 5, 6; Boisd. Ic. 25, 5, 6; Frr. 379. — A variety inhabiting South-Eastern Europe, Asia Minor, and Armenia, in which the black markings are in excess of the white portions of the wings, the insect having the appearance of a black butterfly spotted with white. The hind-marginal and basal black markings are increased in size on all the wings, the inner margin of the fore wings being broadly and continuously black. The under side differs little from the type. Pl. LV., 3.

d. **Turcica** (ab.), Boisd. Gen. p. 25; Stgr. Cat. p. 22. — *Turca,* Millière, Icon. 1860, p. 126. — An exaggerated form of the preceding, in which the wings are almost entirely black.

HABITAT. — The Balkans, occasionally in other parts of the South of Europe. The following is Millière's description of a specimen taken by Donzel near Digne:—" The black colour, which in *Galatea* is only indicated by large patches of different forms, has invaded in *Turca* almost the entire surface of the fore wings; with some difficulty one distinguishes two or three small white oblong blotches placed near the centre of the fore wings. The hind wings are less black than the fore wings, and allow us to distinguish four or five large white patches sprinkled with black. The nervures are very broadly marked with black. The ordinary lunules are very feebly traced. The under side of the fore wings is a repetition of the upper. The hind wings are entirely black, with a greenish tinge, the ordinary patterns being marked in deeper black."

2.—M. Lachesis, Hüb. ii. 3, 3 ; Hüb. 186-7 ; O. i. 1, 247 ; Godt. ii. 19, 1, 2 ; Mill. Ic. 62, 4, 5.

Expands from 2·25 to 2·50 in.   Somewhat resembles *M. Galatea*, but, besides being on the average larger than that species, it has the ground colour white in both sexes.   There is no black patch at the base of the fore wings, and no central black band on the hind wings, which, with the exception of a black spot on the costa, are plain white, with a hind-marginal black band, on which are some ocellated spots, especially well-marked in the female. Under side somewhat similar to that of *M. Galatea*, but lighter ; the hind wings have five distinct black ocellated spots.   Pl. LV., 4.

Times of Appearance.—May and June.

Habitat.—The Spanish Peninsula ; and the South of France, in Provence, Languedoc, Roussillon, and at Vernet-les-Bains, where it is said to replace *M. Galatea*, which in its habit it closely resembles.   It seems to be confined to the above localities, though there are some very closely-allied species both in Asia Minor and the Amur ; they, however, are considered distinct from *Lachesis*.

Larva. — " Fusiform, the last segment without a forked tail, and entirely pubescent, like the larva of its congener, *Cleanthe*.   Its general colour is flesh-tint, with the ordinary lines of a pale red. The transverse marks showing on each segment are terminated by a longitudinal line, straight and continuous, placed beneath the line of the stigmata.   The ventral surface is paler than the dorsal, and is not marked with lines.   The head is yellow, globular in shape, with the eyes and mandibles marked with brown.   The legs are light brown ; the stigmata large and black.   It lives, like its congeners, on different species of grass ; the *Lamarekia aurea*, L., however, seems especially to form its food." — Mill. Icon. 1865, p. 92.

VARIETY.

Cataleuca (ab.), Staud. Cat. p. 22.—An aberration occurring in Central Spain in which the dark markings are absent from the under side of the hind wings ; it thus corresponds to the aberration *Leucomelas* of *M. Galatea*.

3.—**M. Larissa,** H.-G. 896-9; Frr. B. 73, 1; Boisd. Ic. 28, 4, 6; Dup. i. 26, 1-4.

Expands from 1·50 to 2·0 in. Marginal fringes black and white. All the wings with the ground colour creamy white; bases broadly dark brown; hind margins broadly dark brown or black, with a row of white spots. Fore wings with two large confluent black spots in the centre of the costa. Hind wings with two or three indistinct ocellated spots in the black border, formed by white rings enclosing black ones, with blue or white pupils. Under side: Fore wings somewhat as in the under side of *Galatea,* the dusky basal patches seen on the upper surface not appearing. Hind wings with a central band of clear white spots edged with dark brown or black; hind-marginal band much fainter than above, and with five ocellated black and white spots. Pl. LVI., 1.

Time of Appearance.—June.

Habitat.—The Balkan Provinces, and Asia Minor.

Larva.—Unknown.

VARIETY.

Hertha, H.-G. 900-3; Boisd. Ic. 28, 1, 3; Tr. x. 1, 39.— On the under side this variety resembles the typical form described above; but above it is much lighter and more resembles *Galatea* on the one hand, and *Iapygia* on the other. The bases have not the broad dusky patches seen in the type, but, as in the allied species, have black markings enclosing large white spots; the hind margins are without the broad dark bands, but have them somewhat as in *Galatea,* the white spots being triangular. It may be distinguished from *Galatea* by the blue centres of the hind-marginal eyes, by the white central band on the under side of the hind wings, and by the average size of the insect being greater. The broad hind-marginal band on the under side of the hind wings will distinguish it at first sight from the next species.

Habitat.—Dalmatia and Greece; not occurring so far eastward as the type. Pl. LV., 5.

2 H

4.—**M. Iapygia,** Cyr. Ent. Neap. i. T. iii. 5, 1787; Esp. 105, 3, (1789).

CLOTHO, Luc. Expl. Alg. Pl. 2, 4.

ATROPOS, Hüb. 192, 3.

CLEANTHE, Boisd. Ic. 26, 1-3; Dup. i. 25, 5, 6; Mill. Ic. 62, 1, 3.

Expands from 1·90 to 2·25 in. Marginal fringes black and white. Wings white, with a yellowish tinge in the male; bases dusky; nervures black. Fore wings with a central black spot, from the lower part of which a thick black streak descends to the anal angle, when it merges into the black hind-marginal border, which contains a row of large white spots; between the central black spot and the base is a wavy black line occupying the centre of the discoidal cell. Hind wings almost as in *Galatea,* but there is no hind-marginal black border, its place being taken by a series of black lines arranged in triangles. The ocellated spots are very distinct, especially in the female; there are five of these, two being placed near the costal edge, and three near the anal angle; they are white, with black rings enclosing white pupils, and without any tinge of blue. The basal patches enclose a large white blotch similar to those of *Galatea.* Under side: Fore wings with markings corresponding to those above, but lighter and more slender, the hind-marginal band being reduced to a wavy black line. Hind wings with a yellow central band bordered with dark brown; hind margin slightly yellow, with a zigzag black line, internal to which are spots corresponding to those above, but yellowish, bordered with dark brown. and with white pupils in dark rings. Pl. LVI., 2.

TIMES OF APPEARANCE.—June and July.

HABITAT.—The South of Italy, Sicily and the North of Africa, South-Eastern France, Central Spain, Hungary, South Russia, and Persia.

LARVA. — The following description is taken from Millière, Icon. 1865; Pl. 62, i. 3:—"More elongated than that of its congener *Galatea,* less fusiform, and entirely pubescent, the head and the legs included. The segments are distinct. The body is of a citron-yellow, somewhat inclining to green. The five usual lines are fine, well-marked, and extend from the second to the eleventh segment. The vascular line is darker than the ground colour, and marked with white on each side; the subdorsal and stigmatal lines are yellow, and larger than the first. The stigmata are black, visible to the naked eye, and encircled by faint red. The ventral surface, which is not so bright as the rest of the body, does not show any lines. The head is tolerably large, globular, and separated from the first segment; it is reddish yellow, with the eyes marked with brown; the mandibles are reddish. This larva was first discovered in Spain by Dr. Staudinger; it lives on different grasses, but particularly on *Lamarckia aurea*, L., a plant peculiar to the southern parts of Europe. To undergo its metamorphosis the larva hangs itself head downwards to a stalk of grass, and three or four days are enough for it to complete its transformation. The pupa is plump, and of a bright brown colour and dull surface; it is finely striped with red on the wing-cases and thorax; the covering of the eyes and the top of the head is brown. The imago emerges in fifty or sixty days."

Obs.—This species, in consequence of its widely-extended habitat, presents several local varieties or races. The typical form originally described is that found in Italy and Sicily, and is smaller than the Franco-Spanish form *Cleanthe* of Boisduval; the male figured on Pl. LVI. represents *Cleanthe*, which is the commoner form of the species; it is very slightly darker, besides being a little larger than the Italian specimens ("*Cleanthe* vix nom. conserv.," Staud. Cat. 23). A third form is the Russian and Hungarian var. *Suwarovius*, Hbst. viii. p. 13, T. 182, 5, 7. This is somewhat larger and paler than the Italian or French forms, both sexes resembling the female figured on Pl. LVI. The flight of this species is said to be very rapid. M. Maurice Girard, speaking of *Cleanthe*, says, "Elles semblent toujours pressées, comme si elles avaient de longues courses à faire" (Traité Elementaire d'Entomologie, Macro-lepid. p. 210, 1882).

5.—**M. Syllius**, Hbst. viii. p. 15, T. 182, 8, 9 (1796); O. i. 1, 254.
Psyche, Hüb. 198, 9 (1799); Godt. ii. 19, 3, 2.
Occitanica, Esp. 96, 3, 4.

Expands from 1·80 to 2·20 in. All the wings white, with a black hind-marginal border enclosing a row of white spots. Fore wings with several white spots on the black border near the apex, the border being widest at this part; in the centre of the wing, touching the costa, is a large black spot enclosing one or two white ones; the base dusky. Hind wings with a black spot on the costa, having the appearance of being continued from a similar spot on the fore wings; near the anal angle are three indistinctly marked black eyes with darkish rings. Nervures of all the wings black above. Under side: Fore wings similar in markings to the upper side, but the nervures are reddish brown along the hind margin. Hind wings similar to the upper side, but the nervures are marked with reddish brown; the black markings also have a tinge of the latter colour; there are five submarginal ocellated spots, not very distinctly marked, having blue centres enclosed in reddish and yellowish rings. Pl. LVII., 1.

Time of Appearance.—June.

Habitat. — Piedmont, the South of France, and the Spanish Peninsula.

Larva.—The following is translated from Millière, Icon. 1873, p. 276:—"At this time (May 10th to 15th, *i. e.*, when it is full grown) it is rather elongated, sensibly tapering posteriorly, pubescent, of a yellowish flesh-colour, with well marked and continuous lines. The vascular line, which begins and ends in an acute point, is large, of a clear brown, and bordered narrowly with white on each side. The subdorsal line is Naples-yellow, and is edged with green above and below; the lower edging extends along the sides down to the stigmatal line, which is fine and clear. The head is globular, pubescent, and of an indistinct green, with two black ocelli on each side. The legs are flesh-coloured. . . . . . The stigmata are very small, white, encircled with black. About a third part of the larvæ collected had a different colour from the type. This constant variety has the ground colour bright green, and an entirely black vascular line. M. Guenée has pointed out to me that the same thing occurs in the larva of *Galatea.* This *Melanargia* lives on a large kind of grass very common in the neighbourhood of Cannes, and peculiar to dry and stony calcareous districts—the *Brachypodium pinnatum.* The pupa is rather elongated, plump and smooth, yellowish in colour, and finely striped with dark brown."

VARIETY.

Ixora (ab.), Boisd. Ic. 27, 3, 4.—Differs from the type in the absence of the ocelli from the under side of the hind wings.

6.—M. Pherusa, Boisd. Ic. 26, 4-6, i. p. 141; Dup, i. 45, 1, 2.

Expands from 1·80 to 2·20 in. Greatly resembles the last above, especially in the markings of the fore wings, which are almost identical, though perhaps not quite so dark. The hind wings are lighter at the base than in *Syllius,* and the submarginal spots are smaller, with very faintly defined lightish centres. Under side: Fore wings as in *Syllius,* but lighter, the black markings finer, and not so distinctly marked with brown. Hind wings with the nervures dark brown, and only indistinctly marked with reddish; submarginal spots indistinct, and only sometimes with blue centres. Pl. LVI., 3.

Time of Appearance.—June.

Habitat.—Sicily.

VARIETY.

Plesaura, Bell. Ann. S. Fr. 1860, pl. 12, 1, 2.—An aberration corresponding to the form *Ixora* of *Syllius*—that is, wanting the ocelli, and sometimes entirely without the submarginal black spots above.

Larva.—Unknown.

7.—M. Arge, Sulz. Abg. G. 1776, T. 16, 8, 9; Esp. 27, 1, 70, 1, 111, 2; O. i. 1, 251; Dup. i. 24, 5, 6.

Amphitrite, Hüb. 194-5 (1799); Boisd. Ic. 27, 1, 2.

Expands from 1·75 to 2·20 in. All the wings creamy white, black at the base, and with a wavy hind-marginal black band. This species differs from the last two in having fewer black

markings on the fore wings above, and in the absence of any
brownish red line along the nervures beneath. Fore wings with
an irregular hollow spot near the centre of the wing touching the
costa, and between this and the base another smaller blotch ; near
the apex is a blue-centred ocellated spot in the female ; there are
two, and these are more distinct than in the male. Hind wings
with a submarginal row of five black spots with blue ocelli, two
placed near the costa and three near the anal angle ; there are two
faintly marked central lines traced round the discoidal cell. Under
side : Fore wings white, with markings similar to those above, but
finer ; one or two ocellated spots are placed near the apex. Hind
wings with the black discoidal lines stronger than above, and
several other black lines running across the wing internal to them;
nervures black ; between them, at the marginal ends, are some light
reddish brown streaks ; submarginal row of eyes well marked, yellow,
with blue ocelli enclosed in reddish brown rings. Pl. LVII., 2.

TIME OF APPEARANCE.—June.

HABITAT.—Calabria, Sicily, Spain, and Portugal.

LARVA.—Unknown.

8.—**M. Ines**, Hffsgg. Ill. Mag. iii. p. 205 (1804); O. i. 2, 237 ;
Boisd. Icon. 27, 5, 6 ; Dup. i. 24, 1, 4.
THETIS, Hüb. 196-7 (1799), (Nomen preocc.).

Expands from 1·80 to 2·0 in. Very close to the last three
species, but altogether darker, and the markings more pronounced.
Wings yellowish white, marked with brownish black ; nervures
black, but not very strongly marked. Fore wings with a broad
black hind-marginal border, in which are two rows of white spots,
those of the internal row being the largest ; in the centre of the
costal margin is a black blotch containing a white spot ; between
this and the base an angulated black mark ; basal portion of inner
margin dusky ; near the apex are one or two ocellated spots, almost
obliterated by the black band. Hind wings with a black hind-
marginal border, having the usual row of white lunules ; the submarginal spots are deep black and large, with blue ocelli ; the
discoidal black line strongly marked, merging into a line of dusky
shading, which extends from the base. Under side : Fore wings

whiter than above, except at the apex, where the ground colour is yellowish; black markings as above, but lighter; near the apex are one or two ocellated spots, of a pinkish red, with white centres; on the costal margin, at its basal end, is a row of minute black spots, not found in the other species. Hind wings yellowish white; nervures distinctly black; discoidal lines forming a reticulated pattern about the basal portion of the wing; submarginal ocellated spots very large and distinct, each one formed by an external black ring enclosing a yellow one, and inside this a reddish pink one containing a blue centre. Pl. LVII., 3.

TIMES OF APPEARANCE.—May and June.

HABITAT. — The South of Spain. I have received it from Huelva, but it occurs at other places near the sea-coast, and on the coast of Africa.

LARVA.—Unknown.

Genus 2.—**EREBIA**, Dalm. Sverig. Handl. 1816. p. 58; Boisd. Gen. Ind. Meth. p. 26; Westw. Gen. Dal. p. 376. MANIOLA, Schrank. Faun. Boica, ii. 1, p. 152; Meig. Eur. Schmett. 1, p. 104 (Nomen vetustius).

CHARACTERS. — Larva, in the few cases where it is known, fusiform, bifid at the anal extremity, green or yellowish in colour, with darker dorsal or lateral stripes; feeding on various species of grasses, chiefly of the genera, *Poa, Festuca, Panicum,* &c.

Pupa.—Wing-cases long; body plump, thickest in the middle, tapering to the tail, and ending in a blunt flat spike; back of thorax rounded; head- and eye-piece prominent; not attached by the tail, but placed in an upright position among grass near the ground.

Imago.—Head of moderate size. Antennæ variable in length, terminated by an oval club distinct from the shaft, and flattened. Palpi separated, covered with close silky hairs. The eyes smooth. Wings more or less hairy at the base; the nervures not dilated at the base, with the exception of the subcostal nervure of the fore wings; fore wings generally rounded; hind wings sometimes but not always denticulate; colour dark brown or black, with brown or reddish marginal bands, generally enclosing eye-like spots. The fringes of the wings are usually unicolorous greyish white; but some, as *E. Ligea* and *Euryale*, have them chequered with grey and black.

This generic group is well marked by the general colouring of the wings, and by the habit of the species. They are all inhabitants of regions which are more or less cold, generally on account of altitude, but sometimes of latitude, as in the case of *Erebia Embla* and *Disa*, which are found in Lapland. In countries which are quite flat or but little elevated above the sea-level the genus *Erebia* is entirely absent; it is poorly represented on islands, however large and mountainous they may be—England and Scotland, for example, though possessing considerable tracts of elevated country and by no means insignificant mountains, have at present only two species, *E. Æthiops* (the Scotch Argus) and *E. Cassiope* (the Mountain Ringlet). In Ireland, though that island is mountainous in parts, *Erebia* is absent. Corsica, too, a large island situated comparatively near to the great central alpine ranges of Europe, which may be said to be the head-quarters of the genus, and itself possessing mountains on whose summits the snow often remains throughout the summer, does not number a single *Erebia* amongst its lepidopterous fauna.

About sixty species are known, twenty-nine of which inhabit Europe; others are found in Siberia, the Amur, the Himalayas, Arctic America, Colorado, Chili and Patagonia, several in South Africa, and one in Madagascar. Though some species occur in the plains and on the lower grounds, as *E. Æthiops, Ligea, Medusa,* &c., by far the greater number of the European species are found in mountainous regions at various elevations, as far as the verge of the unmelting snow. Many species only occur at particular elevations, so that the *Erebiæ* are more or less distributed in zones upon the sides of the mountains. The best plan, therefore, for those who wish to obtain a considerable number of species is to reside at some place having a certain elevation, and to work upwards and downwards from that station, thus saving the labour of making continual ascents in search of particular species. Zermatt and Chamounix are perhaps as good as any other places in Switzerland to choose for this purpose. All the species occur during the summer season, from the beginning of June to the end of August; those that inhabit mountain regions are single-brooded.

The great general similarity between the species of this genus would cause anything like lengthy descriptions to involve considerable repetition; in describing the *Erebiæ*, therefore, we have principally to notice the expanse of the wings, their colour, the colour and extent of the submarginal bands, and the size and number of the spots, and whether they are ocellated or not. The females are usually larger than the males, and with light and more distinct markings beneath. The life-history of all but a few species is entirely unknown; this is perhaps accounted for by the fact that so few entomologists who visit the places where the imagines abound are able to stay long enough to watch the earlier stages; to this may be added the extreme difficulty or impossibility of obtaining proper food-plants for the larvæ in localities far distant from their real habitat. There is a wide field of work in alpine regions open to those who wish to bear their share in taking away from entomological literature the reproach of having to write "Larva unknown" against the majority of the species of *Erebia*.

1.—E. Epiphron, Kntz. Btr. iii. 131, 6, 7 (1783); O. i. 1, 258; H. S. 92, 94; Frr. 544, 1, 2.

Expands from 1·25 to 1·50 in. Marginal fringes dark brown. Wings brown, darkest at the base; submarginal band marked with about equal intensity on all the wings, narrow, and generally broken up into four or five spots on each wing, fulvous in colour, each fulvous patch containing a distinct black spot, which sometimes has a white centre. Under side similar to the above, but paler, both in the ground colour and the fulvous bands.

TIMES OF APPEARANCE.—June and July.

HABITAT. — Elevated meadows and mountain-slopes in Germany, Silesia, and the nort-east of France (the Vosges and Rhine Provinces). It is less widely distributed than the variety *Cassiope*, described below.

LARVA (of *Cassiope*). — Pale green, with longitudinal lines of a darker colour; a lateral stripe of a white colour runs along the line of the spiracles. Feeds on various grasses, principally on *Poa annua*. Pl. LVIII., 1.

a. **Cassiope**, Fab. Mant. 42; Hüb. 626-7; O. i. 1, 261; Godt.
ii. 15, 1, 2; Frr. 20, 1, 2. — Smaller than the German type, and
paler in colour, especially as regards the fulvous bands; the black
spots are smaller, less evenly arranged, and without the white
pupils in the female; the submarginal fulvous bands are much less
distinct, especially on the hind wings, where the black spots also
are sometimes entirely wanting. Pl. LVIII., 2.

TIME OF APPEARANCE.—June to August.

HABITAT. — The Alps of Switzerland and Piedmont; the
Pyrenees; mountains in Hungary; also in the North of England,
and in Scotland, in marshy hollows at a great elevation. In the
Vosges it occurs simultaneously with the typical *Epiphron.*

b. **Nelamus**, Boisd. Gen. p. 26. — An Alpine form occurring in
Switzerland. It has the black spots absent from the fulvous bands
on all the wings.

c. **Pyrenaica.** — Larger than *Cassiope,* with large ocelli on all
the fulvous bands. Inhabits the Pyrenees.

2.—E. **Melampus**, Fuess. Verz. p. 31, fig. 6 (1775); Esp. 103, 1;
        O. i. 1; Boisd. Ic. 35, 5, 6; Frr. 19, 1, 2.
    JANTHE, H. S. 624-5.

Expands from 1·16 to 1·25 in. Dark brown. Submarginal
fulvous band of fore wings broken up into five or six spots of an
almost oval shape, and separated only by the nervures, which are
strongly dark brown or black; in the centre of each spot is a very
small black dot. Hind wings with three or four fulvous spots,
smaller than those on the fore wings, each with a black dot.
Under side : Dark brown, rather lighter than above; fulvous spots
small and distinctly separated. Pl. LVIII., 3.

TIMES OF APPEARANCE.—July and August.

HABITAT. — Elevated meadows in the Pyrenees, the Alps of
Switzerland, and South-Eastern France. It is a local species, but
common where it occurs.

LARVA.—Unknown,

3.—**E. Eryphile**, Frr. 187, 3, 4 (1836); Meyer-Dür. Tgf. p. 154. Tristis, H. S. 387-90.

Expands from 1·30 to 1·50 in. All the wings dark krown; fore wings with the submarginal band broken into six spots of a bright fulvous colour, the two upper ones having each a black dot in the centre. The hind wings with three or four fulvous spots, without black dots. Under side : Fore wings tinged with fulvous towards the base, so that the basal edge of the submarginal band becomes somewhat indistinct; its outer edge is, however, well marked, the marginal portion of the wing being dark brown. Hind wings dark brown in the male, and with a submarginal band of a lighter colour; this band is more fulvous in the female, and the ground colour is lighter. Pl. LVIII., 4.

Times of Appearance.—July and August.

Habitat. — Elevated mountain meadows in Switzerland and Styria. Rare and local; by some supposed to be a variety of *Melampus*.

Larva.—Unknown.

4.—**E. Arete**, Fab. Mant. 42; Hüb. 231-2; O. i. 1, 301; Bell. Ann. Soc. Fr. 1858, Pl. 11, ii.

Expands from 1·25 to 1·30 in. Fringes yellowish. All the wings brown. Fore wings with a fulvous band, divided by the nervures into five or six divisions; there are no black dots. Hind wings brown, with the fulvous altogether absent or only faintly marked. Under side : Fore wings fulvous, lighter towards the hind margin, and more or less shot with green; near the apex are two small black spots. Hind wings reddish brown in the male, and light yellowish brown in the female; sometimes there are some black dots placed parallel to the hind margin. Pl. LVIII., 5.

Time of Appearance.—July.

Habitat. — Alpine regions in Styria and Carinthia. Another rare and local species.

Larva.—Unknown.

5.—E. **Mnestra**, Hüb. 540-3 (1802); Esp. 120, 3, 4; O. i. 1,
    264; Boisd. Ic. 35, 1-4; Düp. i. 34, 3, 4; Frr. 19, 3;
    H. S. 96.

Expands from 1·20 to 1·40 in. Fringes dark brown. All the
wings brown. The male has a band of fulvous spots on the fore
wings, and four fulvous spots on the hind wings. The female has
a very broad fulvous band on the fore wings, less sharply defined
on the basal than on the outer edge, and showing two small black
dots at the part where it is closest to the apex; the base of the
wing is darker than the hind-marginal portion; hind wings brown,
darkest at the base. Under side : Fore wings fulvous in the centre,
the basal and hind-marginal portions dark brown; hind wings dark
brown, with four fulvous spots in the male, but not in the female.

TIMES OF APPEARANCE.—July and August.   Pl. LVIII., 6.

HABITAT.—Elevated meadows in most of the alpine regions of
Central Europe. It is common but local in Switzerland. I have
taken it on the Furka Pass in August; it is to be taken in France
in the departments of the Hautes—and Basses--Alps, in which
localities seventeen species of *Erebia* are said to exist. (Girard,
p. 229).

LARVA.—Unknown.

6.—E. **Pharte**, Hüb. 491-4; Esp. 120, 4; O. i. 1, 259; Boisd.
    Ic. 35, 7, 8; Dup. 1, 24, 1-2; Frr. 20, 3; H. S. 95.

Expands from 1·20 to 1·40 in. All the wings brown. Fore
wings with a band of five fulvous spots; hind wings with four
spots in the male, generally not well defined, and as a rule absent
in the female. Under side : Fore wings with the fulvous band
distinct; hind wings with the row of fulvous spots in both sexes.
Pl. LVIII., 7.

TIMES OF APPEARANCE.—July and August.

HABITAT. — The Alps of Switzerland, in elevated meadows;
also in the South-Eastern Departments of France, in the same
localities as the last.

7.—E. **Œme**, Hüb. 530-3 (1803); Esp. 120, 2; O. i. 1, 270; Düp. i. 34, 4-8; Boisd. Ic. 34, 5, 8; Frr. 31, 1, 2.
CŒCILIA, Esp. 120, 1.

Expands from 1·40 to 1·70 in. Fringes dark brown. Antennæ proportionably rather longer than in the preceding species. All the wings dull brown, without any fulvous bands; bases nearly black, and the nervures darker than the colour of the wings; the fore wings have two small dull red spots near the apex, sometimes these have white pupils; there are one or two similar spots on the hind wings, generally very indistinct above, but more plainly visible beneath. Under side almost as above, but the red spots are somewhat plainer. Pl. LVIII., 8.

TIMES OF APPEARANCE.—June and July.

HABITAT. — The Alps of Switzerland, and Savoy; in France, principally in the South-Eastern Departments; in Auvergne, but rare, in July and August at Puy de Sancy (Girard).

VARIETY.

Psodea, Frr. 121, 3, ii. p. 44; H. S. 165-5; O. i., 1, 171.— *Spodia*, Staud. Cat. p. 24. — This form of the species is found in the alpine regions of Austria and Styria, and differs from the type in being of a larger size, and in having the ocelli larger and more defined.

8.—E. **Manto**, Esp. 70, 2, 3 (1781), ii. p. 106; Hbst. 210, 8, 9, 211, 1, 2; Bkh. i. 100, 245.
PYRRHA, W. V. p. 167; Fab. Mant. 42 (1787); Hüb. 235-6, 616; O. i. 1, 267; Godt. ii. 15, 3, 4; Frr. 31, 3, 4.—*Œme*, var. Esp. 120, 1.
ERINA, Fab. E. S. 237.

Expands from 1·25 to 1·35 in. All the wings dark brown. Fore wings with a submarginal row of reddish brown spots; in the female these are fewer and paler, and have one or two small black dots. Hind wings with three or four reddish brown spots in the male, plain brown in the female. Under side: Brown; fore wings in the male with a reddish brown band; hind wings with a few reddish brown spots as above. Female much paler than the male;

fore wings with a light yellowish brown band, in which are one or two black dots; hind wings brown, with a submarginal row of yellowish spots, and some others near the base of a similar colour. Pl. LIX., 1.

TIMES OF APPEARANCE.—July and August.

HABITAT. — The Alps of Switzerland, in elevated meadows; the Pyrenees, Mont Dore in Auvergne, Cacadogne in the Vosges (Girard).

LARVA.—Unknown.

9.—**E. Ceto**, Hüb. 578-9; O. i. 1, 272; Godt. ii. 16, 1, 2; Frr. 37, 1, 2; H.-G. 1002, 3.

Expands from 1·30 to 1·60 in. All the wings dark brown; submarginal brown spots tolerably well marked in both the fore and the hind wings; two or three of the spots on the fore wings and all those on the hind wings have black dots in the centre; in the female these black dots have minute white centres. Under side the same as above, but paler, and entirely without basal markings. This species differs from the last in being of larger size, in having the submarginal spots distinctly dotted with black, and in the greater similarity between the sexes. P. LIX., 2.

TIMES OF APPEARANCE.—July and August.

HABITAT. — Elevated alpine meadows in Switzerland, North Italy, the South-Eastern Departments of France, and the Pyrenees.

VARIETY.

Phorcys, Frr. 193, 2, iii. p. 4.—A variety, figured by Freyer, in which the submarginal spots on the under side are white, with dark brown centres.

10.—**E. Medusa**, Fab. Mant. 40; Hüb. T. 45, 103-4; O. i. 1, 273; Frr. 43, 1; Godt. ii. 15, 5, 6; Meyer-Dür. Tgf. p. 163.

LIGEA, Esp. 7, 2 (1777, sed non *Ligea*, Linn.)

MEDEA, Bkh. i. 74, 235 (1788); Hbst. 208, 3, 43.

Expands from 1·60 to 1·75 in. All the wings dark brown, with submarginal rows of light brown spots, forming a band and enclosing round black spots with very distinct white centres.

Under side similar to the upper surface, but paler; the fulvous spots are distinct, and do not coalesce as above. This species may be distinguished from the last by the greater regularity and distinctness of the fulvous spots, which, though they are coalescent, appear as distinct fulvous rings surrounding the black ocelli; the ground colour of the wings is lighter, and in the male shows a tendency to greenish reflections in certain lights. The black ocellated spots are more numerous in the female than in the male; there are generally six on the fore wings in that sex, the two nearest the apex being the largest. Pl. LIX., 3.

TIMES OF APPEARANCE.—May and June.

HABITAT. — Elevated woods and occasionally in wooded plains in Central and Southern Germany, Belgium, and Switzerland; in France it appears to be found in almost all the departments to the extreme east, especially in the Vosges and the neighbourhood of the Jura; it also occurs in Scandinavia and in South Russia.

LARVA.—Pubescent, clear green, with a dorsal band and three lateral lines of a stronger green; the dorsal band is bordered on each side by an edging of greenish white; besides this there is a line of the same colour along the legs, which are green, as also is the head and the caudal points (Girard). Lives on *Panicum sanguinale*, and is full fed in May.

VARIETIES.

a. **Psodea,** Hüb. 497-9; Boisd. Ic. 34, 3, 4; Düp. 1, 40, 1, 2.— *Eumenis*, Frr. 85, 4, 5. — The ocelli are larger and more numerous than in the type.

HABITAT. — Eastern Hungary, South Russia, Bulgaria, and Asia Minor.

b. **Uralensis,** Staud. Cat. p. 24. — The ocellated spots are smaller and less numerous than in the type, and the hind wings are banded with a lighter colour.

HABITAT.—The Ural Mountains and Siberia.

c. **Polaris,** Staud. Cat. p. 24.—*Embla*, H. S. 382-3. — Smaller than the type, and darker.

HABITAT.—North Lapland.

d. **Hypomedusa,** Meyer-Dür. Tgf. p. 163. — ? *Lefebvrei*, H. S.

280-2. — Somewhat resembles *E. Œme.* Smaller than the type, with the ocellated spots very few and inconspicuous.

Habitat. — Austria and the North of Switzerland, in elevated alpine pastures.

11.—E. **Stygne**, O. i. 1, 276 (1807); H. S. 90, 91.
  Pirene, Hüb. 223-4.—*Pyrene*, Frr. 43-2.
  Irene, Hüb. Text. p. 37.
  Lefebvrei, H. S. 88-89 (*Pyrene*, Linn., et *Irene*, Fab. alii erant papiliones).

Expands from 1·20 to 1·60 in. The male has all the wings dark brown; fore wings with a submarginal deep fulvous band extending from near the costa to a point rather less than two-thirds of the distance between the costal and inner margin; this fulvous band encloses three or four black spots, the two nearest the apex being larger than the rest, and containing bluish white pupils; hind wings with the submarginal band broken into four spots; the three nearest the anal angle containing each a black spot, sometimes with a minute white centre. Under side: Fore wings dark brown, with a bright fulvous band having the same extent as above, but unbroken; the three ocellated spots are very distinct. Hind wings dark brown, with a slight indication of a lighter submarginal band; near the anal angle is a very minute ocellated spot. Female lighter than the male; fore wings with the submarginal band reaching nearly to the inner margin, and enclosing five black spots with bluish white centres; the 1st, 4th, and 5th being very small, the 2nd and 3rd large and coalescent; hind wings with a continuous submarginal band enclosing three black spots of equal size, and with white centres. Under side: Fore wings the same as above, but paler; hind wings brown, paler towards the hind margin, with three submarginal black spots enclosing small white pupils. Pl. LX., 1.

Time of Appearance.—July.

Habitat. — Mountain meadows, but not at a great elevation, in Switzerland, the Pyrenees, the South of France; also in the Vosges, and in Auvergne at Mont Dore, where it is common; it occurs rarely in Central France, in the Department of Creuse

(Girard). This is probably one of the most common and widely distributed species of the genus, and is perhaps often passed over by collectors on account of its general resemblance to *E. Æthiops*.

LARVA.—Unknown.

12.—E. Nerine, Frr. 13, 3, 4; Boisd. Ic. 31, 6, 7; Tr. x. 1, 49; Dup. i. 35, 5, 6; H. S. 69, 74.

Expands from 1·90 to 2·20 in. All the wings dark brown. Fore wings with a well-defined submarginal fulvous band extending nearly to the inner margin, enclosing two large coalescent black spots near the apex, each having a conspicuous bluish white pupil; the third spot, which is placed nearer to the anal angle, is much smaller, but has a distinct white centre; hind wings with a distinct submarginal fulvous band containing three black spots of equal size, and with bluish white centres. Under side: Fore wings fulvous, the hind margin and costa brown, the ocellated spots as above; hind wings mottled with dark brown and greyish white, with an indistinct dark central band.

The female is lighter in colour than the male above, and beneath there is a nearly white central band on the hind wings, which are altogether lighter than those of the male; the outline of the wings is more dentated than in the other sex. Pl. LX., 2.

TIMES OF APPEARANCE.—August and September.

HABITAT. — A very local species, being only found in alpine meadows at a great elevation in South-Eastern Germany and the Tyrol.

VARIETIES.

a. **Reichlini**, H. S. Corr. Inst. Nr. p. 5; Spr. Stett. Ent. Zeit. 1865, 243.—Differs from the type in having the fore wings blackish brown instead of fulvous beneath, and in having a small ocellated spot above the two large ones near the apex; the hind wings are darker, and without brownish white markings. Pl. LX., 3.

HABITAT.—The South of Bavaria, and the Tyrol. (Perhaps a local form of *E. Evias*).

b. **Morula**, Spr. Stett. Ent. Zeit. 1865, p. 248. — Smaller and darker than the type, the under side being of a uniform colour.

HABITAT.—Elevated meadows in the Tyrol.

13.—E. **Evias**, Lef. Ann. Soc. Linn. Paris, 1826, 488, Pl. 10;
    Boisd. Ic. 31, 3-5; Tr. x. 1, 44; Dup. i. 37, 1, 2, p. 326.
    BONELLII, Hüb. 892-5 (1827); Frr. 73, 1, 2 ; Butl. Cat. 83.

Expands from 1·50 to 1·75 in. All the wings dark brown,
with distinct and broad submarginal bands, having a tendency to
be broken into separate spots, each portion enclosing a white-
centred round black spot; the submarginal bands are reddish
fulvous in the male, and yellowish in the female. The fore wings
have five ocellated spots, two larger than the rest; above these,
near the apex, a small one, and below them two small ones; the
hind wings have four spots of nearly equal size. Under side: Fore
wings with a broad submarginal fulvous band enclosing the spots,
and above and distinct from the ground colour, which is dark
brown; hind wings dark brown, lighter along the hind margins;
there is a submarginal row of four round black spots with white
pupils. Pl. LX., 4.

    TIME OF APPEARANCE.—July.

    HABITAT. — Elevated alpine meadows, &c., in the Valais,
Central Spain, Piedmont, and France, in those Departments which
are in proximity to the Pyrenees. The species is not one of wide
distribution, and is only found at a considerable elevation.

    LARVA.—Unknown.

14.—E. **Melas**, Hbst. 210, 4, 7, viii. p. 191 (1796); O. i. 1, 277 ;
    Godt. ii. 17, 1, 2 ; Boisd. Ic. 33, 3, 4 ; Dup. i. 39,
    1-4 ; Frr. 49, 2, 61, 1 ; H. S. 65-8.
    MAURUS, Esp. 107, 3, 4 ; 110, 4.

Expands from 1·40 to 1·60 in. Male blackish brown. The
fore wings have two coalescing black spots with white pupils near
the apex, and a smaller one near the centre of the wing. Hind
wings with a submarginal row of three black spots with white
centres. Female somewhat paler. Fore wings with a faint trace
of a submarginal fulvous band. All the spots are larger than
in the male, and there are four on each wing instead of three.
Under side : Fore wings dull fulvous, dullest at the base ; the apex
grey, this colour extending for a short distance along the hind

margin; the black spots wth white centres, as above. Hind wings
dark grey, with a slight bronze reflection, mottled by numerous
narrow dark brown lines, which cross the wing in a direction
parallel to the hind margin; submarginal white-centred spots as
above, but not so large. Pl. LXI., 1.

TIMES OF APPEARANCE.—June and July.

HABITAT. — The mountains of Hungary and the Eastern
Carpathians, at a considerable elevation. It also occurs in Greece.

LARVA.—Unknown.

VARIETY.

Lefebvrei (ab.), Boisd. Ind. p. 23; Boisd. Ic. 33, 1, 2;
Dup. i. 35, 3, 4; Tr. x. 1, 47; H. S. 375; Staud. Hor. 1870, 65.
—Differs from the type in being more of a uniform colour and
darker beneath, and is the Spanish form occurring at great elevations in the Pyrenees and the mountains of Castile.

15.—**E. Glacialis**, Esp. 116, 2 (1800).—*Alecto*, Frr. 49, 4; Boisd.
Ic. 6, 7; H. S. 173-4.

Expands from 1·20 to 1·60 in. Wings dark brown; fore
wings with a dull fulvous band reaching only two-thirds of the
length of the hind margin. Hind wings in the male uniform dark
brown; in the female sometimes with a trace of a submarginal
band. Under side: Fore wings fulvous, the basal portion dull;
hind margin and costa dark brown; hind wings uniform dark
brown. Pl. LXI., 2.

TIME OF APPEARANCE.—July.

HABITAT.—The highest parts of the Alps, on the verge of the
snow-line, especially in the south-west of Switzerland and the
eastern alpine parts of France. Girard remarks of this species
that it is found at a higher elevation than any other *Erebiæ*, and is
both difficult and dangerous to capture on account of the precipices.
He mentions that M. Fallou has taken a worn specimen on August
10th near the summit of the Gornergrat, at an altitude of more
than 10,000 feet.

LARVA.—Unknown.

252 SATYRIDÆ.

VARIETIES.

Alecto (ab.), Hüb. 515-6 (1802); O. i. 1, 279; Frr. 149, 3.—
? *Atratus*, Esp. 104, 1.—*Persephone*, Esp. 121, 5, 6.

Expands from 1·50 to nearly 2 in. Specimens are frequently
found of a larger size than the type, and seldom as small as the
smaller specimens of *Glacialis*. All the fulvous markings are
nearly or quite obsolete, the insect being of a uniform dark brown
above, and the hind wings beneath nearly black. Pl. LXI., 3.

TIME OF APPEARANCE.—June to August.

HABITAT.—The Alps of Switzerland, at a lesser altitude than
the type.

Pluto (ab.), ♂, Esp. 121, 1 (1805); Butl. Cat. 77.—*Alecto*,
Boisd. Ic. 32, 4, 5; Dup. i. 38, 1, 2.—*Tisiphone*, Esp. 122, 5.—A
very dark form of the male, in which the wings are entirely black
above and beneath; scarcely distinct from the preceding.

16.—**E. Scipio**, Boisd. Ic. 30, 1-6, p. 152; Dup. 1, 38, 5, 6;
H.-G. 980-3.

Expands from 1·70 to 2 in. Male dark brown; all the wings
with a reddish fulvous submarginal band; fore wings with two
large coalescing black spots near the apex with conspicuous white
centres; beneath these are two small ones, with minute white dots
in their centre; hind wings with three small black spots of equal
size, and with minute white centres; marginal fringe brown.
Under side: Fore wings reddish fulvous; costa and hind margin
dark brown, the two large coalescing spots as above, only one
smaller one beneath them; hind wings dark blackish brown, with
a slightly lighter submarginal band, on which are three inconspicuous black spots. Female lighter brown, fulvous bands less
tinged with red, and the spots much more conspicuous; those on
the hind wings with light blue centres; marginal fringes whitish.
Under side: Fore wings fulvous, costa and hind margin grey, two
white-centred spots beneath the two large upper ones; hind wings
grey, very slightly mottled with light brown, and with one or two
inconspicuous dark spots. Pl. LXI., 4.

TIME OF APPEARANCE.—July.

HABITAT. — Elevated alpine meadows only in France in the following localities in the department of Basses Alpes : the neighbourhood of Digne and Barcelonette, the meadows of Ozglosse, Larche, Malmorte, &c. (Girard).

17.—E. Epistygne, Hüb. Verz. p. 62 (1816) ; Boisd. Ic. 31, 1-2 ; Hüb. 855-8 ; Tr. x. 1, 45.

    STYGNE, Hüb. 639 (post 1807); Frr. 49, 1. (*Stygne*, O. (1807) alia est species).

Expands from 1·50 to 1·70 in. Wings dark brown. Fore wings with a broad submarginal band of a light yellowish brown, divided into six by the nervures ; the three upper spaces contain each a round black spot with a white pupil ; the fourth has a small blind spot in the female, but is without a spot in the male ; the fifth space has a small black spot, which in the female has a minute white central dot ; the sixth space is fulvous, and generally has a black dot in both sexes, but is often without ; between the submarginal band and the base is a yellowish mark on the costa. Hind wings dark brown, with a deep fulvous submarginal band divided into six parts by the nervures ; four or five of these spaces contain white-centred black spots. Under side : Fore wings deep rich brown, tinged with dusky towards the inner margin : costa and hind margin edged with grey ; near the apex are two or three white-centred black spots and an obscure brown spot near the anal angle ; near the centre of the wing is an irregularly-triangular dusky mark. Hind wings dark brown, crossed by numerous greyish white lines, having a direction parallel to the hind margin ; two of these are somewhat broader than the rest, and enclose a central band of brown slightly darker than the rest of the wing ; near the hind margin is a row of three or four very indistinct black dots. Pl. LXI., 5.

    TIMES OF APPEARANCE.—March and April.

    HABITAT.—Only in the south-east corner of France, in the provinces of Dauphiné and Provence, the chief locality being the dry hills in the neighbourhood of Aix. It is easy to take, on account of its dull heavy flight, and continues on the wing only for a short time.

Obs.—This species, as will be noticed, differs greatly from any other *Erebia* both in habits and appearance ; in the latter it reminds one somewhat of the genus *Pararge* rather than *Erebia*, whilst, as regards the former, its restricted range and early appearance are very remarkable.

Larva.—Unknown.

18.—E. **Afra**, Esp. 83, 4, 5 (1783); O. i. 1, 275; Frr. 37, 4 ;
　　　　Boisd. Ic. 34, 1, 2 ; Dup. i. 25, 1, 2.
　　　Phegea, Bkh. i. 101 ; Hüb. 500-1.

Expands from 1·35 to 1·50 in. Wings dull brown. Fore wings greyish towards the apex and along the hind margin ; near the apex is a small round black spot with a white centre ; submarginal band represented by a row of five round black spots with shining white centres, and surrounded by faint rings of fulvous, the two top spots being larger than the rest and coalescent. Hind wings with a row of spots similar to those on the fore wings, but of equal size. Under side : Fore wings dark mahogany colour, the nervures greyish white, the apex and hind margins are tinged with grey, only the two largest eye-like spots are prominent, the rest being very small ; but they are all surrounded by yellowish white rings. Hind wings dark brown, the nervures nearly white ; submarginal row of spots very evenly placed, one occupying each interneural cell ; they are round and black, with minute but brilliant white centres, and surrounded by rather uncertainly-defined yellowish white rings. Pl. LXII., 1.

Times of Appearance.—June and July.

Habitat.—The steppes of South Russia (Sarepta) and the Caucasus, and Siberia.

Larva.—Unknown.

VARIETY.

Dalmata, Godt. Ency. Méth. p. 350. — A form of the species described by Godart as inhabiting Dalmatia. Larger than the type and less variegated beneath.

19.—E. Lappona, Esp. 108, 3 ; Thnb. Diss. Ins. Succ. ii. p. 37, t. 5, 6, 7.

Manto, Fab. E. S. 231 (1793); Hüb. T. 45, 107-8, 512-14 ; O. i. 1, 296 ; Godt. ii. 17, 7, 8 ; Frr. 85, 1, 2 ; Wallengr. Rhop. Scand. p. 57. (*Manto*, Esp. alia est *Erebia.*

Castor, Esp. 67, 2 ; Schn. Syst. B. p. 108.

Expands from 1·35 to 1·70 in. Dull brown, slightly bronze in some lights. Fore wings with a broad dull fulvous band, not reaching for more than half the length of the hind margin, and containing a row of four or five black spots of nearly equal size ; hind wings with a submarginal row of three or four black spots in very inconspicuous fulvous rings. Under side : Fore wings fulvous, the costa and hind margin grey ; the black spots described above are seen again beneath, but are smaller. Hind wings grey, with the slightest tinge of violet, speckled minutely with brown, and crossed by two wavy dark brown lines, one near the base and the other across the centre of the wing ; between the central line and the hind margin is a row of four black dots. Pl. LXII., 2.

Times of Appearance.—July and August.

Habitat.—The higher Alps of Switzerland and the Pyrenees, always on the verge of the snow-line. Mountains in Scandinavia, Lapland, and the Altai. A local species in Switzerland on account of the great altitude necessary for its existence, but common where it occurs.

VARIETY.

Pollux, Esp. 67, 3. — An aberrant form in which the hind wings are not banded beneath.

20.—E. Tyndarus, Esp. 67, 1 (1781); O. i. 1, 299 ; Frr. 80, 1, 2 ; H.-G. 969-974.

Cassioides, Esp. 103, 2, 2.—*Neleus*, Frr. 80, 3, 4.

Herse, Bkh. i. 94.

Cleo, Hüb. 209-12 ; Godt. ii. 17, 5, 6.

Dromus, Fab. E. S. 224 (1793).

Expands from 1·16 to 1·40 in. Wings dull brown, shot with greenish bronze. Fore wings with an indistinct dull fulvous submarginal band, on which are placed, near the apex, one or two small black spots, sometimes with white centres. Hind wings with faint traces of a fulvous band, but often quite plain. Under side : Fore wings fulvous, with a broad hind-marginal band of grey ; near the apex are one or two small black dots ; hind wings grey, slightly tinged with violet, a few light brown lines running across from the costa to the inner margin. Pl. LXII., 3.

Time of Appearance.—June to the middle of August.

Habitat. — The Alps of Switzerland, the Pyrenees, Italy, Hungary, South-Eastern and South-Central France, especially in Auvergne. It occurs at a great elevation, nearly as high though at the same time lower, than the last species. I have taken it on the summit of the Furka in the middle of August.

VARIETIES.

a. **Cæcodromus** (ab.), Gn. et Vill.—An aberration in which the fulvous bands and ocelli are absent.

b. **Dromus**, H. S. 168-9, 275, vi. p. 8. — *Cassioides*, St. Cat. p. 11.—A variety in which the fulvous bands are broader and redder than in the type, and the ocellated spots larger.

Habitat.—The Pyrenees ; the Caucasus and Armenia.

c. **Hispania**, Butl. Cat. 86, pl. ii. 7 (rect. *Hispanica*). — Nevadensis, Staud. Cat. — Larger than the type, the ocellated spots larger, and the hind wings without dark bands beneath.

Habitat.—Sierra Nevada.

d. **Ottomana**, H. S. 376, 379-80, vi. p. 8. — Very like the preceding variety, but somewhat larger.

Habitat.—Mountains in Greece, Asia Minor, and Southern Armenia.

21.—E. **Gorge**, Esp. 119, 4, 5 ; Hüb. 502-5 ; O. i. 1, 294 ; Frr. 79, 3.

Æthiops minor, Esp. 112, 1, 2.

Expands from 1·16 to 1·45 in. Wings dark brown, with submarginal fulvous bands broadest in the female, which is also lighter. Fore wings with two or three, and the hind wings with three black spots with white centres; these spots are most clearly defined in the female; two of them are near the apex. Under side: Fore wings fulvous; costa and hind margins dark brown; the two apical eyes are distinct. Hind wings dark blackish brown in the male, lighter in the female, the basal portion darkest; there is a submarginal band of grey on which are several black spots, most distinct in the female, and outside this is the dark brown hind-marginal border. Pl. LXII., 4.

TIMES OF APPEARANCE.—July and August.

HABITAT. — The Alps of Switzerland and the South-East of France; also in the Pyrenees. It always occurs at a great elevation, having been taken as high as the Gornergrat, and above the Hospice of St. Bernard.

LARVA.—Unknown.

VARIETIES.

a. **Erynis** (ab.), Esp. 121, 3.—*Gorgone* (var.), H. S. 175. — The ocelli are absent from all the wings.

b. **Triopes**, Spr. Stett. Ent. Zeit. 1865, p. 248. — Has three white central black spots at the apex of the fore wing, instead of two.

HABITAT.—The Upper Engadine and the Tyrol.

c. **Gorgone**, Boisd. Ic. 29, 5-8, 1, p. 150; H. S. 75, 76.— Larger than the type; the male has the under side of the hind wings less raised; the female has the nervures of the hind wings whitish.

HABITAT.—The Pyrenees, at a considerable elevation.

22.—**E. Goante**, Esp. 116, 1; O. i. 1, 293; Frr. 79, 1, 2; H. S. 77-9.

SCÆA, Hüb. 253-4.

Expands from 1·45 to 1·60 in. Brown, with broad sub-marginal fulvous bands, very bright in the male, lighter in the female. Fore wings with two apical white-centred spots, and one or two smaller ones nearer the anal angle; hind wings with three spots of equal size, and with bluish white centres. Under side: Fore wings fulvous, the portion corresponding to the fulvous band

2 L

above lighter than the rest, the white-centred spots as above, the
hind margin dark brown; hind wings dark blackish brown in the
male, lighter in the female; a lighter, almost white band runs
across the centre of the wing, and there are other light markings
towards the base and along the hind margin; the submarginal row
of spots is plainly visible.

TIMES OF APPEARANCE.—July and August.

HABITAT.—The Alps of Switzerland, and in the South-East of
France (Digne, &c.). It does not appear to be found at any great
elevation. Chamounix and Zermatt, where it is common in August
(Fallou). I have found it very common in the Maderaner-Thal,
Canton Uri, during the middle of August. Pl. LXII., 5.

LARVA.—Unknown.

23.—E. **Pronoë**, Esp. 54, 1 (1780); Bkh. i. 28; O. i. 1, 290;
    Frr. 73, 3, 4.

    ARACHNE, Fab. Mant. 40 (1787); Hüb. 215-7; Godt. ii.
    16, 7, 8.

Expands from 1·40 to 1·75 in. Brown, with a somewhat
bronzy appearance. Fore wings with a dull and rather indistinct
submarginal fulvous band; on this are two black apical spots with
pupils, placed close together, but hardly coalescent; below these is
another smaller spot towards the anal angle; hind wings with a
submarginal row of three white-centred black spots surrounded by
dull fulvous rings. Under side: Fore wings fulvous, with the spots
as above, slightly mottled with grey towards the apex; hind wings
dark brown, with a whitish grey basal patch, and a broad wavy
submarginal band of the same colour; inner and hind margins also
mottled with grey. Pl. LXIII., 1.

TIMES OF APPEARANCE.—June and July.

HABITAT. — Alpine meadows in Switzerland, at a moderate
elevation; also mountains in Italy and Hungary, in South-Central
France (principally at Mont Dore, in Auvergne), in the Pyrenees,
in the Caucasus, Asia Minor, and Armenia.

VARIETIES.

a. **Pitho,** Hüb. 574-7.—Darker than the type; the fulvous band of the fore wings almost wanting, and only the two apical spots being present; hind wings uniform dark brown. Under side: Fore wings dark brown, with a triangular reddish patch containing the apical spots; hind wings darker than in the type, the grey bands being more strongly defined, and the hind and inner margins dark brown, without grey mottling. Pl. LXIII., 2.

HABITAT.—Switzerland, at higher elevations than the type.

b. **Pyrenaica,** Staud. Cat. p. 26. — Smaller than the type, the ocellated spots larger, and the under side more variegated.

HABITAT.—The Pyrenees.

24.—E. Neoridas, Boisd. Ind. p. 23; Boisd. Ic. 29, 1-4; Tr. x. 51; Dup. 1, 36, 5, 6; H. G. 984, 7.

Expands from 1·40 to 1·50 in. Hind margin of fore wings rather straighter than in the preceding species. All the wings velvety dark brown; fore wings with a bright fulvous submarginal band, broad and well defined, containing three black spots with white centres, two near the apex, and one below slightly smaller; hind wings with a bright fulvous submarginal band about half the length of the hind margin, and occupying a central position, containing three white-centred black spots of equal size. Under side: Fore wings dark brown, nearly black, the fulvous band and spots as above, apex grey; hind wings greyish brown, with a broad whitish grey submarginal band, wavy and sharply defined along its basal outline, its hind-marginal edge less distinct; near the anal angle is a small black spot; the inner margin is tinged with grey.

TIMES OF APPEARANCE.—July and August. Pl. LXIII., 3.

HABITAT. — Only the South of France, in mountainous and alpine regions. M. Girard gives the following localities (*loc. cit.*, p. 228):—Isère, Lozère, Mont Cenis; Saint Jean, in Dauphinie; pastures in the Basses Alps, Barcelonnette, Larche, Clermont-Ferrand, Royat; on the Plateau of Gentioux, in Creuse; at Lioran; and at Murat, in the Department of Cantal.

LARVA.—Unknown.

25.—E. **Zapateri**, Oberth. Ann. Soc. Esp. iv. p. 379, t. 17, f. 1, 2 (1875).

Expands from 1·25 to 1·30 in. Hind margin of fore wings straight. All the wings brown, somewhat shot with green; fore wings with a broad brownish yellow submarginal band, containing only two white-centred black spots near the apex; hind wings uniform brown, without spots or markings. Under side: Fore wings fulvous towards the base, darker towards the inner and hind margins, submarginal band and spots as above; hind wings light greyish brown, with a very faint indication of a lighter submarginal band; ocellated spots none. Pl. LXIII., 4.

Habitat.—Mountains in Spain.

Obs. — I have been enabled to figure this species through the kindness of Mr. F. D. Godman, F.R.S., who received it from Dr. Staudinger with the locality-label marked "Albarracin," a place in the south of the Province of Arragon.

26.—E. **Æthiops**, Esp. 25, 1 (1777); Hbst. 209, 3, 4.

　　Medea, Hüb. 220-2; O. i. 1, 281; Frr. 55, 1, 2.

　　Blandina, Fab. E. S. 236 (1793), Steph. et Auct. Brit.

Expands from 1·50 to 1·80 in. Wings dark brown; fore wings with a broad submarginal band, bright fulvous in the male, lighter in the female, containing three large white-centred black spots of equal size, and generally one smaller one below the two nearest the apex; hind wings with a submarginal fulvous band somewhat narrower than that on the fore wings, and containing four small white-centred black spots; all these spots are larger in the female, and in that sex there are sometimes five spots on the fore wing. Under side: Fore wings brown, tinged with fulvous towards the base; fulvous band as above, but with only three spots, becoming narrower towards and seldom reaching as far as the inner margin; hind wings grey, with a central broad band of brown, and another narrower one along the hind margin; the grey band formed in the space between these two frequently contains some indistinct and minute ocelli; marginal fringes never spotted with white. Pl. LXIII., 5.

Time of Appearance.—July to September.

Habitat.—Woods and plains in Central Europe, not ascending to any great elevation in the mountains. It is common in Switzerland, South and South-East Germany, North-East and Central France, Livonia, North-Eastern Turkey, the Caucasus, the Altai, North-East Siberia, and the Amur. Common but local in Britain, being confined to the north, its range extending from Yorkshire to the North of Scotland.

Larva.—Light green, with a brownish or dark greenish dorsal stripe: lateral stripe whitish, bordered by dark brown; head reddish brown. Feeds on various species of *Poa* in the autumn, and again after hybernation.

VARIETY.

Leucotænia, Staud. Cat. 26; Frr. 55, 3, 4. — A variety in which there is a whitish band in the centre of the hind wings beneath.

Habitat.—The South of Switzerland, Dalmatia, &c.

27. **E. Ligea**, Linn. Syst. Nat. x. 473; Faun. Suec. 239; Hüb. 225-8; O. i. 1, 283; Godt. ii. 13, 1, 2; Frr. 67. Alexis, Esp. 44, 1, 2, 54, 2.

Expands from 1·60 to 1·90 in. Marginal fringes brown and white; hind margins slightly dentate. All the wings dark brown, with fulvous submarginal bands almost exactly like those in *E. Æthiops*. Under side: Fore wings coloured as in the preceding, but the fulvous band is the same width throughout its length, and does not narrow towards the inner margin; hind wings dark brown, with a submarginal row of four very conspicuous ocellated spots in fulvous rings; internal to these and near the costa is a very strongly marked white dash, often extended so as to form an irregular and interrupted white band, and sometimes there is another less distinct white band near the base. Pl. LX., 6.

Times of Appearance.—August and September.

Habitat. — Central Europe, at a moderate elevation in the mountains, and in wooded plains; also in Scandinavia and North Russia, Bulgaria, the Caucasus, and Siberia. It is admitted by

Westwood and others as a British insect on account of its reported capture in the Isle of Arran, whence it is called in this country "the Arran Brown."

LARVA. — Greyish or yellowish green; dorsal stripe darker green or brown; a pale yellow lateral stripe runs immediately above the spiracles. Feeds on *Poa* in the autumn, and again after hybernation. Figured after Hübner on Pl. LXXVI., 3.

VARIETIES.

a. **Adyte**, Hüb. 759, 760. — Smaller than the type, possibly intermediate between this species and *E. Euryale.*

HABITAT.—Lapland.

b. **Livonica**, Teich. Stett. Ent. Zeit. 1866, p. 133. — Differs from the type in having the hind wings of an uniform brown colour beneath, without any marking.

HABITAT.—Livonica.

28.—**E. Euryale**, Esp. 118, 2, 3; O. i. 1, 286; Hub. 789-90: Frr. 61, 3, 4, &c.; H. S. 97-101; Godt. ii. 13, 3, 4.

PHILOMELA, Esp. 116, 4.

Expands from 1·30 to 1·50 in. Marginal fringes brown and white. Hind margins slightly dentate. Wings dark brown; fulvous bands and spots very much as in the preceding species. Under side : Fore wings as in *E. Ligea,* but darker; hind wings dark brown, with a faint submarginal greyish band containing two or three very inconspicuous ocelli without fulvous rings; white dash near the costa only faintly indicated, sometimes replaced by grey; basal portion of wing dark brown. Pl. LXIV., 1.

TIMES OF APPEARANCE.—July and August.

HABITAT. — Elevated meadows in the Alps of Switzerland, Silesia, Hungary, the Pyrenees, the Appennines, and South-Central France, especially in Auvergne.

LARVA. — According to Freyer, pale yellowish brown, with a brown dorsal stripe, and a white lateral stripe edged with brown. On grasses in June.

VARIETIES.

a. **Ocellaris**, Staud. Cat. p. 26. — The upper surface has the fulvous band broken up into reddish spots with black centres.

HABITAT. — The Styrian and Carinthian Alps, and Eastern Siberia.

b. **Euryaloides,** Tengstr. Cat. p. 5, 295. — Has the ocelli almost entirely absent from the fulvous bands.

HABITAT.—Finland.

29.—**E. Embla,** Thunb. Diss. Ent. ii. (1791) p. 38; Frr. 416,
    3, 4; O. i. 1, 287; Mén. Schrk. p. 35.
    DIOXIPPE, Hüb. 538-9; Dup. i. 36, 1, 2.
    ETHUS, Fab. E. S. 217.

Expands from 1·75 to 1·80 in. Marginal fringes grey, with dark brown spots. All the wings dark brown, with a submarginal row of three or four fulvous circles, more or less separated, in which are placed black spots, often with white pupils; the spots on the fore wings are larger than those of the hind wings, and the first spot is larger than the rest and sometimes contains two pupils; beneath, the fore wings are the same as above in colour and markings, but with some ashy coloration towards the apex; the hind wings beneath are obscure dark brown, speckled with ashy grey along the hind margin, and with two or three black spots corresponding to the ocelli of the upper surface; there is a white spot near the costa, and another in the middle of the wing; also a faintly marked grey central band. The hind wings are much more rounded than is usual in this genus. Pl. LXIV., 2.

TIMES OF APPEARANCE.—June and July.

HABITAT.—Peat marshes in the northern parts of Norway; Swedish Lapland (Tornea, Lulea, Storland, Jukkasjarvi, Jockmock, &c.—Wallengren), Livonia, North Russia, and Eastern Siberia.

LARVA.—Unknown.

30. – **E. Disa,** Thunb. Diss. Ent. ii. p. 37; Frr. 416, 1, 2.
    GEFION, Esp. 108, 2.
    GRIELA, Fab. E. S. 236; Hüb. 228-9.
    EMBLA, Boisd. Ic. 32, 1-3; Dup. i. 36, 3, 4.

Expands from 1·75 to 1·85 in. All the wings dark brown. Fore wings with a submarginal row of four or five black spots surrounded by reddish fulvous rings; there are no white pupils in the male, and only sometimes in the female; all the spots are much about the same size, but the foremost two are approximated. Hind wings unicolorous dark brown, without any markings. Under side : Fore wings the same as above, but the spots are somewhat smaller, and there is some slight greyish coloration near the apex and along the hind margin; hind wings greyish brown, with a broad central brown band dentated in outline; there are no white markings. The marginal fringes are dirty white, spotted with dark brown. Pl. LXIV., 3.

TIME OF APPEARANCE.—July.

HABITAT. — Marshy places in Lapland, in Umea, Tornea, &c., flying over the flowers of *Rubus chamæmorus* (Wallengren).

LARVA.—Unknown.

Genus 3.—**ŒNEIS**, Hüb. Verz. bek. Schmett. p. 58 (1816).
CHIONOBAS, Boisd. Lcc. Lep. Amer. Sept. p. 214 (1833);
Westw. Gen. D. L. p. 381 (1851).

CHARACTERS.--Larva unknown in all but two species. Scudder figures and describes the various states of *Œ. Semidea;* in this case it is fusiform, with the anal extremity scarcely if at all forked, and has conspicuous dorsal and lateral stripes of a dark green colour. The larva of *Œ. Semidea* is described by Scudder as feeding on lichens; that of *Œ. Bore* is believed by Sandberg to feed on grasses.

Pupa (of *Œ. Semidea*).—Compact and rounded, with the head obtuse and regularly rounded ; abdomen cylindrical, tapering very regularly and rather rapidly to the apex, which is blunt; ventral surface somewhat flattened. Mode of pupation not properly ascertained ; but probably the pupa lies upon the ground and in crevices of rocks, and is not suspended.

Imago.—Antennæ rather short, with a gradually formed club. Eyes naked. Wings medium size, with the nervures of the fore wings very little dilated at the base. Discoidal cell of fore wings closed ; the colour of the wings is brown, lighter than the general colour of the last genus, and they usually have a submarginal band of a still lighter colour containing black spots, which are often ocellated. The hind wings beneath are sprinkled with white and brown, and often have white lines following the course of the nervures; the marginal fringes are dark brown, chequered with white; the hind margins of the fore wings are entire, those of the hind wings hardly at all dentated. The fore wings are rather long and narrow in comparison to the hind wings, and both pairs are somewhat thinly clothed with scales.

There are about twelve species of this genus at present known, and they are almost confined to the Polar regions, or to places at a great elevation in lofty and extensive mountain ranges of the Palæarctic and Nearctic regions. The only exception to this rule of habitat is Œ. Tarpeia, Pall., which inhabits the dry steppe-lands of South Russia and Siberia; of the four remaining European species one inhabits the Alps of Switverland, viz., Œ. Aëllo, Hüb., and the rest are only found within the Arctic circle, in Lapland and Siberia. Many species occur in Labrador, and others are found on the Rocky Mountains, in California and in Vancouver's Island; one species occurs in the Himalayas; and two are found in South America, in Chili and Patagonia.* The genus Œneis appears to come in as a natural link between Erebia and Satyrus, the first-described species, Œ. Jutta, very much resembling the Polar Erebiæ, whilst the majority of the species are a good deal like the genus Satyrus in their colour and markings; the habits of all, however, remind one of Erebia more than of any other genus.

1.—Œ. Jutta, Hüb. 614-5; Boisd. Ic. 38; Dup. i. 40, 3, 5; H. S. 116-8; Wallengren, Scandin. Rhop. p. 46. BALDER, H. S. 384-6.

Expands from 1·80 to 2·0 in. All the wings brown, nearly as dark as in Erebia Embla, with a submarginal row of light orange spots, some of them enclosing black centres; the black centre is nearly always present on the fore wing in the spot nearest the apex, and on the hind wing in that nearest the anal angle. The size and

---

* The occurrence of these species affords another instance of the analogy between a large part of the lepidopterous fauna of Temperate South America, and that of the Palæarctic and Nearctic regions. Does the fact of this analogy point to the existence of the remnant of an ancient Antarctic fauna, or have Nearctic forms been extended downwards by means of the great chain of the Andes?

extent of the orange spots is very variable; sometimes they are very small and separate, and in other specimens they form a rather broad submarginal band. The male has some black scales near the base of the fore wings beneath the median nervure; these are absent in the female, which is larger and paler than the other sex. Under side: Fore wings light brown, speckled with grey at the apex; there is always one black spot present near the apex, and sometimes one or two more below this; hind wings smoky brown, mottled with light grey, the hind-marginal half lighter than the basal portion; there is an angular dark central band and a black spot near the anal angle, which is sometimes ocellated. Marginal fringes greyish white and black. Nervures not differing from the ground colour of the wings. Pl. LXIV., 4.

TIME OF APPEARANCE.—July.

HABITAT.—Marshy places and river-banks in Norway, Sweden and Lapland, Livonia, North Russia, Labrador, Siberia, and the Amur.

LARVA.—Unknown.

VARIETY.

Balderi, Hüb. Zutr. Exot. 981-2; Butler, Cat. 160; Balder; Dup. i. 49, 4, 5.—Differs from the type only in being smaller, and with fewer ocellated spots.

2.—ŒE. **Aëllo,** Hub. 519-21; Esp. 115, 1; O. i. 1, 199: Boisd. Ic. 36, 1-3; Dup. i. 31, 1-3; H. S. 125-6.
NORNA, Hüb. 141, 2.

Expands from 1·60 to 2·0 in. Wings brown, the male with some darker scales below the median nervure of the fore wings. All the wings with a broad light brown submarginal band extending nearly throughout the entire length of the hind margins, and containing a variable number of black spots; the fore wings sometimes have only one near the apex, sometimes one, two, or three more below this; frequently one or more of these spots contains a white dot; the hind wings have one ocellated spot near the anal angle, and a much smaller one sometimes appears above it; there are also occasionally two or three minute white spots. Hind margins darkish brown, with the fringes very distinctly brown and white.

Under side : Fore wings light brown, the spots as above, the costa indistinctly speckled with brown; the apex is greyish brown, speckled with white. Hind wings speckled with white and brown, the brown colour predominating over the basal half of the wing and along the hind margin, thus forming a whitish submarginal band; the nervures and nervules are all strongly marked with white. Pl. LXV., 1.

TIME OF APPEARANCE.—The beginning of July.

HABITAT. — The Alps of Switzerland and the Tyrol, at a considerable elevation, being taken on the verge of the snow-line. It is found in alternate years near the highest part of the Simplon Pass ; at Chamounix, at the "Jardin"; and at the Grand Mulets on the ascent of Mont Blanc, and many other localities of a like nature.

LARVA.—Unknown.

3. Œ. Tarpeia, Esp. 83, 1, 2; O. i. 1, 203; Hüb. 779-82; Frr. 427, 3, 4; H. S. 61-64; Fab. Mant. 32; Dup. i. 31, 6, 7.
CELIMENE, Cramer, Pap. Exot. pl. 375, 5, 7.

Expands from 1·60 to 1·80 in. All the wings light fulvous, with the hind margins darker; fore wings with a submarginal row of distinct black spots entirely without ocelli; hind wings with their basal half slightly darker than the rest, and with a submarginal row of four spots similar to those on the fore wings. Under side : Fore wings light brown, speckled along the costa and hind margins with darker brown and whitish; submarginal spots distinct; hind wings smoky grey, the neuration very distinctly marked with white; a narrow interrupted band of white runs from the centre of the costa nearly to the inner margin; midway between the commencement of this and the base is a rather square costal white spot; the submarginal black spots are distinct. Marginal fringes dirty white. Pl. LXV., 2.

TIME OF APPEARANCE.—May to July.

HABITAT. — The Steppes of South-East Russia ; the Kirghis-Steppes ; South-Eastern Siberia, and the Altai.

LARVA.—Unknown.

4.—Œ. **Norna**, Thnb. Diss. Ins. ii. (1791), p. 36, 11 ; Wallengr.
    Scand. Rhop. p. 41, 43 ; Esp. 108, 4 ; Hüb. 763-6 ;
    O. i. 1, 201 ; Boisd. Ic. 36, 4-6 ; Dup. i. 31, 4-5.
    Celæno, Hüb. 152-3.

Expands from 1·50 to 1·70 in. The male has all the wings
pale brown above, with a pale fulvous marginal band, broad and
continuous, reaching very nearly to the hind margins. Fore wings
with two black ocellated spots, one near the apex, and another
between the second and third median nervules ; the hind wings
have one ocellated spot between the second and third submedian
nervules. The female has the wings lighter than the male, the
marginal band being less distinct ; but the ocellated spots are
larger, and sometimes there is a third spot on the fore wings
between the first and second submedian nervules. Under side :
Fore wings pale brown, and almost the same as in Œ. Aëllo. Hind
wings also marked and coloured nearly the same as in Œ. Aëllo ;
but the neuration is not marked with white as in that species, and
the ocellated spot is not so distinct ; moreover, the marginal fringe
is not distinctly marked white and dark brown, but is narrower
and greyish. Pl. XLIV., 5.

TIMES OF APPEARANCE.—June and July.

HABITAT. — River-banks and dry grassy places in many
localities in Swedish Lapland, in Umea, Tornea, and Lulea ; in
Norway at Dowre (Wallengren).

LARVA.—Unknown.

VARIETIES.

This is a very variable species as regards the number of the
ocellated spots ; Wallengren gives no less than six varieties ; it
seems, however, sufficient to mention the two named in Staudinger's Catalogue.

a. **Hilda**, Quens. Act. Holm. 1791, p. 272 ; Schn. N. Mag. iv.
414 ; Stett. Ent. Zeit. 1861, 358.—*Norna,* var. *c,* Wallengr. p. 42.
—Only one ocellated spot on each wing.

b. **Fulla**, Ev. Bull. Mosc. 1881, ii. 614 ; H. S. 615-6. — The
spots are absent from the fore wings, the hind wings having one.

5.—**Œ. Bore**, Schn. N. Mag. p. 415 (1792) ; Esp. 100, 1 ; Hüb. 134-6; O. i. 1, 205; Boisd. Ic. 37, 1 ; Dup. i. 32, 1, 2. Norna, Quens. Act. Holm. 1791, T. 10, 1, 2 ; Thnb. Diss. Ins. ii. p. 36.

Expands from 1·50 to 1·70 in. The male has the fore wings pale brown, with a greyish tinge, without any markings, except that in good specimens the disco-cellular nervule forms a fine dark line ; hind wings with a submarginal band slightly lighter than the rest of the wing. Female lighter brown, with a greyish brown hind-marginal border to all the wings ; fore wings with a faintly marked central wavy line or narrow band, the disco-cellular nervule showing more distinctly than in the male ; hind wings with a broad central band of smoky brown. Under side : Fore wings like those of *Œ. Norna*, but of a duller brown, and entirely without black spots ; hind wings greyish white, with a broad central band of grey and brown, bounded on each edge by a distinct dark brown wavy margin ; base mottled with dark brown, and the hind margin suffused with the same colour throughout its whole length. Neuration brown in the typical form. Pl. LXV., 3.

Time of Appearance.—July.

Habitat.—Dry sandy places in North Lapland ; it also occurs in Siberia and in Labrador, being circumpolar.

Larva.—Pastor Sandberg has described a larva which he has taken in abundance on *Festuca ovina*, which grows on the sandbanks by the Jacob River, in the extreme north of the Scandinavian Peninsula, close to the Varanger Fiord, where the Norwegian and Russian territories join. In this northern locality the butterfly is extremely common. He did not succeed in rearing the larvæ, some on account of their being unfortunately ichneumoned, others dying during hybernation. Schøyen * quotes his description of the larva as follows :—" Clothed with fine hair, light yellowish brown, with darker lines along the body, namely, a narrow interrupted dorsal line, and a broader one along each side. Head small in proportion to the body, greyish yellow, with six slender dark longitudinal lines and black eyes. Body nearly spindle-shaped, white,

* Schøyen & Sandberg, Tomsö Mus. Aarsh. iv. p. 81.

tapering at the extremity, back arched, anus bifid. Surface of the skin hard and coarse. Legs 3 + 1 + 1 = 5 pairs. Length about 4 centimetres. Habits sluggish." Schøyen adds—" There is scarcely any doubt that the larva really belongs to this species. It is found on the same grassy sandbanks where the butterfly is abundant, and it seems quite sufficient that Sandberg collected it in July, 1881, along with many specimens of the butterfly which were flying around the grassy banks of Jacob River; and he also observed a crippled specimen, incapable of flight, which must have been hatched on the spot."

VARIETY.

Taÿgete, Hüb. Zutr. Pap. i.; Möschl. Wien. Mts. 1863, 213. —*Boötcs*, Boisd. Ic. 37, 4-6; Dup. i. 32, 3, 5; H. S. 391-2. — *Calaïs*, Scudder, Rev. 1865, No. 3; Möschl. Stett. Ent. Zeit. 1870, 124. — In this variety the hind wings are more varied in colour on the under side than in the type, the contrast between the light and dark bands being greater. The neuration is distinctly marked with white, as in *Œ. Aëllo*, &c. This form occurs in Lapland as an aberration, but in Labrador, Greenland, and Kamschatka it exists as a true variety or local race.

Genus 4.—**SATYRUS**, Latr. Enc. Méth. ix. p. 11 (1819); Westw. Gen. Diurn. Lep. p. 388 (1851).
HIPPARCHIA, Fab. Ill. Mag. vi. p. 281 (1807). Nomen vetustius et restituendum?
ERICICOLES et RUPICOLES, Dup.

CHARACTERS. — Larvæ smooth, tapering towards the anal extremity, which is bifid. Usually brown or grey in colour, with dorsal and lateral longitudinal lines. They generally feed at night on various kinds of grasses.

Pupæ.—Smooth, fusiform, rounded towards the head, and tapering towards the anal extremity to a point. Usually not suspended, but lying in a cavity in the ground made during the larval state, sometimes protected by a sort of loose cocoon.

Imagines.—Large or middle-sized butterflies, usually difficult to take on account of their habit of settling on rocks and tree-trunks in inaccessible positions; but the flight is not strong or sustained like that of the larger *Nymphalidæ*. The colour of the wings is brown, varying from light fulvous to nearly black; the wings have broad submarginal bands of a lighter colour, sometimes white. There is always at least one ocellated spot near the apex of the fore wing, and generally one near the anal angle of the hind wing. The under side of the hind wings nearly always has a characteristic coloration and pattern of grey and whitish blended in a most pleasing manner, like that seen in *Satyrus Semele* (the Rock Underwing); this pattern is protective, from its assimilating to the objects on which the butterflies settle. The antennæ are rather long, with distinct clubs of an elongated ovate shape, rounded at the apex. Palpi not longer than the head, covered with stiff hairs placed close together at their base, having the last articulation very short, conical, and more or less pointed. The eyes are smooth. The anterior legs are short and slender, very hirsute in the male, in the female more sparingly covered with hair, and have the tarsi single-jointed. The anterior wings are rather pointed at the apex, the hind margins are scarcely if at all dentate. Hind wings rounded, and nearly always slightly dentate along the hind margins. The costal nervure very much dilated at the base, the median sometimes very strongly, sometimes only slightly so; the submedian not sensibly dilated. The males of this genus have a thickened patch of scales beneath the median nervure of the fore wings.

This genus contains about forty known species. Europe proper possesses less than half of these, and they occur principally in the southern and south-central parts. Only one species, *S. Semele*, L., occurs in Britain, and only one or two more in the immediately adjoining countries. Armenia, Asia Minor, Syria, North Africa, and generally the more southern parts of the Palæarctic region furnish many species; several are found in the Himalayas, Northern India, and Japan. The Nearctic region has about eight species, and several more are found in Chili.

1.—S. Hermione, L., M. L. U. p. 281 (1764); Schiff. S. V.
  p. 169; Schn. N. Mag. p. 471; Hüb. 122-4; O. i, 1,
  173; Esp. 70, 4.
  Hermione major, Esp. 8, 3.

Expands from 2·25 to 2·65 in. The male has the wings dark
brown, with a slight reddish tinge; fore wings with a brownish
grey submarginal band, sometimes divided into streaks by the
nervules, which are broadly marked with dark brown; near the
apex is a broad black spot with a white centre, and another
is situate between the second and third median nervules; hind
wings with a dull white submarginal band, the inner edge of which
is more clearly defined than the outer; there is a small white-
centred spot near the hind margin, between the second and third
median nervules. Female larger and more strongly marked than
the male, the submarginal band of the fore wings being yellowish
white, darker towards the apex; the hind wings have the band
pure white, or sometimes with a slight bluish or violet tinge;
there is a row of black lunules between this and the hind margin.
The fringes in both sexes are black and white. Under side: Fore
wings brown, shot with greenish bronze, with some darker markings
near the costa, mottled with grey at the apex; the submarginal
band yellowish white in both sexes, dentated along the internal
margin, and with a white-pupilled ocellus near the apex; hind
wings grey, mottled with black, with two narrow black central
bands, and a broad white submarginal band, speckled with brown,
and suffused along its hind-marginal border; the white-pupilled
spot as above. Pl. LXVI., 1.

Times of Appearance.—July and August.

Habitat.—Rocky woods in Central and Southern France (it is
taken at Fontainebleau and round Paris); Germany and Switzerland (especially in the rocky woods by the sides of lakes); South
Europe generally (excepting the central and southern parts of the
Peninsula); and Asia Minor.

Larva. — Reddish grey, with two dorsal streaks placed close
together, and of a lighter grey than the ground colour; lateral
streaks of a similar colour, bordered with black and white. It is

found in May, feeding on *Festuca* and other grasses. It is a night-feeder, and is obtained by searching for it on the stems with a lantern.

2.—**S. Alcyone**, Schiff. S. V. p. 169; Schn. N. Mag. p. 471; Hüb. 125-6; O. i. 1, 176; Boisd. Ic. 40, 5, 6; Dup. i. 27, 1. 2.

HERMIONE MINOR, Esp. 8, 2.

Expands from 2·0 to 2·25 in. General appearance somewhat resembling that of *S. Hermione*, but the expanse of the wings is less, and the sexes do not differ so markedly as in that species. The male has the submarginal band of the fore wings much more distinct than in *Hermione*, and yellowish white in colour and with two ocellated spots; hind wings with a band of a similar colour, without or with a very indistinct spot towards the anal angle. Female resembles the male in colour and markings, but both are more distinct. Under side very like that of *Hermione*, but the hind wings have a very distinct white submarginal band, suffused along its hind-marginal edge: base of the wings very dark brown.

TIMES OF APPEARANCE.—July and August. Pl. LXV., 4.

HABITAT. — Rocky places in Central and Southern Europe; also sandy places where pine-trees grow. It is possibly only a variety of the preceding (conf. *Statilinus* and *Allionia, Actæa* and *Cordula*).

LARVA.—Not described.

3.—**S. Circe**, Fab. Syst. Ent. 495 (1775).

PROSERPINA, Schiff. S. V. p. 169, T. 1, fig. 9, a, b (1776); Esp. 39, 2; Hüb. 119-21; O. i. 1, 167.

VELLEDA, Rott. Nat. vi. p. 17 (1775).

Expands from 2·25 to 3·25 in. Wings dark brown, nearly black; marginal fringes black and white. Fore wings with a sub-marginal band of five or six large white spots placed close together; near the apex is a round black spot, sometimes with a white centre in the female; hind wings with a broad white band originating

2 N

from the costa, where it has the appearance of a prolongation of the band on the upper wing, and reaching to the anal angle. The abdominal space is lighter than the ground colour of the rest of the wing. Under side : Fore wings mottled with brown and grey, with the band of white spots as above, the upper one containing a white-centred black spot; between this band and the base, in the discoidal cell, is a black spot, with a white one on each side of it. Hind wings mottled with grey ; the central white band as above, but somewhat narrower, and sharply defined along its basal edge ; between this and the base is a broad white streak originating at the costa and terminating at the median nervure ; between this point and the white band is an oval grey blotch. The sexes are similar in colour and markings ; but the female is as a rule much the larger, and has the markings more strongly expressed. Pl. LXVI., 2.

TIME OF APPEARANCE.—From the middle of June to the end of August, according to the altitude of its locality.

HABITAT.—Central and Southern Germany, and Switzerland ; the South of France, and South Europe generally (excepting Andalusia and Sicily); Asia Minor, and Armenia. It is generally found on rocky hill-sides and dry bushy places, settling on rocks and tree-trunks, and is very difficult to capture. It is the finest and largest species of the family *Satyridæ*, and indeed is to be placed among the largest European butterflies, some female specimens reaching nearly three and a half inches in expanse.

LARVA. — Smooth, of a yellowish brown colour, with a lateral stripe of bluish or greenish grey ; there is a dorsal stripe of the same colour, edged with whitish and dark brown ; head striped with blackish brown ; stigmata black ; ventral surface dark brown, with a narrow whitish stripe between the stigmata and the legs ; anal extremity bifid, and yellowish brown.

PUPA. — Smooth and fusiform, very pointed at the anal extremity ; dark reddish brown, the wing-cases spotted with yellow. The larva feeds at night on various grasses, *Lolium perenne*, &c. ; in the daytime it hides under stones. The pupa is not suspended, but rests in a slight hollow in the ground. The larva is full fed in May. Pl. LXXVI., 4.

4.—S. Briseis, L. Mus. Lud. Ul. 276; Syst. Nat. xii. 770; Hüb. 130-1; O. i. 1, 170; Godt. i. 7, 1; Frr. 481; H. S. 180, 1.

Expands from 1·75 to 2·50 in. Wings dark brown. Fore wings with a submarginal row of six white spots placed in the spaces between the nervules; the first of these spots has in the centre a round black spot, sometimes with a white pupil; a similar spot is found in the fourth white spot, whilst the third sometimes also bears a black dot; the costa is yellowish white. Hind wings brown, with a central white band, which is not very sharply defined along the margins; there are some very indistinctly marked marginal lunules. Under side: Fore wings brown, with a very broad yellowish white submarginal band, on which are the two black spots seen above; the whole area of the discoidal cell is whitish grey, and just at the junction of its middle with its outer third is a black semilunar spot with its convex side turned outwards. Hind wings dark grey, with an indistinct central band of a lighter colour; hind margins also somewhat lighter. Marginal fringes of all the wings light brown, and in some lights iridescent. Head and thorax dark brown, abdomen somewhat lighter. Antennæ brown, with black clubs. The female has a much greater expanse of wings than the male. Pl. LXVI., 3.

Times of Appearance.—July and August.

Habitat. — Germany, Central and Southern France, Switzerland, the South of Europe generally, North Africa, Asia Minor, Armenia, and Persia. It frequents dry hill-sides, grassy places, and woods where there is much herbage; thus, like many others of the *Satyridæ*, being very varied in its habitat.

Larva. — Grey, with three dorsal lines of a darker and lateral stripes of a lighter colour; ventral surface lighter, the head rather redder grey. It feeds on various grasses in May, being full-fed by the beginning or middle of June.

VARIETY.

a. Pirata (♀), Esp. 100, 3; Hüb. 604-5.—Differs from the type above merely in the colour of the light bands, which are dull ochre-yellow instead of white; beneath, the hind wings are much less varied, being much lighter at the base; the central light band is absent, and the whole wing is more grey in tone than in the type.

HABITAT.—The South of France, Asia Minor, and Armenia.

5.—S. Anthe, O. i. 1, 169; Boisd. Ic. 40, 3, 4; Dup. i. 27, 3, 4.
   PERSEPHONE, Hüb. 589-90 (*Persephone,* Fab., alius erat
   Papilio).

Expands from 2·50 to 2·70 in. All the wings dark brown. Fore wings with an oblong white blotch near the apex, turning upwards at its inner end till it touches the costa; in the middle of this is a round black spot; below this are four white spots placed one above the other, the second being oblong and enclosing a large black spot; hind wings with a broad submarginal band of dirty white, with the margins rather indistinctly defined. Fringes white and brown. Under side: Fore wings light brown, varied with whitish and black, the black spots as above; in the discoidal cell are some black streaks; hind wings mottled with grey, white, and black, the neuration white; near the anal angle is a black spot. Pl. LXVII., 1.

TIME OF APPEARANCE.—?

HABITAT.—Mountains in South Russia, Siberia, and Armenia.

LARVA.—Unknown.

VARIETY.

Hanifa, Nord. Bull. Mosc. 1851, ii. T. 9, 1-3; H. S. 477-8. —This variety corresponds to the variety *Pirata* of the last species, having in the female those portions of the wing which are white in the type coloured yellow. Pl. LXVII., 2.

6.—S. Autonoë, Esp. 86, 1-3 (1783); Hüb. 137-8; O. i. 1, 177;
   Boisd. Ic. 41, 5, 6; Dup. i. 28, 3, 4; Frr. 283, 1, 2;
   H. S. 127-30.

Expands from 1·50 to 1·75 in. All the wings dark brown; fore wings with a submarginal band of dull yellowish brown, on which are two large round black spots with white centres; hind wings with brown submarginal band differing little in colour from the rest of the wing; near the anal angle is a small black spot.

Under side : Fore wings light brown, the submarginal band only a little lighter than the rest of the wing, the black spot near the apex very distinct, the second spot smaller than above; near the apex are some grey and white markings; hind wings mottled with brown and white; a white angulated line, bordered on its basal edge with dark brown, runs across the centre of the wing; there is a similar one parallel to it, between it and the base; and a third, submarginal, line; between this and the central line is a row of faint white spots, and a black spot near the anal angle; the neuration of the hind wing on the under side is entirely white. Marginal fringes brown and white. Pl. LXVII., 3.

TIME OF APPEARANCE.—?

HABITAT. — South-Eastern Russia, the Kirghis Steppes, the Caucasus, and the Altai.

LARVA.—Unknown.

7.—S. Semele, Linn. Syst. Nat. x. 474; Faun. Suec. 275; Esp. 8, 1; Hüb. 143-4, 826-7; O. i. 1, 197.

Expands from 1·50 to 2·50 in. Brown; fore wings with a broad submarginal patch or band of light brown, indistinct in the male, but strongly marked in the female, containing two large round black spots with white pupils. Hind wings with a somewhat similar patch or band, containing a small white-centred black spot near to the anal angle. Under side: Fore wings light fulvous in the male, with two black spots as above, and with some grey and white costal marks; in the female with a broad submarginal band of light brown, as on the upper side. Hind wings mottled with grey and brown, the basal half being darker, and divided from the submarginal portion by a zigzag dark brown line, external to which is sometimes a broad indistinct whitish band. Hind margins of hind wings slightly dentate. Marginal fringes white and brown. Pl. LXVII., 4.

TIME OF APPEARANCE.—July to September.

HABITAT. — Rocky and stony places, heaths, and chalk hills throughout Europe (except the Polar regions), including the British Isles; North Africa, Asia Minor, and Armenia; Madeira, and the Canaries.

LARVA. — Light brown, with a dark olive-brown dorsal stripe, and lateral stripes of brown edged with white. Feeds on various species of grasses, such as *Triticum repens*, &c., in the autumn, and again in the spring, after hybernation.

VARIETY.

Aristæus, Bon. Descr. p. 177, T. ii. 1; Tr. x. 1, 30; Frr. 397, 1.—This variety has the brown markings much more intense than in the type, sometimes approaching to fulvous; the under side of the hind wings is also more strongly marked with grey and white.

HABITAT.—Corsica, Sardinia, and Sicily.

OBS.—*S. Semele* acquires a darker and more intense character, as regards coloration and markings, as we approach the south than that seen in northern specimens. The present variety is an extreme form, and perhaps may be considered as a subspecies, like many others peculiar to the above named islands. Pl. LXVII., 2.

8.—S. Græca, Stgr. Hor. 1870, p. 70.

PELOPEA, var. GRÆCA, Stgr. Cat. 1871, p. 28.

Expands from 2·0 to 2·25 in. Wings dusky brown, powdered with grey towards the base. The male has an indistinct submarginal band of light orange, much broken, and enclosing two black spots with white pupils; hind wings with a broad submarginal band, narrowly edged with whitish internally. Under side: Fore wings brownish grey, with a yellowish submarginal band enclosing spots as above; external to this the hind margin is bluish grey; hind wings grey, with a white central band, edged internally by a dark grey line, and bounded externally by a row of faint yellowish brown blotches; hind margin grey. The female has the hind margins of the fore wings straighter than in the male, the submarginal band broader and more distinct, the basal part of the wing darker. The hind wings also have the submarginal band broader, and deeper in colour. The coloration of the under side differs from that of the male in being darker and more distinct; the central white band is wanting. Pl. LXVIII., 1.

TIME OF APPEARANCE.—June.

HABITAT.—Mountains in Southern Greece.

Larva.—Unknown.

Obs.—Dr. Staudinger, who first described this species, now considers it specifically distinct from the Asiatic *S. Pelopea*, of which he at one time thought it a variety; he has supplied me with the specimens figured in this work.

9.—**S. Amalthea**, Friv. Magy. Ac. 1845; Stgr. Hor. 1870, p. 68.
Pontica, Frr. 475, 2, 3 (Jan. 1846); Butl. Cat. 51.
Anthelea (♀), H. S. 303-4.
Telephassa, var. *b.* Amalthea, Stgr. Cat. 1871, p. 28.

Expands from 2·0 to 2·25 in. Wings in both sexes dark brown, with white bands; the marginal fringes are brown and white. The male has the hind margin of the fore wings somewhat concave, and the apex rather pointed; the submarginal white band is narrowed towards the apex, and is widest at the centre of the wing; it encloses a large round black spot at its widest part, and another smaller one is placed external to it near the apex; in the discoidal area is a well-marked oblong black spot or dash; the hind wings are slightly dentate, and have a slight angular projection; near the anal angle there is a central white band, fading off into an orange-colour along its hind-marginal edge; near the anal angle is a small black spot. In the female the hind margin of the fore wings is straight; the submarginal white band is broader than in the other sex; it is dilated at its apical end so as to include the black spot, which is larger than in the male, and sometimes has a white centre; the discoidal black spot is wanting; the hind wings are more rounded than in the male, and are without angular projections; the white and orange band is broader, and the spot near the anal angle longer. Under side: Fore wings dull white, the costa, apex, and hind margin brownish grey; the two black spots are very conspicuous; hind wings mottled with brownish grey, with a central white band, and between this and the hind margin an indistinct orange one. Pl. LXVIII., 2.

Time of Appearance.—June.

Habitat.—Greece, and the southern part of Turkey.

Obs.—This is by some considered to be a variety of the Syrian *S. Telephassa*, Hüb., which resembles it in all respects, except that in *Telephassa* the markings are entirely fulvous instead of white. The variety *Anthelea*, Hüb., found in Asia Minor, is intermediate between the two, having the male with white and the female with fulvous markings. These forms do not occur in Europe.

10.—S. Hippolyte, Esp. i. 2, p. 164; Hbst. 211, 3, 4; O. i. 1,
      206; B. Ic. 42, 1, 2; Dup. i. 28, 5, 6; Frr. 278,
      1, 2; H. S. 80, 3.
    AGAVE, Esp. 84, 4 (sed *Agave*, Cr., alius erat *Papilio*);
      Hüb. 139-40; Kirby, Man. Eur. Butt. 57.

Expands from 1·50 to 1·75 in. Fore wings light brown, with an interrupted submarginal fulvous band containing two round black spots, sometimes with small white pupils; hind wings with a continuous submarginal fulvous band, containing a very small black spot near the anal angle. Under side: Fore wings yellowish brown, with a narrow dark submarginal line; internal to the black spots, which are placed as above, is an angulated reddish line; along the costa are some black streaks; hind wings pale yellowish brown, with three or four indistinct wavy dark brown lines crossing the wing; near the centre is an indistinct light band; the neuration is somewhat lighter than the rest of the wing. Pl. LXVIII., 3.

TIME OF APPEARANCE.—June.

HABITAT.—The Ural Mountains; also mountains in Andalusia (Sierra Nevada).

11.—S. Neomiris, Godt. Enc. Méth. p. 827 (1822); Boisd. Ic.
      42, 6-8 (recté *Neomyris*).
    IOLAUS, Bon. Descr. T. 3, 1 (1824); Frr. B. 67, 2; Tr.
      x. 1, 27.
    MARMORÆ, Hüb. 814-7; Butl. Cat. 51 (nomen vetust?).

Expands from 2·0 to 2·25 in. The male has the basal portion of all the wings dark brown; fore wings with a rather indistinct submarginal orange band, commencing at the inner margin and fading off into the ground colour at about the second median nervule; the hind margin is dark brown, and near the apex is a small black spot with a white centre; hind wings orange for two-thirds of their area, the hind margin being broadly dark brown, the same colour as the basal portion; near the anal angle is a small dark brown spot. The female is larger and paler than the male, the fore wings have a more distinctly defined submarginal orange band, the spot near the apex is larger, and has a distinct white centre; between the third and fourth median nervules is a second black spot; the hind wings are much the same as in the male, but the orange portion is paler along its internal edge, in some specimens being nearly white; the spot near the anal angle has a distinct white centre. The marginal fringes in both sexes are brown and white. Under side : Fore wings with the basal half dark brown, the rest light orange, except at the apex and along the hind margin, which are brown mottled with white; the spots seen above are still more distinct. Hind wings mottled with brown and white, the basal part being darkest; there is a central white band, and in the female the neuration is more or less marked with white. Pl. LXIX., 1.

TIME OF APPEARANCE.—July.

HABITAT.—Corsica and Sardinia, frequenting mountains.

LARVA.—Unknown.

OBS.—This species appears to be very distinct from its congeners, and affords another example of a good species of butterfly unknown to any other part of the world but these Mediterranean Islands.

12.—**S. Arethusa,** Esp. 69, 3, 4 (1781); Hüb. 154-5; O. i. 1, 208; Godt. i. 7.

Expands from 1·60 to 1·75 in. All the wings dull brown; in the male with a submarginal row of orange spots, separated by brown lines along the course of the nervules, but forming a more or less distinct band; the fore wings have a large round black spot near the apex; the hind wings have one sometimes near the anal angle. Female similarly marked, but paler; the fore wings have two black spots, and the submarginal band on the hind wings is very indistinct, being often entirely absent. Both sexes have the marginal fringes brown and dull white, and the hind wings have their hind margins somewhat dentate. Under side: Fore wings orange, paler in the female; along the costa are some dark brown streaks; the apex and hind margins are brownish grey, and the black spots are distinct and have white centres; hind wings mottled with brown and grey, being lighter in the centre; the neuration is not conspicuous, being of the same colour as the wing pattern; along the hind margin is a row of indistinct dark brown spots. Pl. LXIX., 2.

TIMES OF APPEARANCE.—July and August.

HABITAT.—South Germany, Central and Southern France, Switzerland, the South of Europe, Armenia, and the Altai. It occurs in France as far north as Fontainebleau and Versailles, but it is not found in Britain or in any of the more northern parts of Europe. It frequents limestone or chalky districts, dry and rocky woods, &c., often settling on the ground.

LARVA.—Unknown.

VARIETIES.

a. **Erythia**, Hüb. 591-2; O. i. 1, 209.—Differs from the type merely in having the under side of the hind wings paler, and with whitish central band.

HABITAT.—South Eastern Europe.

b. **Dentata**, Stgr. Cat. p. 29 (1871); *Erythia*, Stgr. Hor. 1870, 1. —Has the submarginal spots larger than in the type, and distinctly dentate in shape. The hind wings have the neuration marked with white on their under side; they also have a distinct white central band, and generally a darkish dentate line between it and the hind margin.

HABITAT.—South Eastern France. Pl. LXIX., 3.

c. **Boabdil**, Ramb. Faun. Andal. Pl. 12, 1, 2; H. S. 474-6.— Very dark above, with hardly any fulvous markings, some specimens being almost unicolorous. Beneath the colour is lighter than in the type, the hind wings being even more variegated with white than in var. *Dentata*, and the neuration being, as in that variety, marked with white.

HABITAT.—Mountains in Andalusia.

13.—S. Statilinus, Hufn. Berl. M. ii. p. 84 (1766); Rott. Naturf.
vi. p. 13 (1775); O. i. 1, 184; Frr. 499, 2, 3;
H. S. 177.
FAUNA, Sulz. Abg. G. T. 17, 8, 9 (1776); Esp. 29, 1, 63, 7;
Hüb. 507-9; Godt. i. 7.
ARACHNE, Esp. 95, 2, 3.

Expands from 1·50 to 2 in. Dark brown. Male: Fore wings
with a somewhat lighter submarginal band, or at least with faint
indications of one; on this are two round large black spots, and
between them two white dots; hind wings lighter in the centre,
with a submarginal row of white dots, and with a black spot near
the anal angle. Female somewhat lighter brown; the black spots
on the fore wings are sometimes surrounded by yellow rings, and
there are often some yellow submarginal spots on the hind wings.
Under side: Fore wings with yellow rings surrounding the black
spots; the spot nearest the apex has a white centre; internal
to these is a white streak, bounded internally by a black line, hind
margin grey; hind wings brown, with a central white band,
bounded internally by a wavy dark brown line; between this and
the base is another similar line, the hind margin is lightish grey,
and there are one or two submarginal light spots. The hind
margins of the hind wings are dentate; the submarginal fringes of
all the wings white. Pl. LXIX., 4.

TIMES OF APPEARANCE.—July and August.

HABITAT.—Central Europe (but not Great Britain); in the
South of Europe it is rarer, being generally replaced by the variety
Allionia. It frequents rocky woods, though not usually in chalky
or limestone districts. In France it is found as far north as
Fontainebleau, and used to occur in the Bois de Boulogne.

LARVA.—Undescribed as far as I am able to discover. Girard
quotes Maurice Sand to the effect that it " feeds on grasses in woods
during June, and that it is easy to procure by sweeping at
night" (p. 213).

VARIETIES.

a. **Allionia,** Fab. Spec. 83 (1781); Cyr. Ent. Neap. I., i, ii., 13
(1787); Esp. 105, 4; O. i. 1, 181; Tr. x. 1, 30. *Fauna*, Hüb.

145, 6, 510, 11; *Fidia* var. 52, 4. *Statilinus* var. *Martiani*, H. S.
190, 1; vi. p. 15.—Expands from 1·75 to 2·50 in. Larger and
paler than the type; the female has the black spots distinctly
ringed with yellow. Under side: Hind wings much paler and
more uniform in colour than in the type; the white band indistinct
or, as in the figure, altogether absent, but the dark lines across the
wings are rather more conspicuous by contrast to the lighter
ground colour; near the hind margin are some whitish or yellow
spots. Pl. LXX., 1.

Times of Appearance.—June and July.

Habitat.—The South of Europe and Asia Minor.

b. **Fatua**, Frr. 415, 3, 4; H. S. 192, 3; Stgr. Hor. 1870, p. 72.
—Very close to *Allionia*, but the under side of the hind wings is
much more uniform in colour, being almost entirely one shade of
grey, without any white band or central dark streaks; the hind
margin is somewhat more dentate than in *Allionia*. Pl. LXX., 2.

Times of Appearance.—June and July.

Habitat.—Eastern Greece, principally in the neighbourhood
of Athens; also Asia Minor.

Obs.—Dr. Staudinger gives *Fatua* specific rank, but I am
inclined to agree with those who consider it merely a local race of
S. Statilinus, and do not think it sufficiently distinct from *Allionia*
to be considered a distinct species. I have a series, taken last year
in the South of Spain, all in one locality, containing several
forms intermediate between *Allionia* and *Fatua*.

14.—**S. Fidia**, Linn. Syst. Nat. xii. 770; Esp. 49, 3; Hüb.
147, 8; O. i. 1, 179; Godt. ii. 11, 34.

Expands from 1·75 to 2 in. Very similar above to the last
species, especially to var. *Allionia*, but there are some whitish
spots on the fore wings internal to the two black ones, and the hind
wings sometimes show a trace of a whitish central band; the hind
margin of the hind wings is somewhat more dentate than in the
last. Under side: Fore wings much the same as in the last
species, but somewhat lighter in the ground colour, and with more
strongly marked lines, especially near the base and on the costa.
Hind wings white and grey, the inner marginal portion of the

wing being almost entirely white, whilst the hind marginal part is dark grey; a wavy black line runs across the centre of the wing; internal to this is a blackish streak, and near the base is another black line; the central black line has a broad whitish band accompanying it throughout its whole length; there is a sub-marginal narrow black line, and near the anal angle is a small black spot with a white centre. Marginal fringes of the fore wings brown and white, those of the hind wings white. Pl. LXX., 3.

TIME OF APPEARANCE.—July.

HABITAT.—The South of France, Spain, and Italy; frequenting hilly calcareous districts.

LARVA.—Elongate, fusiform, and smooth, except the head and the last segment, which are furnished with short white hairs. In colour it is reddish grey, marked with longitudinal lines; the dorsal line is dark brown, very fine on the first and last segments, but thicker in the middle; the subdorsal is fine and nearly black, thickest on the last segment, it has a broad edging of yellow throughout; the stigmatal line is broad and yellow, edged with brown below; ventral surface and legs light brown. It feeds on grasses of different kinds. Millière says he has found it most often on *Piptaterum multiflorum* (Millière, Icon. ii. 413).

PUPA.—Obtuse, the thorax with a slight projection; it is brown, with the stigmata darker. It is not suspended, but lies free on the ground.

15.—S. **Dryas**, Sc. Ent. Carn. p. 153 (1763); Esp. 40, 1, 2; Schn. Syst. B. p. 100; Butl. Cat. 61.
PHÆDRA, Linn. Mus. Ul. p. 280 (1764); Syst. Nat. xii. 773; Hüb. 127-9; O. i. 1, 186; Godt. 1, 7; Frr. 373.

Expands from 1·75 to 2·25 in. All the wings dark brown; fore wings with two large round spots, each having a light blue centre; hind wings with their hind margins somewhat dentate, fringes brownish white; between the third and fourth median nervules, near the anal angle, is a small black spot with a bluish centre. Under side: Fore wings paler than above, the black spots are surrounded by yellow rings; hind wings light brown, with basal and central streaks of greyish white. Pl. LXX., 4.

TIMES OF APPEARANCE.—From the beginning of July to the end of August.

HABITAT.—Central Europe (excepting Denmark, Holland, Belgium, and the British Isles), North and Central Italy, Bulgaria, South Russia, and Asia Minor. It is usually found in calcareous districts, and I have found it common in the rocky woods on the borders of the Swiss lakes, and similar situations. In France it occurs as far north as Rouen.

LARVA.—Reddish grey or flesh-colour, with a dark brown dorsal stripe edged with white; lateral stripes dark grey; stigmata black; head reddish, with dark brown longitudinal lines. It is found in woods in June on *Avena elatior;* at the end of June it becomes a rounded brown-coloured pupa, not suspended, but placed on the ground in a kind of small earthy cocoon (Girard).

16.—**S. Cordula**, Fab. Ent. Syst. 226 (1793); Hüb. 619-20; O. i. 1, 190; Godt. ii. 12, 3, 4; H. G. 969-70; H. S. 176.
    HIPPODICE, Hüb. 718-19.
    ACTÆA var., Esp. 85, 4.

Expands from 2 to 2·50 in. Male dark brown. Fore wings with a round black spot near the apex, and another similar one below it near the anal angle, both with bluish white centres; hind wings with one or two submarginal black spots near the anal angle. Under side: Fore wings with a yellow ring surrounding the spot near the apex; hind wings dark grey, with a white central band, and with a dark wavy submarginal line; the nervules are marked with whitish. Female lighter than the male, the black spots enclosed in a light fulvous band on the fore wings; hind wings with a trace of a fulvous band. Under side: Fore wings light fulvous, with the black spots as above; hind wings grey, with a lighter central band, the neuration being lighter than the ground colour; near the anal angle are two black spots. Pl. LXX., 5.

TIMES OF APPEARANCE.—June and July.

HABITAT.—The Alps of the Valais, mountains in the South of France, and the Ural.

LARVA.—Unknown.

OBS.—Generally reckoned an Alpine variety of the following.

17.—S. Actæa, Esp. 57, 1, a, b (1780); Hüb. 151, 2, 610, 11; O. i. 1, 193; Godt. i. 7.

Expands from 1·75 to 2 in. Dark brown, darker than the last. Male with a bluish white spot at the tip of the fore wing; hind wings dark brown. Under side: Fore wings dark brown, with a black spot near the apex, having a bluish white centre, and surrounded by a yellow ring; below this is another black spot, and between the black spots two small white ones; hind wings dark grey, crossed by two white bands, and with a black spot near the anal angle. Female somewhat lighter, with an extra spot on the fore wings near the anal angle, that near the apex surrounded by a yellow ring. Under side: Fore wings fulvous, with a large black spot near the apex pupilled with white; below this is a white spot, and sometimes a black one. Pl. LXXI., 1.

TIMES OF APPEARANCE.—June and July.

HABITAT.—The South of France, Spain, and Italy; frequenting dry wooded hills.

LARVA.—Unknown.

VARIETIES.

a. **Podarce**, O. i. 1, 195; Esp. 123, 1, 2; H. S. 49-52; Frr. 463, 3, 4; Stgr. Hor. 1870.—Smaller than the type; the under side of the hind wings with white veins.

HABITAT.—Mountains in Portugal, and in Syria.

b. ♀ **Peas**, Hüb. 132, 3. ♀—All the wings above with a broad ochreous band; hind wings beneath whitish, with a central band of dark grey.

c. **Bryce**, Hüb. 149, 50, 724-7.—About the same size as the type, but the fore wings have a black spot with a white centre in the male as well as in the female; the hind wings are less variegated beneath, being lighter, and the light bands less conspicuous.

HABITAT.—South Russia, the Caucasus, and Altai.

d. **Virbius**, H. S. 45-48; Frr. 463, 1, 2; Stgr. Hor. 1870, p. 75. —Smaller than the type. The fore wings with two ocelli; hind wings beneath unicolorous.

HABITAT.—South Russia.

Genus 5.—**PARARGE**, Hüb. verz. bek. Schmett. p. 59 (1816).
   SATYRUS, Latr. Consid. Gen. p. 355 (1810); ? Sect. typ.
   LASIOMMATA, Westw. Brit. Butt. p. 65 (1840); Gen.
      D. L. p. 385 (1851).
   VICICOLES et RAMICOLES, Dup.

CHARACTERS.—LARVA elongate, pubescent, with two anal projections, generally green. PUPA short, thick, with two small angular projections and two cephalic points; usually suspended by the anal extremity. IMAGO.—Head hairy, with a frontal tuft. Eyes prominent, hairy. Palpi with the front of the basal and second articulations thickly hairy, the terminal articulation short. Antennæ moderately long, straight, and generally ringed with white, terminated by a distinct compressed pyriform club. Thorax hairy, of moderate size. Fore wings with the costal margin moderately arched, the apex rounded, hind margin entire; costal and median nervures dilated at the base, the median less so than the costal; discoidal cell reaching beyond the middle of the wing. Hind wings subovate, with the hind margins moderately dentate; inner margin not incised near the anal angle. In colour the wings are brown, varying in the different species from pale dull brown to bright fulvous; there is always at least one spot near the apex of the fore wings, generally ocellated; the hind wings on their under side are lighter in colouring than the rest, and have a submarginal row of black spots surrounded by yellow rings.

The butterflies of this genus are of moderate size; some of the species frequent the neighbourhood of human habitations, such as gardens, lanes, road-side hedges, &c. (such as *S. Megæra*); others, such as *P. Egeria* and *Achine*, are found in woods, generally preferring rather damp and shady places. Only two species occur in the British Islands.

1.—**P. Roxelana,** Cr. Pap. Exot. Pl. 161, c. f. (1782); Fab. Ent. Syst. 227; O. i. 1, 217; Hüb. 680-3; Boisd. Icon. 43, 1-3; Dup. i. 30, 1-4; Frr. B. 109, 2.

Expands from 1·80 to 2·20 in. The male has the fore wings dull brown, with an irregular patch of dull fulvous in the centre; near the apex is a small black spot. Hind wings dull brown, with a submarginal row of yellow rings, usually three in number; the neuration is rather darker than the rest of the wing, the hind margin is dentate. Under side : Fore wings fulvous, costa, apex and hind margin brown, the apical spot has a yellow ring. Hind wings brown, with two narrow wavy reddish lines running across the centre ; there is a narrow wavy whitish band edged with brown running close to the hind margin, and a submarginal row of black spots surrounded by yellow rings, and having white centres; two of these are placed close together near the costa ; the next two are very much smaller, and pushed out of their proper places, as it were, by a triangular white blotch ; then come three more, that nearest the anal angle containing two white dots. Female usually larger than the male : fore wings more suffused with fulvous, the costa and hind margins broadly dark brown ; there is a yellowish white blotch nearly in the centre of the costa, and outside this two more, having the black apical spot between them ; the neuration is lined with dark brown. Hind wings as in the male, but with a marginal dark line. Under side as in the male, except that the fore wings have white costal blotches. The marginal fringes in both sexes are white and brown. Pl. LXXI., 2.

Times of Appearance.—June and July.

Habitat.—South Eastern Hungary, Turkey, and Asia Minor.

Larva.—Unknown.

Obs. — The neuration of the fore wings of this species is exceptional in appearance, owing to the thick black scales which follow the course of the median, and to the curve taken by the submedian,

2.—P. Climene (recté *Clymene*), Esp. 85, 1-3 (1783) ; Hüb. 165-6.
CLYMENE, O. i. 1, 215 ; Boisd. Ic. 43, 4, 6 ; Dup. i. 29, 45 ;
Frr. B. 109, 1 ; H. S. 102-3.

Expands from 2 to 2·25 in. Wings dull brown. Fore wings
with an orange blotch near the centre, and some light streaks near
the costa ; near the apex is a small black spot surrounded by
a yellow ring, and below it a white dot. Hind wings dull brown,
with one or two submarginal spots in the female surrounded by
yellow rings. Under side : Fore wings fulvous, the apex and hind
margin dull brown. Hind wings grey, with asubmarginal row of
six or seven black spots in yellow rings. Pl. LXXI., 3.

TIME OF APPEARANCE.—July.

HABITAT.—South Russia, in the large forests on the shores of
the Volga (Millière).

LARVA.—Fusiform, moderately pubescent ; green with a slightly
bluish tint, with the longitudinal lines indistinct ; the legs well
developed, with the last segment ending in a forked point ; the
head distinct from the first segment ; the head is rather large,
rounded, ciliated, of one colour, surmounted by two prominent
points, reminding one somewhat of the head of the larvæ of the
genus *Nemoria ;* the mandibles are reddish, and the eyes brown ;
the three usual lines are straight, continuous, brighter than the
ground colour, finely powdered with dark green, but at the same
time not very distinct. Like all the larvæ of the *Satyridæ*, that of
Climene lives only on grasses.

The above is taken from Millière's description (Ic. III., 183) ;
he adds that in all probability it is suspended as a pupa ;
he further remarks that the structure of the head of this larva is
exceptional, and may necessitate a new genus for this species.

3.—P. Mœra, Linn. Syst. Nat. x. 437 ; Faun. Suec. 275 ;
Esp. 6, 2 ; 68, 3 ; Hüb. 174-5 ; O. i. 1, 231.
ADRASTA, Dup. i. 46, 1, 2.

Expands from 1·50 to 1·75 in. The male is brown, with a
broad submarginal fulvous band divided by brown streaks, following

the course of the nervules; there are one or two fulvous patches between this and the base; near the apex is a large black spot bipupilled with white. Hind wings with a fulvous submarginal band, divided like that on the fore wings; it contains two or three black spots with white centres. The female has the fore wings fulvous, the basal portion, costa and hind margin being brown; from the costa a brown streak runs outwards towards the hind margin, forming a triangular fulvous space, with the apex directed downwards; in this space near its base is placed the bipupilled black spot, one or two other shorter brown streaks run from the costa to the median nervure; hind wings as in the male, but sometimes with a fulvous tinge. Under side: Fore wings in both sexes greatly resembling the upper side of the female fore wings, just described, except that the apical spot is surrounded by light yellowish brown, and the costa and hind margins are lighter. Hind wings grey, the basal half crossed by zigzag lines of a brown colour; there is a submarginal row of seven black spots, having white centres, and enclosed in brown and yellow rings; the two near the anal angle are smaller than the rest, and coalescent. Pl. LXXI., 4.

TIME OF APPEARANCE.—April to August.

HABITAT.—The whole of Europe, except Denmark, Holland, Great Britain, and South Russia; it is also common in Asia Minor and Syria. On the Continent this is a common species, frequenting much the same kind of localities as *Megæra*. It has been reputed British, but erroneously.

LARVA.—Bright green, with darker dorsal and subdorsal lines, the stigmatal line being yellow; head and legs green. It feeds on grasses in April and in June.

PUPA.—Green or greenish black, very slightly angular, slightly bifid, with two dorsal rows of tubercles of a yellow or brown colour; it is generally suspended on walls, palings, and similar places.

VARIETY.

a. **Adrasta**, Hüb. 836-9; Tr. 10, 1, 36. *Mæra* var., Esp. 49, 1.
—Slightly larger than the type, deeper in colour, and more suffused with fulvous; the under side of the hind wings is darker

grey, strongly speckled with brown. It is found with the type in cold and elevated localities, in the mountains, &c. Pl. LXXI., 5.

4.—**P. Hiera,** Fab. Gen. 262; Hüb. 176; O. iv. 135; Boisd. Ic. 44, 1-3; Dup. i. 46, 3, 4; Wallengren Rhop. Scand. p. 25.

Expands from 1·50 to 1·60 in. Wings dark brown. Fore wings with a fulvous patch near the apex; extending about half-way down the wing it encloses a black spot with a white centre, and occasionally there are two others smaller; a few dark lines cross the wing from the costa; hind wings dark brown, with a submarginal row of black spots with white centres, and surrounded by fulvous rings. Under side very similar to that of *Mæra*, but generally rather darker. Pl. LXXII., 4.

Like the last, this is a somewhat variable species.

TIME OF APPEARANCE.—May to the end of July.

HABITAT.—Alpine regions in Switzerland, but not at very great elevations; also Scandinavia, Bulgaria, South Russia, and Armenia.

OBS.—This species is intermediate between the last and the next; some specimens greatly resemble small examples of *Mæra*, but on the whole are duller and have the spots proportionately larger.

LARVA.—Undescribed.

5.—**P. Megæra,** Linn. Syst. Nat. xii. 771; Esp. 6, 3, 68, 4; Hüb. 177-8; O. i. 1, 235.

Expands from 1·75 to 2 in. Wings fulvous. Fore wings with three blackish brown wavy lines running across them from the costa towards the inner margin; hind margins dark brown, the basal portion suffused with the same colour; near the apex is a round black spot with a white centre. Hind wings fulvous, the hind margins dark brown; a submarginal row of white-centred black spots, four in number, is placed between two dark brown lines; the basal portion dusky. Under side: Fore wings as above, but the bases are lighter, and the apical white-centred spot is surrounded by a yellow ring. Hind wings brownish grey, with several dark brown zigzag lines running across from the costa to the inner margin, and with a submarginal row of black spots with white centres, and surrounded by yellow rings. The female is paler in colour than the male, and has both the central and hind-marginal bands narrower; it is also somewhat larger. Pl. LXXII., 5.

TIME OF APPEARANCE.—In this country there seems to be two or three broods in the season, the first appearance being at the end of April; in southern countries it is on the wing throughout the year.

HABITAT.—The whole of Europe, except the Polar regions, North Africa, Asia Minor, and Armenia; generally distributed over the British Isles. (The Wall-butterfly.)

LARVA.—Pubescent, apple-green, with five longitudinal lines of dark green, and yellow stigmatal line; head rounded, green, with fine black hairs; the thoracic legs brown, posterior pairs green. On grasses.

PUPA.—Green or greenish black, slightly angulated, with two dorsal rows of yellow or white tubercles. Suspended.

VARIETIES.

a. Var. **Lyssa**, Boisd. Ic. 44, 4, 5, i. p. 222; H. G. 914-17.—Differs from the type in the ground colour of the hind wings, which, instead of being brownish grey, is bluish or ashy grey; this difference in colour is well marked.

HABITAT.—Dalmatia, the Balkans, and Asia Minor.

b. Var. **Tigelius**, Bon. Descr. T. i. 2 (1824); Boisd. Ic. 45, 1-3; Dup. i. 30, 5-7; Frr. B. 68, 1. *Parameyœra*, Hüb. 842-4; Butl. Cat. 124 (*nomen vetustius?*).—Smaller and lighter than the type. In the female the dark bands that enclose the submarginal spots in *Megœra* are wanting; the spots themselves have bluish centres. Pl. LXXII., 6.

HABITAT.—Corsica and Sardinia.

6.—**P. Egeria**, Linn. Syst. Nat. x. 473; xii. 771.
MEONE, Esp. 95, 1 (1789); Hüb. 179-80; O. i. 1, 240.

Expands from 1·70 to 2 in. Wings dark brown. Fore wings spotted with fulvous, and with a small white-centred black spot near the apex. Hind wings with a submarginal row of fulvous spots, each enclosing a black one with a white centre. Under side : Fore wings as above, but paler, and the spots are somewhat larger. Hind wings brown, varied with fulvous and purplish red ; there are some faint submarginal light spots. Pl. LXXIII., 2.

Time of Appearance.—Throughout the summer, there being two principal broods in the year.

Habitat.—Southern and South-western Europe, North Africa, and Syria.

Larva.—Dull green, with yellowish white dorsal and lateral stripes. Feeds on *Triticum repens* and other common grasses.

Pupa.—Greenish, somewhat angular, with a double row of small dorsal tubercles.

Obs.—The above is a description of the southern form of the species, which is doubtless the original Linnean typical form. In Northern and Central Europe the form *Egerides*, Stgr., about to be described, entirely replaces it ; but in some places the species exhibits seasonal dimorphism, giving the northern var. *Egerides* as the spring and typical *Egeria* as the summer brood.

VARIETY.

a. **Egerides**, Stgr. Cat. 1871, p. 30. *Ægeria*, Esp. 7, 1 ; Hüb. 181, 2 ; Frr. 403. *Egeria* aut *Ægeria* autorum anglicorum.—The size and shape of the type, but the ground colour is darker and the light spots smaller, and instead of being fulvous they are light straw-colour or yellowish white. The under side resembles that of *Egeria*, the fulvous parts being replaced by whitish yellow. Pl. LXXIII., 3.

Time of Appearance.—April, July, and August.

Habitat.—Common throughout Northern (except the Polar region) and Central Europe, including the British Isles. (The Speckled Wood-butterfly.) It occurs as the spring brood in many localities in Central France, &c. ; also in the South-East of Europe. I have seen it stated that specimens exactly resembling British ones have been taken near Constantinople.

7.—P. Achine, Sc. Ent. Carn. p. 156, 1763 ; Butl. Cat. 122.
DEJANIRA, Linn. Mus. Ul. p. 282 (1764) ; Syst. Nat. xii.
774 ; Esp. 9, 2 ; Hüb. 170-1 ; O. i. 1, 229 ; Frr. 391.

Expands from 1·70 to 2·20 in. All the wings dull brown,
somewhat paler in the female than in the male ; marginal fringes
dark brown and pale yellow. Fore wings with a submarginal row of
round black spots, each enclosed in a ring of light yellow ; there
are generally five of these, the fourth from the costa being larger
than the rest. Hind wings with a somewhat similar row of spots,
the two central ones being much larger and more distinct than the
remaining ones, which are only faintly defined ; internal to the
rows of spots on both wings is a faint light streak. Under side :
Fore wings lighter than above, with a light yellow submarginal
band, on which are placed the black spots. Hind wings with all the
black spots distinct, and with white centres ; encircling the two
upper spots and running internally to the rest is a narrow band of
pure white, and external to the spots a narrow marginal line
of light yellow. Pl. LXXIII., 4.

TIMES OF APPEARANCE.—June and July.

HABITAT.—Moist shady woods in Central Europe (except
Britain), Scandinavia, and South Russia ; the Altai and the Amur.
Its flight is feeble, and it frequently settles on leaves.

LARVA.—Pubescent, green, with three darker dorsal lines, two
double lateral and a stigmatal line ; the head and thoracic legs
yellowish ; the posterior legs and the caudal points green. Feeds
on *Lolium perenne,* and is full grown in May. Pl. LXXVI., 6.

PUPA.—Not suspended, but placed on the ground, like many
of those in the last genus.

Genus 6.—**EPINEPHELE,** Hübner, Verz. bek. Schmett. p. 59
(1816) ; H. S. Schmett. Eur. (1844).
HIPPARCHIA, Fab.
SATYRUS, Lat.
MANIOLA, Schk. ; RAMICOLES et HERBICOLES, Dup.

CHARACTERS.—LARVA pubescent, grey or green, with longitudinal stripes ; head globular ; body shaped as in the last genus,

with two caudal points. They feed on grasses. PUPA slightly elongated, somewhat rounded; suspended by the caudal extremity. IMAGO.—Antennæ tolerably long, club not very distinct. Palpi longer than the head, the articulations hairy. Eyes not hairy. Front legs short and very slender, not hairy. The femora of the rest smooth beneath in the female. Wings rounded; hind wings moderately dentate, slightly incised on their inner margin near the anal angle. The costal and median nervures are equally dilated at their origin; the submedian without any dilatation. The fore wings of the male have generally a patch of dense scales beneath the median nervure. Prevailing colour of wings dark brown, marked with fulvous; there is at least one ocellated spot near the apex of the fore wings.

Various descriptions of country form their habitats, such as meadows, road-sides, heaths, woods, &c.; but they are not found at any elevation in mountainous regions, being rather inhabitants of the plains.

Many of the species are very abundant; *E. Janira* is perhaps one of the commonest species of European Lepidoptera. About thirty species of *Epinephele* are at present known; nearly all belong to the Palæarctic region, but a few inhabit temperate South America, and one occurs in Australia.

1.—E. Narica, Hüb. 704-7; Hüb. Verz. p. 59; Frr. 464.

Expands from 1·60 to 1·80 in. Ground colour of wings brown. Fore wings with the central area light fulvous; the male has a round black spot near the apex, and beneath the median nervure is a patch of black scales; this patch is absent in the female, but there is a black spot near the anal angle. Hind wings uniform brown, the hind margin somewhat dentate, and the fringe whitish. Under side similar in both sexes; fore wings fulvous, the costa and hind margin brown mottled with grey; the apical eye has a white centre; anal eye absent in the female. Hind wings greyish brown, with a narrow white central band, and some whitish mottling near the base, a submarginal indistinct dark brown line; the neuration is marked with white. Pl. LXXIV., 1.

Time of Appearance.—June.

Habitat.—South Eastern Russia and the Kirghis Steppes.

Larva.—Unknown.

2.—E. Lycaon, Rott. Naturf. vi. p. 17 (1775).
> Eudora, Esp. 45, 1 (1777); 69, 1, 2; Hüb. 163-4; O. i. 1,
> 223; Godt. ii. 18, 1-3.
> Janirula, Esp. 113, 1.

Expands from 1·40 to 1·50 in. The male has all the wings dull brown, the marginal fringes being of a similar colour. The fore wings have a black spot near the apex and a patch of black scales beneath the median nervure. Hind wings uniform brown, with faint indications of a narrow submarginal band. Under side: Fore wings fulvous, bordered with light brown apical spot, with a white centre. Hind wings greyish brown, with a slightly lighter central band. Female : Fore wings light brown, with a light fulvous submarginal band, containing two black spots, one near the apex and the other near the anal angle; the former has a white centre. Hind wings light brown, with a very faint lighter band. Under side : Fore wings light fulvous, bordered with light brown ; the spots as above ; hind wings as in the male but lighter. Pl. LXXIII., 2.

Time of Appearance.—From May to July.

Habitat.—Central Europe (but not the British Isles), Spain, France, and South Russia; Western Asia, Siberia, and the Amur.

Larva.—Green, with a narrow white lateral line, below which is a broader one, of a bright yellow or orange colour; head reddish brown and green ; ventral surface and legs green. (Hüb.) On grasses in May.

Pupa.—Green, or brown varied with pink and white streaks. Pl. LXXVI., 7.

VARIETY.

Lupinus, Costa, Fauna Nap. ; Stgr. Hor., 1870, 79. *Rhamnusia*, Fer. 457, 2, 3 ; H. S., 377-8, 427-8.—Larger than the type and lighter in colour ; the male has the thickened patch of scales on the fore wings denser. Somewhat resembles *E. Janira* var.

2 Q

Hispulla, but the female of that form, besides differing in colour from the present, has only one spot on the fore wing.

HABITAT.—South Eastern Europe and Western Asia.

3.—E. **Janira**, Linn. Syst. Nat. x. 475, No. 106, ♂, Faun. Suec.
　　276, No. 1053, ♂ ; Esp. 10, 1, 2; O. i. 1, 219.
　　JUTURNA, Linn., *loc. cit.*, No. 104, Faun. Suec., No. 1052,
　　♀ ; Hüb. 161, 2, ♀ (nomen anterius, recipiendum ?).

Expands from 1·75 to 2·20 in. The male has all the wings dull brown, rather darker towards the base, and with a greeenish bronze iridescence, especially towards the hind margins. The fore wings have a black apical spot, with a white centre, and surrounded by a fulvous ring ; there is a slight trace of fulvous coloration between this, and beneath the median nervure is the usual patch of scales seen in the males of this genus; but this is not so distinct as in the foregoing species. Female larger and lighter than the male ; the apical spot is much larger, and is placed in a well-developed submarginal fulvous band ; and there is generally a tinge of fulvous between this and the base. Hind wings brown with a slightly lighter submarginal band. Under side : Fore wings fulvous, the apex and hind margin from apical spot well marked. Hind wings brown, with a lighter submarginal band ; most distinct in the female. On this band are sometimes one or more small round spots. Pl. LXXIII., 3.

TIME OF APPEARANCE.—June to September.

HABITAT.—The whole of Europe, excepting the Polar and the higher Alpine Regions; it is also found in Armenia and Asia Minor; and, in the forms of varieties, in North Africa, Syria, and the Canaries. In Britain it is the commonest butterfly (the Meadow-Brown), frequenting fields, roadsides, gardens, heaths, &c., throughout the greater part of the summer.

LARVA.—Greeen, with a dark green dorsal line; lateral lines white. Feeds on various grasses in April and May.

VARIETIES.

a. **Hispulla**, Hüb. 593-6; Esp. 119, 1, 2; O. i. 1, 222.—Somewhat larger than the type. The male is darker, and has a more distinct patch of black scales beneath the median nervure of the fore wings; the hind wings are somewhat more dentate, and are more uniform in their colouring beneath. Female has the fulvous spread over the greater part of the fore wings, and the hind wings have a broad submarginal band of the same colour. The under side resembles that of the type, but the markings are stronger and brighter. Pl. LXXIII., 4.

HABITAT.—The South of Europe, North Africa, Syria, and the Canaries; the specimens figured are Spanish.

b. **Telmessia,** Z. Is. 1877, p. 4; H. S. 479-82; Stgr. Hor. 1870, 79.—A somewhat smaller and darker form, occurring in Asia Minor and Turkey. The female greatly resembles *Hispulla*, but is smaller.

OBS.—*E. Janira* is liable to a partial albinism, large white patches occurring on one or all of the wings. Such specimens are not very rare. Esper figures it, also Herrich-Schaffer, 104-5.

E. Janiroides, H. S. 533-4, is not described in this place, as it is not, so far as I can ascertain, a European species, but only occurs in North Africa; it will be referred to at the end of this section.

4.—**E. Nurag,** Ghil. Elenco. p. 83; Dutreux, Stett. e. z. 1855, 78; Stgr. Hor. 1870, 79.

Expands from 1·30 to 1·50 in. Close to *Janira*, but lighter in colouring. Fore wings light fulvous; costa and hind margins brownish grey; running from the inner margin towards the centre of the wing in the male is a dark patch; this is absent in the female. The nervules are marked with dusky brown. Near the apex is a round black spot, with a white centre present in both sexes. Hind wings brown, with a submarginal fulvous band, most distinct in the female. Under side: Fore wings fulvous, the hind margins grayish brown; the apical spot as above. Hind wings greyish brown, with a lighter central band sharply angled along its basal edge, and containing several small round spots. Pl. LXXIII., 5.

TIME OF APPEARANCE.—June.

HABITAT.—Sardinia and Corsica.

LARVA.—Undescribed.

5.—**E. Ida,** Esp. 92, 2 (1784), 102, 3; Hüb. 158-9, O. i. 1, 212; Godt. ii. 18, 4, 5; H. S. 183.

Expands from 1·25 to 1·50 in. Wings fulvous, the costa and hind margins broadly brown; the fore wings have an apical oval black spot, with two white dots; in the male an oblong rather rectangular dark brown blotch reaches from the inner margin to as far as the disco-cellular nervule; this blotch is divided by the nervules, which are fulvous, into five or six portions. Under side: Fore wings fulvous; hind margin broadly dark brown; near the apex is a whitish patch, and internal to and below this a black spot like that on the upper side, surrounded by a pale fulvous ring; across the centre of the wing is a faint reddish streak. Hind wings pale grey, mottled with white, and with a white central band; there are no submarginal spots. The sexes are similar, except that the female has no dark blotch on the fore wing, and is larger in expanse. Pl. LXXIV., 1.

Time of Appearance.—June and July.

Habitat.—Generally common in the South of Europe, such as Spain, Italy, the South of France, &c.; it occurs also in Algeria. It is not found north of the Alps.

Larva.—Grey, with a dark dorsal line; lateral lines reddish grey and white. Feeds on various grasses in April and May.

6.—**E. Tithonus,** Linn. Syst. Nat. xii. ii. 2; Linn. Mantissa, 537 (1771); O. i. 1, 210.

Tithonius, Herbst, viii. t. 189, 5, 8 (1796); Godt. Ency. Méth. ix. p. 542 (1823).

Pilosellæ, Fab. Syst. Ent. 497 (1775).

Amaryllis, Bkh. i. 80 (1788); Godt. i. 7, 2.

Herse, Wien. Verz. 320; Hüb. i. 156-7.

Phædra, Esp. 9, 1, 28, 3 (1777).

Expands from 1·35 to 1·60 in. Much resembles the last above; differs from it in having the wings of a darker and less brilliant fulvous; the fore wings in the male have the central

blotch much larger, sometimes reaching the costa, and not so
distinctly divided by the nervules; the hind wings are not so
dentate, and beneath they are variegated with white, but have
the basal half brown; there is a submarginal brownish yellow
band containing a row of light spots, the hind margin brown.
Pl. LXXIV., 2.

TIME OF APPEARANCE.—July and August.

HABITAT.—Roadsides, heaths, waste places, &c., throughout
Europe, excepting the north-east; Asia Minor and Armenia. It is
common in Britain (the Large Heath).

LARVA.—Greyish green, with a darker dorsal line; lateral lines
light grey; head brown. On grasses in May and June.

7.—E. **Pasiphaë**, Esp. 67, 4 (1781), 97, 1; Hüb. 167-9; O. i. 1,
214; Godt. ii. 18, 6, 7.

Expands from 1·30 to 1·75 in. Wings dark brown; the male
has the fore wings with a submarginal fulvous band; near the
apex is an oval black spot, with two bluish-white pupils. Hind
wings with a similar band, on which are three or four black spots.
Female similar, except that the fore wings are fulvous throughout,
and two of the spots on the hind wings have bluish-white centres.
Under side: Fore wings much as in *Tithonus*, but brighter fulvous,
and the costa and hind margins are darker brown. Hind wings
dark brown, without any white mottling; there is a central cream-
coloured band, and between this and the hind margin a row of five
black spots, surrounded by yellow rings, and with white centres.
Pl. LXXIV., 3.

TIME OF APPEARANCE.—June and July.

HABITAT.—Dry, slightly wooded places in the South of France,
Spain and Portugal, and Algeria.

LARVA.—Light brown, with reddish lateral line and a broad
dark brown dorsal stripe. On grasses in May.

8.—E. Hyperanthus, Linn. Syst. Nat. x. 471; Faun. Suec. 273; Esp. 5, 1, 57, 2; O. i. 1, 225; Godt. i. 7, 3 (*Satyrus*, H.).

POLYMEDA, Hüb. 172-3.

APHANTOPUS HYPERANTHUS, Wallengr, Scand. Rhop. p. 31.

ENODIA HYPERANTHUS, Steph.

MINOIS HYPERANTHUS, Butl.

Expands from 1·60 to 1·75 in. All the wings dark brown, in the male nearly black, in the female paler; there is a submarginal row of black spots on all the wings centred with white, and surrounded by yellowish rings; their number is variable, but there is generally from one to three on the fore wings, and two or three on the hind wings. Under side lighter brown; the fore wings have three black submarginal spots in yellow rings, two of them having white centres. Hind wings with five spots, yellow ringed and white-centred; marginal fringes whitish. Pl. LXXIV., 4.

TIME OF APPEARANCE.—June and July.

HABITAT.—Woods throughout Central and Northern Europe, excepting the Polar Regions ; Piedmont, Dalmatia, Asia Minor, Armenia, the Altai, and the Amur. Generally distributed in the British Isles (the Ringlet).

LARVA.—Reddish or greenish grey, dorsal line darker; two lateral lines of a yellowish colour. In May and June on grasses, especially *Poa annua.*

VARIETY.

Arete (ab.), Mül. Fn. Fr. p. 36; O. i. 1, 288. *Hyperanthus,* var. Esp. 57, 3, 4; Frr. 290-2.—An aberration in which the black spots and yellow rings are entirely wanting on the under side, only the white spots being present. This is the most usual form of *Arete,* but all kinds of intermediate stages occur, and occasionally even the white spots are absent. This aberration is found in the same localities as the type; it is sometimes taken in the British Isles, but is not so common here as on the Continent. Pl. LXXIV., 5.

Genus 7.—**CŒNONYMPHA,** Hübner, Verz. bek. Schmett.
p. 65 (1816); Westw. Gen. D. L. p. 396 (1851).
Cænonympha, Butler Cat. Sat. p. 39 (1868) (nomen
rectius, utendum ?).
Chortobius, Guenée.
Hipparchia, Fab., Ochs.; Erebia, Dalm.; Satyrus,
Lat., Boisd., Zett.; Dumicoles, Dup.

Characters. — Larva more cylindrical, and tapering less
towards the extremities than those of the preceding genera; anal
extremity bifid. The known European species are bright green in
colour, with dark dorsal and yellow lateral stripes. They feed on
grasses. Pupa short, with cephalic projections and dorsal tubercles,
usually green in colour, suspended. Imago.—Antennæ rather long,
with an oblong club. Palpi somewhat longer than the head, the
middle and basal articulations hairy. Eyes smooth. Fore wings
with the costal margin considerably arched, the apex rounded, and
the hind margin entire. All three nervures are dilated at the base.
Hind wings rounded, with the hind margins entire; inner margins
scarcely if at all incised. In colour the wings are brown or fulvous;
the fore wings generally have an ocellated spot near the apex; the
hind wings have a submarginal row of ocellated spots, sometimes
in light rings on the under side, and the majority of the species
have also a marginal line of a dark metallic colour, resembling lead
or bronze.

These butterflies are generally of small size, and frequent
mountains, woods, fields, &c. They are usually local in their
habitats. The genus is represented in Asiatic Russia, the Indian
Archipelago, Australia, New Zealand, and also in California.

1.—**C. Œdipus,** Fab. Mant. 31 (1787); O. i. 1, 315; Godt.
ii. 19, 5.
Geticus, Esp. 102, 2, 107, 5 (1790); Butl. Cat. 39.
Pylarge, Hüb. 245-6, 702-3.
Iphigenus, Herbst, Nat. Schmett. viii. 198.

Expands from 1·25 to 1·50 in. All the wings uniform dark
brown above. Hind wings with a submarginal row of indistinct

black spots. Under side lighter brown; fore wings with one or two small ocellated spots near the hind margin, and with a marginal leaden line. Hind wings with a submarginal row of black spots surrounded by yellow rings, and with silvery centres; these spots approximate closely to one another, and are four or five in number, decreasing in size from the anal angle towards the costa; on the costa, and detached from the submarginal row, is another spot, similar to the largest of the others; there is a marginal leaden line. Pl. LXXIV., 6.

TIMES OF APPEARANCE.—June and July.

HABITAT.—Moist woods and meadows in the South of France, Piedmont, Northern Switzerland, South-East Germany, Hungary, South-East Russia, the Altai, and the Amur. A very local species.

LARVA.—Green, with a dark green dorsal stripe, and light yellow lateral lines. On *Carex* in July and August.

VARIETY.

Miris, Fabr. Ent. Syst. Suppl. p. 429 (1798).—Has the ocellated spots larger than in the type, especially on the under side of the fore wings.

2.—C. Hero, Linn. Faun. Succ. 274; Esp. 22, 4; Hüb. 253-3, 349-50; O. i. 1, 313.

Expands from 1·10 to 1·15 in. Wings dull brown. The fore wings in the male are usually uniform brown. Hind wings with a submarginal row of orange spots, four in number, and with black centres. Under side: Fore wings brown; hind margin reddish, with a row of small black spots and a leaden line. Hind wings brown, with a submarginal row of black spots, five or six in number, surrounded by orange rings and with silvery white centres; internal to these is a white line, and external to them a leaden one. The female is lighter and more strongly marked than the male, and has the black spots surrounded by orange rings on the fore wings.

TIMES OF APPEARANCE.—May and June. Pl. LXXV., 1.

HABITAT.—Moist woods throughout Central Europe, excepting Britain; also in Scandinavia and Livonia.

LARVA. — Unknown. Girard says it is probably green, and ought to be found concealed at the foot of the thick tufts of grasses which abound in the moist and woody places where the imago flies.

3.—**C. Leander**, Esp. 89, 5 (1784); O. i. 309; Boisd. Ic. 45, 7, 8; Dup. i. 33, 5, 7; Frr. B. 110, 1; H. S. 184-5. CLITE, Hüb. 526-7, 747-8.

Expands from 1·0 to 1·33 in. Male dark smoky brown; marginal fringes white. Fore wings with a patch of orange fulvous extending from the base towards the middle of the costa, indistinctly defined externally. Hind wings fulvous along the inner margin, and with two or three small submarginal orange rings. Female: Fore wings light fulvous, smoky brown along the hind margins. Hind wings smoky brown, but lighter than in the male; hind margin fulvous, with a row of black dots. Under side: Fore wings light fulvous, inclining to orange in the male, with a white-centred black spot near the apex; in the female a smaller black spot below it; a narrow white marginal line, bounded internally with a leaden streak and externally by a dark brown one, extends along the entire length of the hind margin. Hind wings orange fulvous, greenish grey towards the base; there is a submarginal row of six black spots with white centres, and a marginal leaden line. Pl. LXXV., 2.

TIME OF APPEARANCE.—June and July.

HABITAT.—Woods in South-Eastern Europe, Asia Minor, and Armenia.

LARVA.—Unknown.

4.—**C. Iphis**, Schiff. S. V. p. 321; Hüb. 249-51; O. i. 1, 310; Godt. ii. 20, 3, 4; Frr. 355. AMYNTAS, Butl. Cat. 40. HERO, Fab. Ent. Syst. 222; TIPHON, Esp. 85, 3, 4.

Expands from 1·0 to 1·25 in. Sexes nearly similar. Fore wings brown, tinged with orange or fulvous, darker along the hind margins, but with a hind-marginal narrow fulvous line. Hind

2 R

wings smoky brown, with a narrow orange hind-marginal line.
Under side: Fore wings as above, but with two small white spots
with black centres near the apex; internal to these is a white
streak. Hind wings brown, with a submarginal row of six black
spots with silvery white centres, and enclosed in light yellow rings;
internal to these are two white patches, forming as it were the
rudiments of the white band seen in the next species; the hind
margin has an orange line, bounded internally by a very narrow
leaden one. Pl. LXXV., 3.

TIME OF APPEARANCE.—June and July.

HABITAT.—Open places in woods in Southern and Eastern
Europe, Asia Minor, and Armenia.

LARVA.—Much resembles that of *C. Hero*, but the lateral line
is white, and the anal segment reddish brown. On grasses in May.

VARIETY.

a. **Iphioides**, Stgr. Berl. E. Z. 1870, p. 101.—The hind wings
have larger ocellated spots beneath, and the white spots are nearly
wanting.

HABITAT.—Central Spain (Castile).

5.—**C. Arcania**, Linn. Faun. Suec. 273 (1761); Hüb. 240-2;
    O. i. 1, 317.
    ARCANIUS, Linn. Syst. Nat. xii. 791 (1767); Esp. 21, 4;
    Godt. i. 8, 2.

Expands from 1·10 to 1·25 in. The sexes are similar. All
the wings blackish brown. The fore wings have the discal area
broadly fulvous. The hind wings have a similarly coloured patch
near the anal angle. Under side: Fore wings fulvous, with a
leaden marginal line, and a small black apical spot surrounded by
a yellow ring. Hind wings pale brown, with a white central band,
broadest at its central portion; external to this is a row of from
three to six black spots in yellow rings, of which three are larger
than the rest, especially that near the costa; hind margin reddish
brown, with a leaden or silvery line. Pl. LXXV., 4.

TIME OF APPEARANCE.—Beginning of June till July.

HABITAT.—Woods in Central and Southern Europe and

Scandinavia, but not in the British Islands, nor in the south of the Spanish peninsula; it is absent also from the south-east of France, but occurs in Asia Minor and Armenia.

LARVA.—Smooth, bright green, with a blackish green dorsal line, and with four lateral lines of pale yellow bordered with a deeper shade of the same colour; the head, legs and caudal points are green. It feeds on different grasses, especially on *Melica ciliata*. In the middle of May it changes to a reddish brown pupa, without angles or tubercles, and rather rounded in shape. It is suspended.

VARIETY.

a. **Darwiniana**, Stgr. Cat. 1871, p. 32. *Arcanues* var., Hüb. 186-7, vi. p. 169.—Somewhat smaller than the type, and has the white band on the under side of the hind wings narrower, and of the same width throughout. Pl. LXXV., 5.

HABITAT.—Alpine regions at low elevations in Switzerland, North Italy, and France. A transitional form between the present species and the next.

6.—**C. Satyrion**, Esp. 122, 2; O. i. 1, 322; Frr. 367, 1, 2; H. S. 289-90.

PHILEA, Hüb. 254-5 (nomen *vetustius* sed *Philea, L. alius* erat *papilio*).

Expands from 0·80 to 1·15 in. Much smaller than the last. The wings above are generally dull brown. The female sometimes is tinged with fulvous towards the hind margins of the fore wings. Under side resembles that of the last, but the ground colour is duller. On the hind wings the white band is broader in proportion, and is of the same width throughout; the ocellated spots are placed in the band, and are well defined; the hind margin is reddish orange, with a narrow leaden line. Pl. LXXV., 6.

TIME OF APPEARANCE.—July and August.

HABITAT.—Elevated mountain meadows in Switzerland, the Tyrol, North Italy, and Central France.

OBS.—Probably an alpine variety of the last, and so described in Staudinger's Catalogue.

LARVA.—Unknown.

7.—**C. Corinna,** Hüb. 536-7 (1880); O. i. 1, 323; Godt. ii. 20, 7, 8; H. S. 285-6.

Norax, Bon. Descr. p. 183; T. ii. 2 (1824); Frr. B. 68, 2.

Expands from 0·80 to 1·15 in. Bright fulvous; hind margins and costa brown. The fore wings have a black spot near the apex. Under side : All the wings with a narrow metallic marginal line. Fore wings fulvous, with a black apical spot, having a white centre, and surrounded by a yellow ring. Hind wings with a narrow white central band, irregular in outline, and not of the same width throughout; there are four submarginal black spots, one of them within the white band; the base is greenish. Pl. LXXV., 7.

Time of Appearance.—June and August.

Habitat.—Dry places in Corsica, Sardinia, and Sicily; and Calabria (Boisd).

Larva.—Green, with a darker dorsal stripe; three lateral lines shading off from dark green to yellow. On *Carex* and on *Triticum cæspitosum* in the spring and summer.

8.—**C. Thyrsis,** Frr. 475; H. S. 297-300 (1846); Stgr. Hor. 1870, 80.

Expands from 1·15 to 1·20 in. All the wings fulvous, with their hind margins dark brown. The fore wings have an apical black spot. The hind wings have three black submarginal spots. Under side : Fore wings as above, but the apical spot has a white centre; internal to it is a black line; the hind margin is brown, with a narrow orange line. Hind wings brownish grey, with a central yellowish white band, a submarginal row of small black spots, and a marginal leaden line. Pl. LXXV., 9.

Time of Appearance.—May and June.

Habitat.—Crete.

Larva.—Unknown.

Obs.—I do not possess a specimen of this species. The figure in this work is copied from H. S. 297, &c.

9.—C. Dorus, Esp. 78, 1; O. i. 1, 320; Godt. ii. 20, 5, 6.
Dorilis, Bkh. i. 93; Hüb. text, p. 42. *Dorion*, Hüb. 247-8.

Expands from 1·10 to 1·25 in. The male has the fore wings dark brown, slightly tinged with fulvous; there is a black spot near the apex surrounded by a light fulvous ring. Hind wings light fulvous, the costal margin dark brown; there is a submarginal row of black spots, having an inward curve. Female similar, but the ground colour of the wings is light fulvous. Under side similar in both sexes. Fore wings fulvous, with a central reddish streak, external to which is a broader light one; near the apex is a black spot, enclosed by a light ring, and with a white centre; the hind margin is yellow, with a silvery line. Hind wings with the basal half brownish grey; across the centre is a narrow white band; external to this a row of fine black spots, with white centres, and surrounded by yellow rings; in the centre of the costa is a larger spot of a similar kind; hind margin yellow, with a leaden or dull silvery line. Pl. LXXVII., 1.

Time of Appearance.—July.

Habitat.—Woods in the South of France, Spain, and Italy. In many localities it seems to replace *C. Arcania*.

Larva.—Unknown.

10.—C. Amaryllis, Cram. Pap. Exot. 391, a. b.; H. S. 188-89; 287-88.
Amarillis, Hbst. 187, 7, 8; Frr. 283, 3, 4; Dup. i. p. 371.

Expands from 1·0 to 1·50 in. All the wings light fulvous. Fore wings with a submarginal row of black spots, very indistinct, except those near the apex and anal angle. Under side: Fore wings light yellowish fulvous, with a submarginal row of black spots, having silvery white centres and yellow rings. Hind wings greenish grey, with a submarginal light orange band, containing a row of spots, similar in character to those of the fore wings; internal to this is a slightly indicated white band. Pl. LXXV., 8.

Time of Appearance.—June and July.

Habitat.—The Ural, the Altai, and the Amur.

Larva.—Unknown.

11.—C. Pamphilus, Linn. Syst. Nat. x. 472; Faun. Suec. 273;
Esp. 21, 3; O. i. 1, 305; Godt. i. 8.
NEPHELE, Hufn. Berl. Mag. ii. p. 78 (1766); Hüb. 237-9;
Bkh. i. 87.
MENACLAS, Poda. Ins. p. 78 (1761).

Expands from 1·10 to 1·25 in. Orange-brown or tawny; hind
margins dark brown, fringes grey. The fore wings have an indistinct
black spot near the apex. Under side : Fore wings tawny, apex and
hind margin grey; near the apex is a black spot, with a white
centre and a yellow ring; internal to this is a light streak. Hind
wings grey, the basal portion.greenish; there is a central whitish
band, and a submarginal row of white dots, sometimes surrounded
by indistinct brown rings. Pl. LXXVII., 5.

TIME OF APPEARANCE.—May to September.

HABITAT.—The entire region of the European fauna, except
the polar regions and the Amur (conf. var. *Lyllus*). It is one of the
commonest butterflies in meadows, roadsides, heaths, &c.; and
extends to some elevation in the mountains. Very common in
Britain (the Small Heath).

LARVA.—Bright green, with a dark green dorsal, and on each
side a lateral, stripe of the same colour, edged with white; head
and ventral surface yellower than the rest; the anal points and
stigmata reddish brown. Feeds on various grasses, especially
Cynosurus cristatus.

PUPA.—Sometimes uniform green, sometimes with three
black lines on the wing-cases, anal point reddish; rounded, and
without angular projections.

VARIETY.

Lyllus, Esp. 122, 1; O. i. 1, 307; Godt. ii. 20, 9, 10; Frr.
499 1; H. S. 430, *Pamphila,* Hüb. 557-8.—Generally somewhat
larger than the type, and brighter in colour; the hind margins have
a narrow black line on all the wings; the fore wings have the apical
spot more distinct than in the type. Under side : Fore wings light
orange, with a central reddish streak descending from the costa
about two-thirds across the wing; apical black spot well defined;

hind margin and apex yellowish grey. Specimens often have an antemarginal leaden line. Hind wings yellowish grey, the basal portion darker than the rest; there is a central lighter streak, and sometimes a few submarginal light dots. Pl. LXXVII., 6.

TIME OF APPEARANCE.—April to August.

HABITAT.—The South of Europe, North Africa, Asia Minor, &c. In many places entirely replacing the type; in some localities existing as a seasonal dimorphic form, as in some parts of South-Eastern France.

12.—**C. Tiphon**, Rott. Naturf. vi. p. 15 (1775); Haw. Lep. Brit. p. 16.

DAVUS, Fab. Gen. 259 (1777); Mant. 28-9; Hbst. 186, 3-6; O. i. 1, 302; Godt. ii. 21, 1, 2.

TULLIA, Hüb. 243-4.

Expands from 1·25 to 1·60 in. All the wings pale fulvous or brownish. Fore wings with an apical black spot in a light ring. Hind wings with two or three similar spots near the anal angle; fringes light brown or whitish. Female with an indistinct lightish streak on the fore wings. Under side: Fore wings dull fulvous; the costa and hind margin greenish grey; from one to three submarginal black spots, with white centres and whitish rings; within this row of spots is a whitish streak. Hind wings dull greenish grey, lighter towards the hind margin, with a central whitish streak beginning at the costa, and a submarginal row of small black spots in white rings, and generally with white centres; internal to these is a narrow broken white band. Pl. LXXVII., 2.

The above is a description of the markings and coloration of the wings in the typical form of this variable species.

TIME OF APPEARANCE.—July.

HABITAT.—Moist mountain meadows throughout Northern and Central Europe, including the British Islands (the Marsh Ringlet). It is local, but common where it occurs.

LARVA.—" The colour of the head and body is apple-green, inclining to olive-green; the head is dull, opaque, and uniformly coloured; the body is striped. There is a narrow medio-dorsal

stripe dark purple-green, bordered on each side by a still narrower
yellow stripe; these three stripes terminate before the anal points;
on each side are two pale yellow stripes, one above and the other
below the pale brown spiracles; the upper of these is bordered
above and below with dark bluish green, and terminates in the
anal flap, which is tinged with pink; the lower terminates before
the anal flap; the ventral surface, legs and claspers are bottle-
green."

PUPA.—" Thorax dorsally convex, with a slight median keel;...
the colour of the wing-cases is pale brown;... the head, thorax,
and abdomen are apple-green, sprinkled with dingy whitish green."
—*Newman, Brit. Butt.*, p. 100. The larva feeds on *Rhyncospora alba*
from June to September, and then hybernates, pupating in the
following May.

VARIETIES.

a. **Laïdion**, Bkh., i. 91.—A form occurring as a variety in
Ireland and Scotland. It has the submarginal spots of the under
side either very small or altogether absent.

b. **Philoxenus**, Esp. 78, 3. *Rothliebii*, Stgr. Cat. i. p. 14;
Newman, Brit. Butt. p. 98, cum figuris.—Differs from the type
principally in having a second spot on the fore wings above, near
the anal angle. The hind wings have a distinct row of five
black spots in light rings. Under side: Fore wings with two or
three submarginal black spots with white centres, and internal to
them a broad white streak. Hind wings brownish grey, with a
central white band, and a submarginal row of six or seven well-
defined black spots with white centres, and surrounded by light
rings, sometimes very large and distinct; hind margins light grey.

TIMES OF APPEARANCE.—June and July, and occasionally later.

HABITAT.—This form seems to be peculiar to the North of
England, principally in Durham, Cumberland, and Yorkshire,
frequenting mosses. Staudinger mentions Southern Holstein as a
locality, but it is extremely probable that it is peculiar to Northern
England. It is probably a distinct species, and is thus given by
Newman, *loc. cit.*, whose description is accompanied by a series of
excellent figures, which agree in every respect with specimens
contained in good British collections. Pl. LXXVII., 3.

c. **Isis,** Thnb. Diss. Ins. ii. p. 31 (1791); Zett. Ins. Lapp. p. 905.
Demophile, Frr. 439, 3, 4 (1844).—Smaller than the type, and
lighter in colour; the ocellated spots either very small or
altogether wanting. Judging from the specimens sent to me
by Dr. Staudinger, it very closely resembles the British variety
Laïdion. Pl. LXXVII., 4.

TIME OF APPEARANCE.—June and July.

HABITAT.—Lapland and Sweden; also the Amur (Ménetriès,
Schrencks' Reisen, p. 34).

OBS.—It should be mentioned that Wallengren's descriptions
of the varieties occurring in Scandinavia seem to be very close to
all the forms of the species described above, except the var.
Philoxenus, Esp.

Genus 8.—**TRIPHYSA,** Zell. Stett. Ent. Zeit. 1850, p. 308.
PHRYNE, H. S. i. p. 90 (1845, sed nomen præocc).

Earlier stages unknown. Small butterflies, having the general
aspect of the last genus. The wings are rounded, and have the hind
margins entire; they are somewhat narrower in the female than in
the male. The sexes differ remarkably in the colouring and in the
neuration of the wings: in the male the colour is dark brown,
whilst that of the female is greenish or greyish white; the male
has only the costal and median nervures dilated at the base, in the
female the submedian has likewise the basal dilatation. The
discoidal cells are closed. The shafts of the antennæ are proportionately shorter than in *Cænonympha;* the clubs are ovate and
flattened. The neuration is marked with silvery white, especially
on the hind wings. All the wings have, beneath, a submarginal
row of black spots, with silvery centres, and with marginal leaden
lines, like those in *Cænonympha.*

This genus belongs rightly to the Siberian and Manchurian
subregions of the Palæarctic fauna. The occurrence of *T. Phryne,*
Pall., in South-East Russia, at Sarepta, and elsewhere, entitles it
to stand as a genus of European butterflies, though it is as truly
Asiatic as *Charaxes* is African. Only one species is known besides
that found in Europe, *T. Nervosa,* Motsch., inhabiting Japan.

(*Sunbecca*, Ev., evidently belongs to the last genus.) It is remarkable that the same white veining of the under side of the hind wings is found in several species of *Satyridæ* inhabiting South-Eastern Russia, &c.: for example, *Œneis Tarpeia*, *Satyrus Autonoë*, *Erebia Afra*, and others: the last especially shows an analogy to *T. Phryne* in this respect.

1.—T. **Phryne**, Pall. Reis. p. 470 (1771); Esp. 89, 3, 4; Hüb. 200-1; O. i. 1, 256; B. Ic. 45, 4-6; Dup. i. 33, 1-4; H. S. 106-7.
    Tircis, Cr. Pap. Exot. T. 373 (1782); Herbst, 183, 7, 8.
    Phryneus, Fab. Mant. ii. 33 (1787); Godt. Ency. Méth. ix. 519.

Expands from 1·0 to 1·25 in. The male is dark brown, with the hind margins of all the wings, as well as the inner margin of the hind wings, whitish. The female is whitish, with the markings of the under side slightly visible above. The under side is nearly similar in both sexes, being brown, with the neuration marked with white. All the wings have a whitish submarginal band, containing a row of round black spots with silvery white centres. The female is somewhat lighter in colour than the male, and has a whitish streak in the discoidal area of all the wings. Pl. LXXVII., 7.
    Time of Appearance.—May and June.
    Habitat.—Meadows in South-Eastern Russia, the Altai, and Eastern Siberia.
    Larva, &c.—Unknown.

ABERRATION.

Dohrnii, Zell. Stett. E. Z. 1850, p. 308; H. S. 641-2.—Has the hind margins whitish in the male.
    Habitat.—Same as the type.

OTHER SPECIES OF *SATYRIDÆ* INCLUDED IN STAUDINGER'S CATALOGUE, &c., AS BELONGING TO THE "EUROPEAN FAUNA."

Genus *MELANARGIA.*

M. Titea, Klüg. Symb. T. 29, 15-18, 1832 ; Ld. Z. B. v. 1855. *Darceti*, Dup. i. 26, 5, 6 (1833).—Expands from 1·75 to 2·30 in. Close to *Lachesis*, but the black markings are more distinct, forming a broad black border, with white spots. All the wings are black at the base. The markings of the under side of the hind wings differ from *Lachesis* in having the central band white, edged with black, and the submarginal row of spots very distinct, and enclosed in light yellow rings. Habitat, Syria.

Var. *Teneates*, Mén. Cat. Rais. p. 252 (1832); H. S. 423-4.— Scarcely differs from the type ; slightly paler in colouring, the black markings being smaller. Habitat, Armenia, Persia, and Asia Minor.

M. Epimede, Stgr. MS. Cat. 1883.—Differs from *Lachesis* in having the neuration of all the wings dark brown or black. The inner margin of the fore wings is dusky thoughout its entire length. The hind wings have the base altogether white, excepting the nervures. Under side white, with the dark markings very distinct. Habitat, the Amur and Japan. Probably a small form of *Halimede*.

M. Larissa var. *Astanda*, Nordm.—Larger and darker than the type. Habitat, Armenia.

M. Halimede, Mén. Schrk. p. 37; T. iii. 6, 7.—Larger than *Epimede*, expanding from 2·20 to 2·30 in. All the markings are darker and broader, especially along the hind margins and the inner margins of the fore wings ; the submarginal ocellated spots on the underside of the hind wings are also very large, whereas in *Epimede* they are very indistinct. I have received a pair of this species from Dr. Staudinger, and am inclined to think his *M. Epimede* is the original *Halimede* of Ménetriès. Habitat, the Amur.

M. Mauritanica, Stgr. MS. Cat. 1883.—A pair of this species has been received from Dr. Staudinger. It greatly resembles

M. Larissa var. *Hertha*, but is to be distinguished at once from it in having the inner margins of the fore wings broadly black; the dark markings are altogether more intense. Habitat, Algeria.

M. Hylata, Mén. Cat. Rais. p. 251; H. S. 425, 6.—Expands from 1·80 to 2·0 in. Creamy white, with a submarginal dark grey zigzag band on all the wings. Fore wings with a grey blotch between this and the base, a blackish line in the discoidal cell, and a black spot near the apex; base dusky. Hind wings with a submarginal row of white-centred spots in yellowish rings, and a narrow greyish band near the base. Under side whitish, like *M. Galatea* var. *Leucomelas*, with the markings of the upper side faintly visible, except a black spot near the anal angle of the fore wings. Habitat, South-Eastern Armenia and Persia.

M. Japygia var. *Caucasica*, Nord. Bull. M. 1851, ii.; T. viii. 3, 6. *Xenia*, Frr. 566, 3, 4.—Larger than the type, and somewhat paler. Habitat, Armenia.

Genus *EREBIA*.

E. Kefersteinii, Ev. Bull. Mosc. 1851, ii. 610; H. S. 617-18 (1852).—Expands 1·50 in. Fore wings, hind margin and costa dark brown, basal portion reddish brown; there is a broad submarginal fulvous band, with four black spots, those nearest the apex being larger than the rest. Hind wings dark brown, with a submarginal row of four or five fulvous spots with black centres. Under side: Fore wings nearly as above, but lighter. Hind wings dark brown, with a submarginal grey band, on which are four black spots in orange rings. Habitat, Eastern Siberia and the Altai.

E. Maurisius, Esp. 113, 4, 5; Stgr. Cat. 1871, p. 24.—Expands 1·30 in. Wings dark brown. Fore wings with an even submarginal band of fulvous, divided into six spots by the nervules; discal portion of the wings fulvous. Hind wings with a submarginal row of five fulvous spots. Under side: Fore wings fulvous, bordered with dark brown. Hind wings dark brown, with a submarginal row of seven light fulvous spots, and a similar basal spot. Habitat, Siberia.

E. Pawlowskyi, Mén. En. iii. p. 145.—"Alis rotundatis, supra fusco-nigris, subtus nigro-ferrugineis, prope marginem exteriorem maculis (anticis 7 quadratis æqualibus, posticis 5 oblongis) flavo-fulvis." Habitat, Eastern Siberia. (Stgr. Cat. p. 24.) Taken first by Pawlowsky near the River Sibagli, in the government of Jakoutsk, at the end of June. Mén. Bull. Acad. Petr. xvii. p. 498.

E. Theano, Tauscher, Mém. Mosc. 1809 ; T. 13, 1 ; Ev. Bull. Mosc. 1851, ii. 611. *Stubbendorfii*, Mén. Bull. Acad. Sc. V. —Expands from 1·20 to 1·35 in. Dark brown. Fore wings with a submarginal band of light fulvous, separated into distinct spots by the neuration ; near the base is a large fulvous spot. Hind wings similarly marked, but the basal spot is smaller. Under side : Fore wings as above, but paler. Hind wings brown, submarginal spots white, and at the base are four white spots of a similar character. Habitat, the Altai.

E. Turanica, Erschoff.—Expands from 1·30 to 1·50 in. Brown, with a submarginal row of light fulvous spots on all the wings ; that near the apex on the fore wings is larger than the rest ; the spots on the hind wings are of uniform size. Under side similar, but the hind wings have a row of conspicuous white spots internal to the fulvous ones. Habitat, Turkestan.

E. Melas var. *Hewitsonii*, Ld. Wien. Mts. 1864, p. 167 ; T. 3, 6, 7.—The wings have a reddish submarginal band, and the ocellated spots are larger and more numerous than in the type. Habitat, Armenia.

E. Hades, Stgr.—Expands from 2·0 to 2·50 in. Purplish black ; marginal fringes black and white. Fore wings with a single bluish white apical spot in the male ; in the female two. Hind wings black, without spots. Under side : Male—Fore wings blackish brown ; apical spots surrounded by a black ring. Hind wings blackish brown, with a submarginal row of white dots. Female—Fore wings brown, hind margin and apex grey ; the apical spots surrounded by a yellow ring. Hind wings grey, with a submarginal row of white spots. Habitat, Turkestan. This beautiful species is described from the specimens in Mr. Godman's collection, received from Dr. Staudinger.

E. Parmenio, Boeber, Nouv. Mém. Mosc. ii. p. 306, pl. 19. H. S., 421-2, 464-6.—Expands from 2·0 to 2·20 in. Dark brown. Fore wings with a large black spot near the apex, containing two smaller white ones, and enclosed in a fulvous ring ; below this are two smaller black spots. Hind wings with a submarginal row of black spots with white centres, and enclosed in fulvous rings. Under side: Fore wings fulvous, with the spots as above; the neuration marked with white. The hind wings have broad white markings, following the course of the neuration. Habitat, Eastern Siberia and the Amur.

E. Oenus, Ev. Bull. Mosc. 1843, iii. 538; T. 8, 5; H. S. 291-2. —Somewhat resembles *Lappona*, but smaller. The fore wings more suffused with reddish fulvous. Under side: Fore wings reddish fulvous; the costa and hind margins dark brown. Hind wings brown, darker at the base, with very faint traces of a submarginal band. Habitat, the Ural. Probably a local form of *E. Lappona*.

E. Melancholica, H. S. 276-9, vi. p. 10. Perhaps intermediate between *Pronoë* and *Neoridas*.—Expands 1·20 to 1·50 in. Dark brown. Fore wings with a broad submarginal fulvous band, on which are three or four black spots with white centres, the two nearest the apex coalescing. Hind wings with a submarginal row of three black spots with white centres. Under side: Male almost as above. Female—Fore wings reddish fulvous; the costa and hind margin dark grey; submarginal band yellowish. Hind wings grey at the base and hind margins, with a central broad dark brown band, external to which is a greyish white one of equal width. Habitat, Armenia and Asia Minor.

E. Sedakovii, Ev. Bull. Mosc. 1847, iii. 70; T. i. 5, 6; H. S. 591-2, vi. p. 11.—Expands 1·50 in. Dark brown. Fore wings with a yellowish submarginal band, containing three black spots with greyish blue centres. Hind wings with a submarginal band of reddish fulvous, narrower than that of the fore wings, and containing three black spots of equal size, and with white centres. Under side: Fore wings as above. Hind wings with a very indistinct submarginal whitish band, containing four black spots with white centres. Habitat, Eastern Siberia and the Amur.

E. Æthiops var. *Melusina*, H. S. 373-4, vi. p. 10.—Larger than the type, and with the submarginal bands very broad and reddish fulvous. Habitat, Armenia.

E. Ajanensis, Mén. En. ii. p. 104. *Eunomia*, Mén. Bull. Acad. xvii. p. 216; Schencks' Reisen, p. 34.—" Alis fusco-bruneis, fascia externa maris feruginea, fœminæ ochracea, singulis ocellis quatuor rarò albo pupillatis; subtus pallidioribus, posticis fascia medea sinuato dentata, extus omnino niveo-marginata." Close to *Ligea*, but smaller; the hind wings, beneath, have a white submarginal band, and three white basal streaks. Habitat, South-Eastern Siberia. July and August.

E. Maracandica, Erch. Lep. Turkest. p. 17.—" Fusca, supra alarum anteriorum plaga magna postica, ocello nigro albo-pupillato notata, posteriorumque plaga postica fulvis; subtus omnibus fuscis, anterioribus ex basi fulvescentibus posterioribus serie transversa punctorum alborum postica insignibus. Exp. 43-48 mm. Habitat, Circa Maracanda in monte Tschupanata, volat mense junio."

E. Cyclopius, Ev. Bull. Mosc. 1844, iii. 590; H. S. 607-8.— Expands from 2·25 to 2·50 in. All the wings are dark brown, the marginal fringes whitish and brown. The fore wings have a large black apical spot, containing two white ones, and enclosed in an orange ring. Under side: Fore wings with the apical spot very conspicuous, and surrounded by a broad yellow ring. Hind wings brownish purple, with two bands of grey. Habitat, the Amur.

E. Tristis, Brem. Bull. Acad. 1861. *Wanga*, Brem. Lep. O. S. p. 20; T. ii. 1.—Expands from 2·50 to 2·70 in. Smoky brown. Female lighter than the male. Wings without any markings, except a black spot near the apex of the fore wings enclosing two white ones, and surrounded by a very pale yellow ring. Under side almost as above; but the hind wings speckled with greyish white. Habitat, the Amur.

E. Discoidalis, Kirby, Richards, Faun. Bor. Amer. p. 298.— Habitat, Eastern Siberia. This species belongs to the Nearctic fauna.

E. Ero, Brem. Lep. O. S. p. 20; T. ii. 2.—" Alæ supra fusco brunneæ; antica, serie submarginali e punctis parvis nigris rufo-cincti composita. Alæ anticæ subtus brunneæ fusco marginatæ ocello duplici apicali; posticæ albido-adspersæ, fascia discoidali obsoletissima, puncto medio et punctis parvis submarginalibus quatuor albis." Very close to *E. Disa*, but smaller and lighter; the fore wings are reddish beneath. Habitat, the Amur, Apfel mountains. July.

E. Edda, Mén. Insect. in Middenhorfs Reisen. in Siberien, p. 58; T. iii. 112.—Habitat, North-Eastern Siberia. (Stgr. Cat. p. 27.)

E. Calmucca, Stgr.—Expands from 1·50 to 1·65 in. Brown, with a slight reddish tinge. Costa and hind margin of all the wings silvery grey (especially in the variety *albo-notata*, which has also the disco-cellular nervule of the fore wings marked with white). Under side: Fore wings chocolate-brown, apex grey. Hind wings silvery grey. Habitat, Kuldja.

E. Radians, Stgr.—Expands 1·25 in. Wings broad antero-posteriorly. Brown, with submarginal reddish fulvous streaks on all the wings radiating towards the hind margin. Under side: Fore wings reddish fulvous, costa and hind margins grey. Hind wings uniform grey. Habitat, Kuldja.

E. Tianchanica, Stgr.—Expands from 1·30 to 1·50 in. Close to *E. Radians*, but has the wings more rounded in outline; the radiating streaks extend over the whole area of the wings, instead of being merely submarginal. Under side: Fore wings as above. Hind wings with a faint submarginal grey band. Habitat, Kuldja.

The above three species are described from specimens sent by Dr. Staudinger. They are not included in the Catalogue of 1871.

Genus *ŒNEIS*.

Œ. Sculda, Ev. Bull. Mosc. 1851, ii. 612; H. S. 613-14.—Expands 1·75 in. Pale fulvous. Fore wings with two black spots, the costa whitish, and the neuration dark brown. Hind wings with a submarginal row of three black spots, that near the anal angle having a whitish centre; there are two narrow central dark parallel bands. Under side: Fore wings as above, but paler, the spots distinct and with white centres, some whitish marks near the apex. Hind wings whitish, spotted with dark brown, with a submarginal row of white-centred spots, and a dark central band. Habitat, Eastern Siberia.

Œ. Nanna, Ménétr., Schrencks' Reisen, 38; T. iii. 5.—This species is nearly of the same shape and size as *Œ. Tarpeia*, Esp. Above it much resembles it in pattern; the base of the wings is dark brown, the remaining portion being dark fulvous; the black spots are smaller and narrower than in *Œ. Tarpeia.* The hind wings beneath are greyish white, with a darker central band; but the neuration is not distinctly white, as it is in *Tarpeia.* Habitat, Amur.

Œ. Urda, Ev. Bull. Mosc. 1847, iii. 69; T. ii. 1-4; H. S. 461-3. —Expands from 1·85 to 2·10 in. Wings subdiaphanous, dull pale brown, the neuration distinctly dark brown, especially in the male. Fore wings with the hind margins dark brown, with two black spots, the apical one white-centred in the male, and both in the female. Hind wings with a submarginal row of black dots in the female. Bases of all the wings dark brown. Under side: Fore wings as above, but paler, and the neuration not broadly marked with brown, apex greyish. Hind wings brown, with a central dark wavy band broadly edged with white. Habitat, Eastern Siberia.

Œ. Crambis, Frr. 440, 3, 4; *Taygete,* H. S. 112, 5; *Also,* Möschl. Wien. Mts. 1860, 342. *Œno,* Scudder, Rev. Nr. 5.— Expands from 1·60 to 1·75 in. Pale brown; the hind marginal portions lighter. The male has some black-centred scales on the fore wings, and a submarginal row of white dots on the hind wings; the female has the fore wings pale brown, and the submarginal dots are absent from the hind wings. Under side: Something like *Bore,* but with a distinct dark central band, edged with white on both sides. Habitat, North-Eastern Siberia and Labrador.

Œ. Semidea, Say. Am. Ent. pl. 30. *Œno,* Boisd. Ic. 39, 4, 6; Dup. 1, 49, 1-3; H. S. 59, 60. *Also,* Boisd. Ic. 40, 1, 2; H. S. 381. *Assimilis,* Butl. Cat. p. 163.—Expands 1·70 in. Pale brown. Fore wings with the basal portion darker than the rest, costa speckled with dark brown and white. Hind wings speckled with dark brown, with a lighter submarginal band without spots, marginal fringes dark brown and white. Under side: Fore wings almost as above. Hind wings speckled with dark brown and whitish, with a central narrow dark band. Habitat, Labrador.

Œ. Mulla, Stgr. Stett. Ent. Zeit. xlii. p. 270.—A species allied to *Œ. Jutta.* Taken by Dr. Staudinger, at Tarbagatai, at the end of June, 1877.

Genus SATYRUS.

S. Hermione var. *Syriaca*, Stgr. Cat. p. 27.—Differs from the type in having the under side of the hind wings less varied in colour, and speckled with whitish. Habitat, Syria, Cyprus.

S. Briseis var. *Pricuri*, Pier. Ann. S. Fr. 1837, p. 304; pl. 12, 6; Butl. Cat. 55.—Intermediate between *Briseis* and the next-mentioned species. Has the band on the fore wings more broken than in the type, and that of the hind wings indented. Habitat, North Africa.

S. Heydenreichi, Ld. Z. B. v. 1853, p. 9; T. i. 2.—Expands from 1·70 to 2·30 in. Close to *Briseis*, but smaller; the ground colour is darker, and the submarginal white bands narrower and more broken. The fore wings have a large discoidal white blotch commencing at the base; the white costa is speckled with dark brown. The hind wings have the white band deeply indented towards the inner margins, and veined by an oval blackish patch towards the costa. Under side: The hind wings are much more varied than in *Briseis*, having large dark brown blotches distinctly marked. Habitat, the Altai.

S. Autonoë var. *Sibirica*, Stgr. Cat. p. 12.—Lighter in colour than the type, and with the bands whiter. Habitat, E. Siberia and the Amur.

S. Semele var. *Mersina*, Stgr. Cat. p. 28.—Has the under side of the hind wings almost uniform grey. Habitat, Asia Minor and Cyprus.

S. Bischoffii, H. S. 307-10, vi. p. 12.—Expands from 2·0 to 2·25 in. Fore wings dark brown, with two black spots and six white ones, more or less ovoid in shape. Hind wings orange; hind margins dentate, and narrowly dark brown. Under side: Fore wings yellowish white, with black spots and streaks; apices brownish. Hind wings light brown, slightly darker towards the base and hind margins. The female differs from the male only in being somewhat larger. Habitat, Amasia.

S. Pelopea, Klug, Symb. Phys. iii., T. 29. 5-8 (1832); Stgr. Hor. 1870, p. 70.—Expands from 1·80 to 2·0 in. Dark brown, with light fulvous submarginal bands, broader and most distinct

on the hind wings. The fore wings have two round black spots
placed within the fulvous band. The hind wings have a small
black spot near the anal angle. Under side : Fore wings yellowish,
with black spots and reddish brown streaks. Hind wings grey,
with a broad submarginal white band shading off, along its hind
marginal edge, into purplish brown. Habitat, Syria and Persia.

a. Var. *Mniszechii*, H. S. 577-9; Frr. 617, 2; Ld. Wien. Mts.
1858, 138.—Expands from 2·40 to 2·50 in. Dark brown; all the
wings with broad orange submarginal bands. Fore wings with two
large black spots with white centres, and between them two white
dots. Hind wings with two white dots near the anal angles, the
innermost surrounded by a black ring in the male. Under side :
Hind wings reddish grey. Habitat, Asia Minor.

b. Var. *Caucasica*, Led. Wien. Mts. 1864, p. 168.—Expands from
2·50 to 2·75 in.—Larger than the last. The hind wings are dusky
grey beneath. Habitat, Armenia.

c. Var. *Mamurra*, H. S. 314-15; Frr. 617. *Pelopea*, H. S. 575-6.
—Smaller and darker than the type. The female very much
resembles that of *S. Græca*, but the under side of the hind wings
is reddish grey. Habitat, Asia Minor (mountains).

S. Telephassa, Hüb. Latr. ix. 3; Klug. Symb. Phys. iii. 29,
1-4; H. S. 305-6.—Expands from 2·0 to 2·25 in. Very like
Pelopea var. *Mniszechii*, above; but has the wings shaped as in
S. Amalthea. The male has a black central patch of scales on
the fore wings, and the female has the orange band of the hind
wings less distinct than in *Pelopea*, and with two white dots near
the anal angle. Habitat, Syria.

Var. *Anthelea*, Hüb. 861-2; B. Ic. 41, 1-4; Frr. 265, 2, 3;
Dup. i. 27, 5, 6; H. S. 178-9.—Resembles *S. Amalthea*, but the
male alone has the hind marginal bands white; in the female they
are fulvous. Habitat, Asia Minor.

S. Beroë, Frr. 415, 1, 2; H. S. 108-11.—Expands from 1·80
to 2·0 in. Brown, with lighter submarginal bands. Fore wings
with two large black spots, having white centres. Hind wings
without submarginal spots. Under side yellowish white, with the
spots as above, and with blackish streaks. Hind wings whitish
grey, with a darker central band; the base and hind margins dark
grey. Habitat, mountains in Asia Minor.

a. Var. *Rhena,* H. S. 571-3.—Male darker and more uniformly coloured than the type. Female with the submarginal bands more distinct and yellowish ; the fore wings with very large black spots. Habitat, Amasia.

b. Var. *Aurantiaca,* Stgr. Cat. 1871, p. 29.—Has the bands on all the wings orange. Habitat, Persia and Armenia.

S. Geyeri, H. S. 301-2 ; Frr. 611, 4.—Expands from 1·50 to 1·75 in. The male has the wings light brown, with lighter submarginal bands, that on the fore wings containing two black spots. Hind wings with a small black spot near the anal angle. Female dusky grey, with whitish submarginal band edged by strongly marked zigzag black lines. Under side : Fore wings yellowish white, with brown or blackish lines and spots. Hind wings brown, with white submarginal band; neuration white. Fringes of all the wings white. Habitat, Armenia and Amasia.

S. Statilinus var. *Sichæa,* Led. Wien. Mts. 1857 ; Stgr. Cat. 1871, p. 29.—Closely resembles the var. *Fatua,* but is more varied, the black spots on the fore wings being surrounded by yellow rings ; the under side is also more brightly and strongly marked. Habitat, Syria.

S. Pisidice, Klug, Symb. Phys. T. 29, 9, 10.—Expands from 2·25 to 2·50 in. Somewhat resembles *Allionia* in colouring, but the black spots on the fore wings are smaller, and are always white-centred. The hind wings have the hind margins deeply indented ; the marginal fringes are pure white. Under side very like that of *Allionia,* but lighter. The hind wings grey, with a central blackish line and a shorter basal stripe, and with a narrow yellow marginal band, as in *Fidia.* Habitat, Syria (Sinai).

S. Parisatis, Koll. Ins. Pers. p. 11. *Macrophthalmus,* Ev. Bull. Mosc. 1857, ii. p. 615.—Expands from 2·0 to 2·25 in. Dark brown. Fore wings with a black spot near the apex, having a white centre ; beneath this two white spots, as in the allied species ; below these a second black spot ; hind margins white along its posterior two-thirds. Hind wings dark brown ;·the hind margin dentate, and broadly white throughout its entire length. Under side light brown, with fine dark streaks. Fore wings with a black spot in a yellow ring, and with a white centre, towards the apex ; the other spots as in *Allionia.* Hind wings with two ocellated spots, similar to those above. Habitat, Persia.

S. Kaufmani, Erschoff. Lepid. Turk. pl. i. 14; Text, p. 19.— "Alis supra fuscis, anteriorum orbiculis nigris albido-cinctis tribus; subtus omnibus griseo-luteis, anteriorum orbiculis nimbo nigrante conjunctis." About the size of *Briseis*. The wings dark brown above. Fore wings with three black spots, enclosed by white rings. Hind wings unicolorous. Under side yellowish grey. The fore wings have the submarginal spots placed on a black band. Habitat, Maracanda in the middle of June.

S. Abdelkader, Pier. Ann. Soc. Fr. 1837, p. 19, pl. i. 5, 6.— Expands from 3·0 to 3·15 in. Dark brown. All the wings with a submarginal row of blue spots. Under side: Hind wings grey, mottled with dark brown; neuration whitish. Habitat, North Africa.

S. Stulta, Stgr.—Size of *Allionia*. Wings dark brown; fringes white. Fore wings with two white spots in the female. Hind wings broadly white in both sexes; in the female with a black spot near the anal angle. Habitat, Samarkand.

S. Dryas var. *Sibirica*, Stgr. Cat. 1871, p. 29.—Has the under side in both sexes uniform brown. Habitat, Eastern Siberia and the Amur.

S. Cordulina, Stgr.—Resembles *Cordula*, but is smaller and duller. Habitat, Samarkand.

S. Actæa var. *Amasina*, Stgr. Cat. p. 13 (1861), p. 29 (1871).— Is smaller than the type, and has the neuration marked with white beneath. Habitat, Amasia.

Var. *Parthica*, Ld. Hor. 1869, p. 83.—Discal portion of the fore wings bright fulvous beneath; the apical spot on the fore wings surrounded by a yellowish ring on the under side; the under side of the hind wings being streaked with white. Habitat, Persia.

(Genus Ypthima, Doub.)

S. Asterope, Klg. Symb. Phys. T. 29, 11-14; Ld. Z. B. v. 1855, T. 1, 6.—Expands from 1 to 1·25 in. Wings dark greyish brown; female somewhat lighter. Fore wings with a large oval black spot near the apex containing two brilliant metallic blue dots, and surrounded by a yellow ring. Hind wings with a spot near the anal angle similar in character but smaller, circular in outline, and with only one blue dot. Under side very similar to the upper side, but lighter, and with the hind wings especially streaked with whitish. Habitat, Syria.

S. *Amphithea*, Mén. Schrk. p. 41, T. iii. 10.—Rather larger than *Asterope*. Wings dusky grey. Fore wings with an obscure black spot near the apex. Hind wings with a similar spot near the anal angle. Under side: Fore wings with a black spot surrounded by a yellow ring, and containing three white dots. Hind wings with three spots of a similar character, but each containing a single white dot. Habitat, Amur.

S. *Baldus*, Fab. E. S. 323; Godt. Ency. Meth. 551.—Expands 1·75 in. Uniformly brown. Fore wings with an apical black spot with two silvery blue pupils. Hind wings with two black spots near the anal angle surrounded by yellow rings, and with silvery blue centres. Under side: Wings indistinctly streaked with greyish white. Fore wings with the apical spot as above, but enclosed in a conspicuous yellow ring. Hind wings with six similar spots varying in size. Habitat, Amur.

S. *Motschulskyi*, Brem. Grey, p. 8; Mén. En. T. vi. 5.— Expands from 1.50 to 1·75 in. All the wings dark fuscous; each with a single black spot, blue-pupilled, and enclosed in a yellow ring. Under side ashy grey, with fine dark streaks. Fore wings with one ocellus, and the hind wings with three; all with blue pupils and yellow rings. Habitat, Amur.

Genus *PARARGE*.

P. Eversmanni, Ev. Bull. Mosc. 1847, iii. T. 2, 5, 6; Erschoff, Lep. Turkest. p. 19.—Expands from 1·60 to 1·70 in. Wings pale fulvous; hind margins irregularly dark brown. Fore wings with a black spot near the centre of the costa. Hind wings with a submarginal row of rounded black spots. Under side: Fore wings pale fulvous, with five dark streaks reaching from the costa; hind margin dark brown, near the centre a black spot, a small apical spot with a white centre. Hind wings grey, blotched and striped with reddish brown; there is a submarginal row of six ocellated spots in yellow rings, bordered on either side by white blotches; fringes brown and white. Habitat, Turkestan, Maracanda.

P. Clymene var. *Roxandra*, H. S. 471-3 (1850); Nord. Bull. Mosc. 1851, ii. 410.—Has the hind wings more varied beneath than in the type, and with a white rather oval-shaped spot between the upper two submarginal spots. Habitat, Armenia.

P. Xiphia, Fab. S. E. 492; H. S. 86, 87.—Expands from 2·15 to 2·25 in.—Resembles *Ægeria* above, but is larger and more distinctly marked, the markings being orange. Under side : Fore wings orange, with a white triangular blotch towards the apex ; hind margin grey and brown. Hind wings warm brown, with a white streak reaching downwards from the costa, a submarginal row of five ocellated spots. Habitat, Madeira.

Var. *Xiphioides*, Stgr. Cat. 1871, p. 30. *Xiphia*, H. S. 84, 85. —Expands from 1·50 to 1·75 in. Smaller than the type, the white streak on the hind wings beneath less distinct or absent. Habitat, the Canaries.

P. Deidamia, Er. Bull. Mosc. 1851, ii. p. 617 ; H. S. 609-10. —Expands 2 in. Dark brown ; marginal fringes brown and white. Fore wings with an apical white-centred black spot in a light yellow ring; internal to this is a faint yellowish streak. Hind wings with a hind marginal row of six ocellated spots, three being more distinct than the rest. Under side: Fore wings with a whitish ring surrounding the apical eye ; internal to it is a white streak. Hind wings with a row of six ocellated spots in light yellow rings. Habitat, Eastern Siberia.

P. Nashreddini, Stgr.—Expands 1·50 in. Brown with a bronze reflection. Fore wings rather pointed towards the apex, with a large apical black spot having a white centre ; in the male a quadrate patch of black scales runs up from the inner margin nearly to the apical spot. Hind wings dark brown, with two submarginal black spots in orange rings, and with white centres. Under side : Fore wings light orange, with light brown and white costal stripes ; a large apical spot as in *Mæra ;* hind margin with a white and a yellow line. Hind wings whitish grey, with yellowish central lines and submarginal ocellated spots as in *Mæra*. Habitat, Amur.

P. Maackii, Brem. Lep. O. S. p. 22, T. iii. 2.—Expands from
2·25 to 2·50 in.   Fore wings dark brown, with an indistinct wavy
light central line.   Hind wings dark brown, with a submarginal
row of four blackish spots in yellow rings, and with white centres.
Under side dark brown, as above.   Fore wings with three or four
small submarginal spots in yellow rings, and with white centres ;
there is a marginal light coloured line.   Hind wings with a row of
six ocellated spots, the first and fifth larger than the rest, the
sixth with two white pupils.   Habitat, Amur.

P. Epimenides, Mén. Schrk. p. 39, T. iii. 8-9.—Expands from
2·20 to 2·50 in.   Wings denticulate, greyish brown.   Fore wings
with some slight markings near the apex, which are pale yellow in
the female, and with a small black spot.   Hind wings with four
submarginal black spots surrounded by yellow rings in the female.
Under side : Fore wings pale yellow, with some white bands near
the apex and in the centre ; the apical spot ocellated and ringed.
Hind wings white in the centre, the basal portion and hind margin
yellowish grey.   There are seven black spots, ocellated and with
yellow rings, more distinct in the female than in the male ; the
hind margin has three brown lines.   Habitat, Amur ; Mount
Bureja in July, Ssoungari in August (Mén.)

P. Epaminondas, Stgr. M. S. Cat. 1883.—Close to *Epimenides*,
but smaller ; the wings are browner and less distinctly marked ;
beneath, the ground colour of the wings is browner ; the hind
wings have a central white band, and the ocellated spots, which in
Epimenides are enclosed in light yellow rings, are in this species
surrounded by rings of an orange-yellow colour.   Habitat, Amur.

(Genus *Pronophila*, Mén.)

P. Schrenckii, Mén. Schr. p. 33, T. iii. 3.—Expands 3 in.
Wings ample, rounded and subdentate, greyish brown above.
Fore wings with a small apical black spot.   Hind wings with a submarginal row of oval black spots, the third smaller than the rest.
Under side yellowish grey, with three brown marginal lines ; the
basal portion brownish grey.   Fore wings with a black ocellated
spot surrounded by yellow.   Hind wings with six large ocellated
spots in yellow rings, discal portion of the wing with several brown

stripes, the spots being placed on a band of a like colour. Habitat, Amur; Mount Bureja in July.

Genus *EPINEPHELE*.

E. Wagneri, H. S. 311-3, vi. p. 16.—Expands from 1·60 to 1·75 in. Fore wings fulvous, bordered with dark brown; in the male there is a submedian patch of blackish scales; in the female there is only a small blotch of fulvous towards the centre; in both sexes there is a conspicuous oblong black apical spot. Hind wings dark brown; hind margin so deeply indented as to give the appearance of having four tails; along the hind margin is a narrow light band. Under side: Fore wings fulvous, with an apical ocellated spot; internal to it a white triangular mark. Hind wings light brown, with a narrow white central band, and with two black spots in yellow rings near the anal angle. Habitat, South-western Armenia and Persia.

Var. *Mandane*, Koll. Ins. Pers. p. 11.—Has the fore wings unicolorous above. Habitat, Persia.

E. Comara, Led. Hor. 1871.—Expands 1·50 in. Fore wings light fulvous, hind margin dark brown; near the apex is a black spot; the male has a patch of black scales below the median nervure. Hind wings dark brown; the fringes white, and the hind margin indented. Under side: Hind wings drab, with a whitish central band; near the anal angle are two black spots in white rings. Habitat, Persia.

E. Dysdora, Led. Hor. 1869, p. 85.—Expands from 1·25 to 1·35 in. Dark brown. Fore wings tinged with fulvous in both sexes. The female has two large black spots on the fore wings. Hind wings beneath with a strongly-marked central line and two spots. Habitat, Astrabad.

E. Amardæa, Led. Hor. 1869, p. 84. — Expands from 1·25 to 1·30 in. Male dark brown; marginal fringes white; hind margin of hind wings subdenticulate. Fore wings with a very indistinct black spot; submedian black scales indistinct. Female lighter, with the central portion of the fore wings light fulvous. Under side: Hind wings grey, with three wavy dark brown lines; a whitish central band and two small black spots near the **anal** angle. Habitat, Persia.

Var. *Naubidensis*, Ersch. Lep. Turk. p. 21 (1874).—**A** small variety inhabiting Turkestan, in which the black apical spots are absent above, and the wings less distinctly marked beneath.

E. Cadusia, Led. Hor. 1869, p. 84.—Expands 1·25 in. Dark brown, very like *Capella*, but is without the spots near the anal angle of the hind wings beneath. The fore wings above have a second very small black spot immediately beneath the apical spot. Habitat, Astrabad.

E. Capella, Christoph.—Expands 1·25 in. Male dark brown ; fore wings suffused with deep fulvous towards the centre ; near the apex is a black spot. Female light fulvous towards the centre of the fore wings, and with an apical spot, as in the male. Under side : Hind wings yellowish grey, speckled with black ; two ocellated spots near the anal angle. Habitat, the Caucasus; Elburz.

E. Kirghisica, Stgr. MS. Cat. 1883. *Naricina*, Stett. E. i. 1870.—Expands 1·30 in. Orange fulvous ; female paler ; hind margin dark brown. Fore wings with two black spots. Under side yellowish drab, lighter in the centre ; there are two black spots near the anal angle. Habitat, Samarkand.

E. Interposita, Ersch. Lep. Turk. p. 2, n. 68 (1874), pl. ii. 16. —Close to *E. Lycaon*, but lighter. The hind wings beneath are grey without spots, with a central band somewhat lighter than the rest of the wing. Female with a large black spot in a yellow ring near the apex of the fore wings above, and with a light yellow indistinct central band. Hind wings blackish towards the base ; the hind margins dentate, with a white fringe. Habitat, Maracanda, in June.

Genus *CŒNONYMPHA.*

C. Hero var. *Perseis*, Led. z. b. v. 1853. *Sibirica*, Stgr. Cat. 1861.—Larger than the type ; the under side with stronger white markings. Habitat, the Altai.

C. Arcanoides, Pier. Ann. Soc. Fr. 1837, 306 ; H. S. 580-1.— Expands from 1·10 to 1·25 in. Wings dark brown above. Fore wings with the inner two-thirds bright fulvous. Hind wings narrowly reddish fulvous at the anal angle, and with a small black spot in a reddish ring. Under side : Fore wings reddish fulvous ; hind margin brown, with a leaden line ; near the apex is a black ocellated spot in a light brown ring. Hind wings dark brown, with a submarginal row of four or five equal-sized ocellated spots ; internal to them is a white band ; hind margin narrowly reddish, with a leaden line. Habitat, Algeria.

C. Saadi, Koll. Ins. Pers. p. 11, 1848. *Iphias*, Ev. Bull. Mosc. 1851, ii. p. 618.—Expands 1 in. Light fulvous above, the hind margin narrowly dark brown. Fore wings with two small black spots. Under side : Fore wings light fulvous, with two leaden marginal lines, and a central dark brown line bordered externally by a white streak; about the centre of the ring are two conspicuous submarginal black spots surrounded by a yellowish blotch. Hind wings yellowish grey, with a central streak similar to that on the fore wings ; there is a submarginal row of grey spots with metallic centres, and a marginal leaden line. Habitat, Persia.

C. Symphita, Led. Ann. Soc. Belg. 1870, p. 44.—Size of *Pamphilus*, and of much the same general appearance above. Under side : Fore wings with a distinct apical black spot. Hind wings with a submarginal row of equal-sized black spots in light rings, extending from the costa to the anal angle. Habitat, Armenia.

C. Rinda, Mén. Schrk. p. 42, T. iv.—Expands 1·50 in. All the wings pale fulvous above, paler than in *Amaryllis;* they are slightly darker towards the base; there are no spots except one indistinct apical spot on the fore wings. Under side pale greenish or yellowish grey. Fore wings with three small black submarginal spots with white centres, and enclosed in light rings. Hind wings darker than the fore wings, quite without spots, but with several hind marginal light blotches. Habitat, the Amur.

C. Nolkenii, Ersch. Lep. Turk. p. 23 (1874), pl. ii. 17.— Wings rich chocolate-brown without spots, margins dark brown. Hind wings suffused with black. Under side : Fore wings with two or three submarginal spots in light rings. Hind wings brown, with an even row of six black spots in yellow rings, and with white centres; hind margin reddish. Habitat, Mount Naubid, in

Russian Turkestan, frequenting grassy places where junipers grow, in June (Ersch.)

C. Sunbecca, Ev. Bull. Mosc. 1843, iii. p. 538 ; H. S. 611-2.— Expands 1·40 in. The wings are white and entirely without black spots; the markings of the under side are, however, so far visible above as to make certain portions of the wing appear greyish, especially the basal portion of the hind wing; the white spots of the under side are also visible above. Under side : Fore wings with the costal and apical portions grey, the rest white ; near the apex is a row of four white spots, and internal to these four more larger ones. Hind wings grey, greenish towards the base ; there is a submarginal row of five small round white spots ; internal to these a central row of six more, very irregular in size ; there are two more oblong white spots near the base. Habitat, Western Siberia.

––––––––––––

The following North American *Satyridæ* may be compared with those of Europe and North Asia.

The genus *Melanargia* is not represented in America.

The following *Erebiæ* are enumerated in Mr. Kirby's Catalogue :—

E. Fasciata, Butl. Cat. Sat. B. M. p. 92.—Arctic America.

E. Vesagus, Doubl. Hew. Gen. D. Lep. t. 64, f. 3 (1851). —Rocky Mountains.

E. Epipsodea, Butl., Cat. Sat. B. M. p. 80, n. 6, t. 2, f. 9 (1868).—Rocky Mountains.

E. Disa, Thub.—This species is circumpolar, occurring in the Rocky Mountains, as well as in North Europe and Siberia.

E. Rossii, Curt. Ross 2nd Voy. App. Nat. Hist. p. 67 (1835).— Arctic America.

E. Discoidalis, Kirb. Faun. Bor. Amer. iv. p. 298 (1837). —Arctic America.

Six species of *Erebia* are found in Temperate South America, chiefly in the Chilian Andes ; and about the same number in South Africa, in the mountains of Caffraria and the Cape ; but the great mountain ranges of Europe and North Asia furnish an overwhelming majority of species.

Nearly as many species of *Œneis* occur in North America as in the Palæarctic region :—

Œ. Chryxus, Doubl. Hew. Gen. D. Lep. t. 64, f. 1 (1851).—Rocky Mountains.

Œ. Nevadensis, Feld. Reise Nor. Lep. iii. p. 489 (1867).—California).

Œ. Gigas, Butl. Cat. Sat. B. M. p. 161 (1868).—Vancouver's Island.

Œ. Californica, Boisd. Lep. Calif. p. 62 (1869).—California.

Œ. Uhleri, Reak. Proc. Ent. Soc. Phil. vi. p. 143.—The Rocky Mountains and Colorado.

Œ. Semidea, Gay, Amer. Ent. iii. t. 50 (1858) ; also Boisd. Ic. t. 37 (1832) ; H. S. 381.—The White Mountains.

Œ. Taygete, Hüb. Calais Scudd. Proc. Ent. Soc. Phil. (1865). —Arctic America.

Œ. Subhyalina, Curt. Ross 2nd Voy. App. Nat. Hist. p. 68.—Appears to be a synonym for *Semidea*.

The genus *Satyrus* is but poorly represented in North America. Those that occur come nearer to the European *S. Dryas* than to the *Semele* group of the genus. They are thus given in Kirby's Catalogue :—

S. Sthenele, Boisd. Ann. Soc. Ent. Fr. 1852, p. 308.—California.

S. Sylvestris, Edw. Proc. Acad. Nat. Sc. Phil. 1861, p. 162.—California.

S. Nephele, Kirb. Fann. Am. Bor. iv. 297 (1837).—Canada.

S. Ariane, Boisd. Ann. Soc. Ent. Fr. 1852, p. 307.—California.

S. Boöpis, Behr. Proc. Calif. Acad. 1864.—California.

S. Pegala, Fabr. Syst. Ent. p. 494 (1775).—The United States.

S. Alope, Fabr. Ent. Syst. iii. 1, p. 229, n. 715 (1793).—The United States.

S. Ridingsii, Edw. Proc. Ent. Soc. Phil. iv. p. 201 (1865).—Colorado.

The genus *Epinephele* is not represented in North America, though ten species occur in Chili.

Cœnonympha is but sparsely represented :—

C. *Pamphilus*, L., occurs in Colorado in the form of the var.
Pamphilioides, Reak, Proc. Ent. Soc. Phil. vi. p. 146 (1866).

C. *Tiphon* var. *Inornata*, Edw. Proc. Ac. Nat. Sc. Phil. 1861,
p. 163.—Canada.

C. *Ochracea*, Edw. *loc. cit.*—Northern States.

C. *Galactinus*, Boisd. Ann. Soc. Ent. Fr. 1852, p. 309.

C. *Californica*, Doubl. Hew. Gen. D. Lep. t. 67, f. 2 (1851).

The two latter species both occur in California, and approach
more nearly than any other species to the Siberian C. *Sunbecca*, Ev.

Fam. 10.—**HESPERIDÆ**, Leach, Samouelle's ' Useful Com-
       pendium,' p. 242 (1819) ; West. Gen. Diurn. Lep.
       p. 505 (1852).

    HESPERIOIDÆ, Wallgr. Scand. Rhopal. p. 248.

CHARACTERS.—Larvæ fusiform, tapering towards the extremities, smooth or pubescent; very rarely, and only in some exotic
species, hairy, and never spiny. The head is large and globular.
The larvæ of the European species feed on low-growing plants,
chiefly those belonging to the orders *Leguminosæ, Malvaceæ,* and
Labiatæ, and on grasses. They are in the habit of rolling themselves up in leaves after the manner of the larvæ of the heteroceous
group Tortrices, spun together in a sort of cocoon.

Pupæ without angles, and tapering at the extremities, especially at the caudal end; they are never suspended or succinct,
like those of the groups of butterflies already described, but are
often enveloped in a slight cocoon made of loosely-spun filaments
of silk.

Imagines, small butterflies, the head large, almost as wide as
the thorax, the antennæ set widely apart from each other at their
base. They have a small tuft of hairs at their base, the shafts are
short, the clubs thick and often arched, and sometimes furnished
with a terminal hook. All the four legs fully developed in both
sexes ; the posterior tibiæ furnished with four spurs in the majority
of the European genera, but in *Carterocephalus* they have only two.

The abdomen stout in proportion to the general size of the insect, the thorax wider than the abdomen and somewhat depressed. The position of the wings during repose is different from that of the butterflies of any of the other families, the hind wings being held in a horizontal position, and the fore wings only half erect; the wings are never closed perpendicularly over the back; the inner margin of the hind wings is not deflected, but is thrown into a slight fold, so that there is no canal for the reception of the abdomen. The flight of these insects is hurried and intermittent, and never steady or sailing like those of other groups.

The *Hesperidæ* are represented in every zoological region and subregion. Mr. Kirby's Catalogue includes nearly a thousand known species and fifty-two genera. The Neotropical region has by far the greatest number of species, and the Nearctic region is richer in species than the Palæarctic to the extent of more than double the number occurring in the latter. Thirty-one species are given in Staudinger's Catalogue as belonging to Europe proper, and besides these there are a great number of marked varieties, many of which are very likely specifically distinct. The European species are mostly sombre in colouring, being usually black and white, fulvous or grey, though the pattern of the wings is somewhat striking in its arrangement in a few species.

The family of *Hesperidæ* is, by the common consent of entomological authors, placed at the end of the Rhopalocera, as being most closely allied to and leading to the Heterocera, with which some have proposed that they should be incorporated.

In their transformations they resemble the genera *Parnassius* and *Doritis* among the *Papilionidæ*, also *Zegris* among the *Pieridæ*; they likewise somewhat remind us of certain species of *Satyridæ* which undergo their pupation on the ground.

Genus 1.—**SPILOTHYRUS**, Boisd. Gen. Ind. Meth. p. 35 (1840).

> ERYNNIS, Schrank, Fauna Boica, ii. 1, p. 152 (1801). Nomen anterius et peferendum.
>
> CHARCHARODUS, Hüb. Verz. bek. Schmett. p. 110 (1816).

CHARACTERS.—Larva short, cylindrical, pubescent, the head large, hollowed out or cleft, the prothorax much contracted.

Pupa more or less rounded anteriorly, and elongate and conical towards the caudal extremity. Enclosed in a slight cocoon.

Imago.—Club of antennæ pyriform, not curved. Palpi widely separated, hairy, the last articulation smooth and short. Head somewhat smaller than the thorax, which is very broad. Abdomen reaching slightly beyond the hind wings. Fore wings with transparent or glassy spots, the costa curved in the male. Hind wings dentate.

The species of this genus are few, and, with the exception of one which occurs in Mexico, are confined to the Palæarctic region.

1.—S. Alceæ, Esp. 51, 3 (1780), i. 2, p. 4.
    Malvarum, Hffsgg. Ill. Mag. iii. p. 198 (1804); O. i. 2, 195.
    Malvæ, Hüb. 450-1 (*Malvæ*, L., alia est species).

Expands from 1·10 to 1·30 in. Dark brown, with a reddish tinge. Fore wings with an obscure dark central band; on the outer half of the wing four transparent white spots. Hind wings brown, with three darker bands. Under side: Fore wings as above, but lighter. Hind wings with a central square white spot and two or three smaller ones towards the base: hind margin with white and brown spots. Pl. LXXVIII., 1.
    Times of Appearance.—June and August.
    Habitat.—Central and Southern Europe (except Denmark and England), North Africa, Western Asia, and Siberia. It frequents places where mallows grow.
    Larva.—Pubescent, dark grey, with two lighter lateral lines; the head is black, and the first segment marked with four yellow points. It feeds spun up in the leaves of different species of *Malvaceæ*. Times of appearance, May and September. Those appearing in the autumn hybernate in the hollow stems of thistles and burdocks (Girard). Pl. LXXX., 1.

VARIETY.

Australis, L. Is. 1847, p. 285.—Smaller than the type, and suffused to a greater extent with reddish brown.
    Habitat.—The South of Europe.

2.—S. Altheæ, Hüb. 452-3. *Malvarum* var., O. i. 2, 197 ; Godt.
ii. 28, 5, 6.
GEMINA, Led. z. b. vv. 1852, p. 50.
MARRUBII, Kirby, Man. Eur. Butt. p. 115.

Expands from 1·10 to 1·30 in. Very closely resembles the
last, but is darker, and has a greenish tinge on the fore wings.
The wings are darker than in *Alceæ* ; the fore wings have four
white subdiaphanous spots. Hind wings with two central white
spots. Under side : Hind wings tinged with green. The clubs of
the antennæ are wholly black, whereas in the last species they
are reddish brown beneath. Pl. LXXVIII., 2.
TIMES OF APPEARANCE.—May and August.
HABITAT.—Central and South-Eastern Europe; the larva
of the type seems to be as yet unknown, though that of the Spanish
variety *Bœticus* is described by Rambur; it may be inferred that
there is some resemblance between the two forms.

VARIETY.

Bœticus, Ramb. Cat. Faun. And. P. C. 12, 3, 4 (1839); Ramb.
Cat. S. And. p. 80.
MARRUBII, H. S. 14, 15.

Smaller than the type and lighter, being of a yellowish or
brownish grey colour.
HABITAT.—South-Western Europe.
LARVA.—Pale grey, with a reddish or yellowish tinge, brown
dorsal and lateral stripes. Feeds on *Marrubium Hispanicum* in
April and in August.

3.—S. Lavateræ, Esp. 82, 4 (1783) ; Hüb. 454-6 ; O. i. 2, 198 ;
Godt. ii. 28, 7.

Expands from 1·25 to 1·30 in. Pale grey, with a greenish
tinge. Fore wings with some white spots near the apex, and
white central band edged internally with dark brown. Hind
wings with central and basal blackish bands; the hind margin

dark grey, with whitish spots. Under side similar to the upper surface, but very pale, and with a greenish tint. Pl. LXXVIII., 3.

TIMES OF APPEARANCE.—April, and from July to the end of August.

HABITAT.—South Germany, Switzerland, Hungary, the South of Europe; the North of Asia Minor, Armenia, and Persia. Frequenting dry calcareous and rocky districts.

LARVA.—Bluish grey, with a brown dorsal stripe, and on each side a pale yellow lateral line edged with dark brown. Feeds on *Stachys recta* in May.

Genus 2. — **SYRICHTHUS**, Boisd. Gen. Ind. Meth. p. 35 (1840).

HESPERIA, Fab. Ent. Syst. iii. 1, p. 258 (1793); Lat. Ency. Meth. ix. (1819).

PYRGUS, Hüb. Verz. bek. Schmett. p. 109 (1816); Westw. Gen. Diurn. Lep. (1852).

CHARACTERS. — Larvæ and pupæ have the same general characters as those of the last genus.

Imago.—Club of the antennæ oval, blunt, and without any terminal hook. Palpi separated, hairy. Head slightly larger than the thorax, Abdomen reaching beyond the hind wings, at least in the male. Wings generally blackish or dark brown, with white spots more or less square in shape; fringes chequered with black and white.

This genus has numerous representatives in Europe; fifteen species are enumerated in Staudinger's Catalogue, the nomenclature of which I implicitly follow. Rambur has separated them into many more species; indeed the genus remains in an uncertain condition as regards its division into species, and is perhaps in this respect the most difficult of all the genera of European butterflies. Only one species, *S. Malvæ*, occurs in Britain.

1.—S. Proto, Esp. 123, 5, 6; O. i. 2, 210; Boisd. Ic. 46, 4, 5; Dup. i. 42, 8; Frr. 361, 5.

Expands from 1·10 to 1·20 in. Dark brown, with a greenish tinge. Fore wings with three brownish white spots near the centre and three smaller ones on the costa near the apex. Hind wings with three whitish central spots and a smaller one near the base. All the wings have a marginal row of faint light spots; the fringes are brown and white. Under side: Fore wings blackish brown chequered with white. Hind wings greenish red, with a basal and a central whitish band; the inner margin dull white. Pl. LXXVIII., 4.

TIMES OF APPEARANCE.—June and July.

HABITAT.—Waste places, &c., in the South of Europe; Algeria, Northern Asia Minor, and Syria.

LARVA.—Yellowish grey, with a reddish dorsal line, and with two red spots on the head. Feeds on *Phlomis* in April and May.

2.—S. Tessellum, Hüb. 469-70; O. iv. 1, 157; Ramb. Faun. Andal. pl. 8, 1, 2.

Expands from 1·25 to 1·30 in. Somewhat resembles *Proto* above, but is larger, and its colouring is darker brown, without any green tinge; the spots are whiter and more distinct, the marginal rows of spots being well defined. Under side brown, with a greenish tinge. Fore wings with two rows of oblong white spots, besides a white discal spot and another on the costa near the apex. Hind wings with a white central band and faintly white hind marginal and basal spot; fringes of all the wings black and white. Pl. LXXVIII., 5.

TIMES OF APPEARANCE.—May and August.

HABITAT.—Meadows in South Russia, Turkey, Asia Minor, and Persia.

LARVA.—Unknown.

3.—S. Cribrellum, Ev. Bull. Mosc. 1841, p. 25; Frr. 349, 1; H. S. 12, 13.

Expands from 1 to 1·10 in. Ground colour blackish brown. Fore wings greenish towards the base; the white spots distinct on all the wings. Hind wings with a central white band, as well as a submarginal row of white spots. Under side: Fore wings white, with a black blotch towards the base, a grey discal spot surrounded

by a black ring, and below this another irregular black ring
enclosing a white spot. Hind wings white, with two greenish
grey bands, one central and one near the base ; the nervules
greenish brown. Fringes not conspicuously spotted beneath.
Pl. LXXVIII., 6.

TIME OF APPEARANCE.—May.

HABITAT.—Meadows in South Russia and Siberia.

LARVA.—Unknown.

4.—S. Cynaræ, Ramb. Faun. And. pl. 8, 4, 5 ; Frr. 349, 2 ;
     H. S. 4-7 ; Ramb. Cat. S. Andal. p. 68.

Expands from 1 to 1·10 in. Brownish black ; very much
resembles *Cribrellum*, but the fore wings are not so distinctly
greenish grey at the base ; the white spots are larger, and are
arranged in two rows, one central and the other submarginal.
Hind wings similar to those of *Cribrellum*, but with smaller spots.
Under side similar to that of the last, but the markings are darker
and more clearly defined. Pl. LXXVIII., 7.

HABITAT.—South Russia, Asia Minor, Armenia, and the Amur.

5.—S. Sidæ, Esp. 90, 3 (1784) ; Hüb. 468 ; O. i. 2, 199 ; Godt.
     ii, 27, 5, 6 ; Frr. 361, 1.

Expands from 1·10 to 1·30 in. Dark brown. Fore wings
with a zigzag row of white spots, a square white discal spot, and
above it two smaller ones near the costa ; about the centre of the
wing, and almost touching the inner margin, is another white
spot. Hind wings brown, lighter in the centre, with a submarginal
row of white dots ; internal to this a brown band edged with black.
Under side : Fore wings blackish grey, with white spots arranged
as above. Hind wings white, with two yellowish orange bands
edged with black, not reaching either to the costa or to the inner
margin. There are traces of a submarginal row of spots.
Pl. LXXVIII., 8.

TIME OF APPEARANCE.—June.

HABITAT. — Woods in the South of France, Italy, South-
Eastern Europe, Western Asia, and Siberia.

LARVA.—Unknown.

6.—**S. Carthami,** Hüb. 720, 723 ; O. iv. 159; Ramb. Faun. And.
pl. 8, 8 ; Frr. 389, 3 ; Ramb. Cat. p. 65.
Malvæ, Esp. 23, 2.
Tessellum, O. i. 2, 205.

Expands from 1·20 to 1·35 in. Wings dark brown; fringes
brown and white. Fore wings with a row of six white spots,
extending in a curved direction from near the apex to the centre of
the inner margin; there is a white discal spot and three placed
close together on the costa. Hind wings with an indistinct light
central band, and sometimes with a submarginal row of spots.
Under side: Fore wings brownish grey, slightly tinged with green
towards the apex, with white spots arranged as above. Hind
wings greenish, with central and basal white bands, and one or
two white blotches near the hind margin; marginal fringes with
more white colouring beneath than above. Pl. LXXVIII., 9.

Times of Appearance.—May and August.

Habitat.—Mountains in Central Europe, Spain, the South of
France, and South Russia; Northern Asia Minor and Siberia.

Larva.—Unknown.

VARIETY.

Moeschleri, H. S. 37, 38, vi. p. 175.—Has the white spots,
above and beneath, larger than in the type. The under side is
almost white.

Habitat.—South Russia (Sarepta).

7.—**S. Alveus,** Hüb. 461-3 ; Ramb. Faun. And. pl. 8, 3 ; Cat.
p. 69 ; Dup. i. p. 259.
Carthami, Hüb. 721-2.

Expands from 1 to 1·10 in. Brown. Fore wings with a
white discal spot and a submarginal row, as in the last species;
there is a small white dot near the centre of the inner margin, and
the usual tripartite costal spot. Hind wings as in *Carthami,* but
even less distinctly marked; fringes brown and white. Under
side: Fore wings grey with white spots, as above, but suffused;
hind margin brownish or greenish. Hind wings white, with

central and submarginal greenish bands, irregular in shape; there is a basal spot of the same colour as the bands. Pl. LXXIX., 1.

Times of Appearance.—May and August.

Habitat.—Central Europe (except Holland and the British Isles), Spain, Italy, South Russia, Asia Minor, Syria, Armenia, and Persia.

Larva.—Unknown.

VARIETIES.

a. **Fritillum,** Hüb. 464, 5; Tr. x. 1, 94; Ramb. Faun. And. pl. 8, 14; Cat. S. And. p. 74. Smaller than the type, and with the under side whiter.

Habitat.—Central and Northern Europe.

b. **Cirsii,** Ramb. Faun. And. pl. 8, 15; Cat. S. And. p. 73; H. S. 33, 34. Smaller than the type; the white spots on the fore wings slightly more distinct; the hind wings with a central and submarginal white band. Under side: Hind wings with the white bands narrower than in the type, the ground colour being of a darker green. Pl. LXXIX., 2.

Habitat.—Central and Southern Europe.

c. **Carlinæ,** Ramb. Faun. And. pl. 8, 11; Cat. S. And. p. 72. Smaller than the type and darker, the spots on the fore wings small and indistinct. Under side: Fore wings grey, with small white and black spots. Hind wings greenish, with white spots, as in the type. Pl. LXXIX., 3.

Habitat.—Central and Southern Europe.

d. **Onopordii,** Ramb. Faun. And. pl. 8, 13; Cat. S. And. p. 72.

8.—**S. Serratulæ,** Ramb. Faun. And. pl. 8, 9; Cat. p. 71; H. S. 18-20; Frr. 621, 3.

Expands from 1·10 to 1·15 in. Brown. Fore wings with a row of small white spots and one or two near the costa; these spots have nearly the same arrangement as in the last species. Hind wings almost unicolorous, with one or two indistinct light spots near the costa. Under side: Fore wings reddish brown, with white spots as above, but larger. Hind wings greenish, with a large white costal blotch, basal and marginal white spots, and another round one near the inner margin. Fringes brown and white. Pl. LXXIX., 4.

TIMES OF APPEARANCE.—June and July.

HABITAT. — Central and South Europe, Central Germany, Armenia, Syria, and Eastern Siberia.

VARIETY.

Cœcus, Frr. 493, 3, 4.—*Serratulæ* var. H. S. 21, 22. Smaller than the type. The white spots are almost entirely wanting above, thus giving the insect a uniform brown appearance. Under side: Fore wings greyish white; hind margins darker. Hind wings greenish grey, with white spots, as in the type. Pl. LXXIX., 5.

HABITAT.—The Alps of Switzerland and Austria.

9.—S. Cacaliæ, Ramb. Faun. And. pl. 8, 6, 7 ; Cat. p. 77 ; H. S. 21, 22.

ALVEUS, B. Ic. 46, 1-3 ; Dup. i. 50, 788 ; Frr. 621, 2.

Expands from 1·5 to 1·20 in. Brown, slightly darker at the base. Fore wings with white spots placed as in the last, but very minute, being mere dots. Hind wings with a slightly perceptible light central band ; fringes brown and white. Under side : Fore wings reddish grey, with a curved row of white spots much larger than those above, but indistinct. Hind wings greenish grey, with white spots arranged almost exactly as in *Serratulæ*. Pl. LXXIX., 6.

TIME OF APPEARANCE.—July.

HABITAT. — A mountain species, inhabiting the Alps of Switzerland and the Pyrenees.

10.—S. Andromedæ, Wallengr. Scandin. Rhop. p. 272.

Expands from 1·10 to 1·20 in. Greyish brown. Fore wings with white spots arranged as in the last species, but much larger. Hind wings with very faint traces of a central and submarginal light band. Under side : Fore wings grey, with white spots.

Hind wings greenish, with a central row of four white spots, two small basal spots, and an indistinct marginal row. The clubs of the antennæ are reddish at the tip. Pl. LXXIX., 7.

Times of Appearance.—June and July.

Habitat.—Mountains in Scandinavia and Lapland; also in the Austrian Alps (Stgr.)

11.—S. **Centaureæ**, Ramb. Faun. And. pl. 8, 10 ; Cat. p. 78 ; H. S. 1-3 ; Wallengr. Scandin. Rhop. p. 265.

Expands from 1·10 to 1·20 in. Very closely resembles the last species, but has the wings greyer, and the white spots on the fore wings larger and more distinct. The hind wings have a narrow submarginal light band, and some faint light markings in the centre ; fringes black and white. Under side : Fore wings dark grey, with white spots occupying the same position as above, but the three lower ones are united to form a short band. Hind wings greenish, with an irregular central white band, and with basal and hind marginal spots. Pl. LXXIX., 8.

Time of Appearance.—July.

Habitat.—Mountains in Central Scandinavia, Lapland, and Labrador.

Larva.—Unknown.

12.—S. **Malvæ**, Linn. Syst. Nat. x. 485 ; Faun. Suec. 285 ; Lew. Ins. T. 46, 4-7 ; Dalm. Förs. p. 202 ; Zett. Ins. Lap. 915 ; Wallengr. Scandin. Rhop. p. 275.
　　　Alveolus, Hüb. 466-7 ; O. i. 2, 2-8 ; Godt. F. 12, 3 ; Ramb. Faun. And. pl. 8, 15 ; Cat. p. 76 ; Frr. 361, 2.

Expands from 1 to 1·10 in. Blackish brown, with white spots variable in their arrangement ; there is generally a row reaching from near the apex to the centre of the inner margin, several more on the costa, and one or two near the base. Hind wings with a row of white spots placed closely together near the costa, and reaching about half-way across the wing, and with a marginal row of white dots. Under side pale brown, with the white spots as above, but larger. Hind wings greenish or yellowish, with one or two large white spots near the costa, some white basal spots, and an indistinct hind marginal row. Clubs of antennæ black. Pl. LXXXI., 1.

Times of Appearance.—May and August.

Habitat.—Dry open places in woods, meadows, heaths, downs, &c. Throughout the greater part of Europe, and in Asia Minor, Cyprus, Eastern Siberia, and the Altai. It is common in England, being the only British species of this genus. (The Grizzled Skipper).

Larva.—Cylindrical, slightly thicker in the middle segments; head larger than the second segment and dark brown or black; the body is green or brown, with darker dorsal and lateral stripes.

Pupa.—Smooth and tapering, without projections, brown in colour, with white spots and lines.

The larva feeds on *Rubus fruticosus, R. Idæus,* and, according to some writers, on *Fragaria vesca.* Both forms of the larva are represented from Hübner, on pl. lxxx. fig. 2, feeding on the leaves of *Rubus Idæus,* the wild raspberry. It appears in April and in September.

VARIETIES.

a. **Taras,** Meig. I. T. 55, 3, *a, b.—Malvœ,* Esp. 51, 2.—*Alveolus,* Hüb. 847-8; Frr. 361-3.—? *Lavateræ,* Haworth (secund. Newman Brit. Butt. 170).—The white spots on the fore wings are very large and confluent, so as to give the wing the appearance of being white in the central portion, with black veins. The hind wings are usually less spotted than in the type, the costal spots being either very small or absent. This aberration occurs throughout the general area of distribution of the type, and there are many forms intermediate between *Malvœ* and *Taras.* Pl. LXXXI, 2.

b. **Melotis,** Dup. i. 42, 1, 2, p. 257 ; Ramb. Cat. p. 76.—*Hypoleucos,* Led. Z. B. v. 1855, p. 193.—Somewhat larger than the type and paler, the under side of the hind wings being almost entirely white.

Habitat.—The Greek Islands and Syria.

13.—S. Phlomidis, H. S. 8, 9, i. p. 153; Frr. 621, 4.

Expands from 1 to 1·30 in. Dark brown. Fore wings with a row of white spots, three being placed together near the costa; below these two more larger ones, and again, below and internal to them, two more, the upper one of which is larger than the rest of the spots; there are three white basal spots and a white semilunar discoidal spot; the basal third of the costa is marked with white; there is a marginal even row of white dots. Hind wings dark brown, with a central band of three white spots and a marginal row of indistinct white dots; fringes dark brown and white. Under side: Fore wings brown, whitish at the apex, and along the costa and at the base; the white spots are arranged as above, but are larger and less distinct. Hind wings yellowish brown, with a central white band and some white basal and marginal spots. Pl. LXXXI., 3.

Time of Appearance.—July.

Habitat.—Turkey, South of the Balkans, Greece, Asia Minor, and Syria.

Larva.—Unknown.

14.—S. Orbifer, Hüb. 803-6; Boisd. Ic. 47, 1, 2; Frr. 362, 1; Dup. i. 42, 5, 6.

Eucrate, Tr. x. 1, 96.

Tesselloides, H. S. 10, 11, vi. p. 36.

Expands from 1 to 1·10 in. Dark brown. Fore wings with white spots arranged very similarly to those in the last species, but the basal spots and the costal white streak are absent, and the discoidal spot is orbicular in shape. Hind wings with a central row of small white spots, and a marginal row similar to that seen in the last. Under side pale brown, with the spots arranged as above. Hind wings olive-green, with a central row of large white spots and a marginal row of smaller ones. Fringes brown and white. Pl. LXXXI., 4.

Time of Appearance.—June.

Habitat. — Mountain valleys in South-Eastern Europe, Western Asia, the Altai, and the Amur.

Larva.—Unknown.

15.—**S. Sao,** Hüb. 471-2 (1801) ; Godt. ii. 28, 3, 4.
  Sertorius, Hffmgg. Ill. Mag. iii. 203 (1804); O. i. 2,
    211 ; Frr. 361, 4.

Expands from 1 to 1·10 in. Much resembles *Orbifer* above, but is rather lighter, and the white spots are smaller on all the wings. Under side : Fore wings as in the last. Hind wings bright reddish brown, with a central row of white spots forming a band, and with basal and hind marginal smaller white spots. Pl. LXXXI., 5.

Times of Appearance.—May and August.

Habitat.—Dry grassy or rocky places in Central and Southern Germany, Belgium, France, Spain, Central and Northern Italy, Dalmatia, and Turkey.

Larva.—Unknown.

VARIETIES.

a. **Eucrate,** O. i. 2, 213 ; Ramb. Cat. S. And. p. 86.—Smaller than the type, and paler in colouring beneath.

Habitat.—The South of Spain, South Italy, and Greece.

b. **Therapne,** Ramb. Ann. S. Fr. 1832, pl. 7, 4 ; Boisd. Ic. 46, 6, 7 ; Dup. i. 42, 9, 10; H. S. 16, 17.—A variety or subspecies peculiar to Corsica and Sardinia. Smaller than the type. Expands from ·85 to 1 in. Wings browner than in the type, the fore wings reddish towards the base, the basal portion of the hind wings being greenish ; the spots are arranged as in the type, but are smaller in proportion, and brownish instead of pure white. Pl. LXXXI., 6.

Genus 3.—**NISONIADES,** Hüb. Verz. bek. Schmett. p. 108
    (1816); Westw. Gen. Diurn. Lep. p. 519 (1852).
  Thannaos, Boisd. Gen. Ind. Meth. p. 37 (1840).
  Erynnis, Schrank.

Characters.—Larva smooth, thickest in the middle ; the head large and contrasting considerably in size with the narrow anterior segment.

Pupa.—Fusiform, with a tubercular elevation on the head, and with the abdomen conical and pointed.

Imago.—Club of the antennæ fusiform and hooked. Palpi separated, hairy, the last joint slender and prominent. The head as large as the thorax, which is thick. Abdomen reaching as far as the border of the hind wings. Wings entire, more or less rounded; fringes not chequered with black. The colour of the wings is very sombre, at least in the two species found in Europe. About forty more species are known, the genus being distributed widely over the earth, but principally found in the tropics; several species inhabit the American States, and two or three South Africa.

1.—N. Tages, L. Syst. Nat. x. 485 ; Faun. Suec. 286 ; Esp. 23, 3 ; Hüb. 456-7 ; O. i. 2, 214.

Expands from 1 to 1·25 in. Fore wings with the discal portion grey; near the costa is a large dark brown spot and another near the inner margin ; there is a submarginal dark brown band, and a marginal band of a still darker colour, having whitish dots. This pattern is most marked in the male; in the female it is less distinct and paler. Hind wings dark brown, with a darker marginal band, on which are dots similar to those on the fore wings, and often some pale spots on the discal portion. Under side almost uniform brown, lighter than above. Fringes of all the wings brown, and without spots. Pl. LXXXI., 7.

Times of Appearance.—April and August.

Habitat.—Dry places throughout Europe, excepting the Polar Region. It also occurs in Western Asia and the Amur. It is a common British species (the Dingy Skipper), and is chiefly found on chalk hills and heaths. Larva tapering towards the extremities, bright green, with a yellow lateral stripe, and a second obscure line above the spiracles. Head brown. Feeds on *Lotus corniculatus* in May and September. Pupa green, tinged with red on the dorsal surface. Pl. LXXX., 3.

VARIETIES.

a. Unicolor, Frr. 505, 1, vi. p. 37 ; Stgr. Hor. 1870, p. 86.— A variety occurring in Greece and Western Asia, in which the wings above are uniform brown without any markings.

b. **Cervantes**, Graslin, Ann. Soc. Fr. 1836, 558, pl. 17 ; Ramb.
Cat. S. And. p. 83 ; Frr. 417, 3.—Somewhat larger than the type
and darker.

Habitat.—Andalusia.

2.—**N. Marloyi**, Boisd. Ic. 47, 6, 7. i. p. 241.
Sericea, Frr. 265, 4 (1838); H. S. 29, 30.
Rustan, Koll. Ins. Pers. p. 11.

Expands from 1 to 1·25 in. Dark brown. Fore wings with
two narrow blackish stripes, and sometimes with a marginal row
of grey spots. Hind wings dark brown, without any markings.
Under side : Fore wings with two small white spots near the
apex. Fringes lighter brown than the wings, and without spots.
Pl. LXXXI., 8.

Time of Appearance.—July.

Habitat.—Dry places in Greece, Turkey, South Russia, Asia
Minor, and Armenia.

Larva.—Unknown.

Genus 4.—**HESPERIA**, Boisd. Ind. Meth. (1840) ; Latr. Ency.
Méth. ix. p. 11 (1819).
Pamphila, Fab. Ill. Mag. vi. p. 287 (1807) ; Westw.
Gen. Diurn. Lep. p. 521, 1852.
Thymelicus, Hüb. Verz. bek. Schmett. p. 113 (1816)
(nomen preferendum).

Characters.—Larva elongate, smooth, tapering at the extremities, especially at the anal end. Head large and globular. In
colour they are generally green or brown, with lighter longitudinal
stripes.

Pupa.—Conico-cylindrical, terminated anteriorly by a short
point, and having a free tube placed anteriorly to contain the
tongue.

Imago.—Head large, thorax very broad, eyes large and prominent ; club of antennæ straight, ovoid, and often with a terminal
hook. Palpi hairy, with the third joint conical or conico-
cylindrical. Abdomen thick and reaching beyond the hind wings.
Hind margin of fore wings rounded. Hind wings with a slight

concavity in the centre of the hind margin. In colour the wings are generally fulvous; the fore wings in the male have a black line immediately below the discoidal cell; this is absent in the females, and the ground colour of the wings is paler than in the other sex. Five out of the seven European species occur in Britain.

1.—H. **Thaumas**, Hufn. Berl. M. ii. p. 62 (1766); Rott. Nat. vi. 4; Esp. 36, 2, 3; Bkh. i. 181.
LINEA, S. V. p. 160; Fab. Mant. 84 (1787); Hüb. 485-7; O. i. 2, 228; Godt. i. 12, 3; Fʀʀ. 631, 1.

Expands from 1·08 to 1·20 in. Both sexes fulvous, but the male is generally deeper than the female; has a broad black streak in the centre of the fore wings, curving downwards towards the inner margin. Hind margins of all the wings blackish in the male, very narrowly so in the female; the neuration is blackish. Under side : Fore wings fulvous, yellowish grey towards the apex. Hind wings greenish, with the inner margin fulvous. Clubs of antennæ reddish brown beneath.

TIMES OF APPEARANCE.—June to the end of August.

HABITAT.—Borders of woods, roadsides, dry fields, and pastures, both in the plains and at some elevation in the mountains; throughout Central and Southern Europe, Scandinavia, and the British Isles, North Africa, Asia Minor, Syria, Armenia, and Persia; also North America.

LARVA.—Green, with yellowish white lateral lines. Head and legs green. Feeds on various grasses, introducing itself between the leaves and the stems. It is full-grown early in June.

PUPA.—Yellowish green, the wing-cases darker; the tube containing the tongue is brownish.

N.B.—The numbers to the figures of this species and that of *C. Palæmon* on Pl. LXXX. are accidentally transposed. The central figure represents the larva and pupa of *H. Linea* (= *Thaumas*); that at the bottom right corner *C. Palæmon*.

2.—H. Ludoviciæ, Mabille, 'Description d'Hespéries, Ann. Soc. Ent. Belg. 1883, p. lxviii.

This is a new species described by M. P. Mabille, *loc. cit.* It seems to me close to the next species, *H. Lineola*, and has been taken at Murat in Auvergne, and in the Val d'Eyna, Pyrenees; also by Fallou in Switzerland.

I have not had an opportunity of seeing any examples, and, as no figure accompanies the description in the Ann. Soc. Ent. Belg., I am unable to figure the species in this work, or do anything further than quote verbatim the original description, as I had to do in the case of *Lycæna Panope* :—

"*P. Lineola*, Och., minor et ei proxima, sed habitu gracili, statura et colore diversa. Intense aut obscure fulva. Alæ anticæ late nigro-marginatæ, cum ramis nigro-scriptis; nervula discoidalis nigra cellulam claudens; lineola nigra, maris indicium, exilis, nervo parallela. Alæ posticæ fulvo-nigrantes, raro fulvæ, marginibus late infuscatis. Alæ anticæ subtus apice nigranti margine interne nigro, parte interiori alæ nigra; posticæ fuscescentes, squamis nigris creberrimis consitæ spatio anali sordide luteo-fusco, cætera ut in *P. Lineola*."

3.—H. Lineola, O. i. 2, 230; Boisd. Ic. 47, 4, 5; Dup. i. 41, 1-3; Frr. 631, 2.

VIRGULA, Hüb. 660-3.

Expands from 1·08 to 1·20 in. Very close to *Thaumas*, which it resembles in size; the wings above are fulvous, but not so bright as in that species; the hind margin is less black, and in the male the central black line is much narrower, sometimes broken into two, and occasionally almost obliterated. Under side : Fore wings fulvous, blackish towards the base. Hind wings yellowish white in the male, greyish white in the female, without markings. Clubs of the antennæ deep black beneath.

TIME OF APPEARANCE.—July.

HABITAT.—The same as *Thaumas*, but it is not so common, and does not occur in Britain.

LARVA.—Bright green, with five yellow lateral lines. Head reddish. It feeds on grasses in June.

4.—H. Actæon, Esp. 36, 4; Hüb. 488-90; O. i. 2, 231; Godt. ii. 27, 3, 4; Frr. 631, 3.

Expands from ·90 to 1·20 in. Wings fulvous, with a slightly greenish tinge. The male has the hind margins blackish, and the subdiscoidal black streak is moderately wider. The female has a crescentic row of pale fulvous spots on the fore wings, the convex side of the crescent being placed outwards; internal to these and near the base is another pale fulvous spot, not always present; the hind wings are rather lighter towards the hind margin than at the base. Under side: Fore wings as above. Hind wings uniformly greenish grey. The clubs of the antennæ are bright fulvous beneath.

TIMES OF APPEARANCE.—June to August.

HABITAT.—Dry places and hills in Central and Southern Europe (except Russia), North Africa, the Canaries, and Asia Minor. In Britain it is very local, being restricted to a few localities in Devonshire, Dorset, and Warwickshire ("The Lulworth Skipper.")

LARVA.—Pale green, with a dorsal line of darker green, edged on each side by a white band; the stigmatal lines narrow and white. The head is prominent and brownish yellow. It feeds at night on grasses.

5.—H. Sylvanus, Esp. 36, 4; Hüb. 482-4; O. i. 2, 226; Frr. 646, 2, 696, 2.

Expands from 1·20 to 1·40 in. The male is fulvous; the hind margins of the fore wings and all the marginal portions of the hind wings brownish fuscous. The fore wings have the discal black line very distinct. Under side: Hind wings yellowish, with some very indistinct central markings; the inner margin tinged with fulvous. The female is somewhat larger than the male; the wings are brownish. Fore wings without the sub-discoidal black streak, with a submarginal row of large fulvous spots; the basal portion is fulvous towards the costa. Hind wings with a row of fulvous spots forming a band. The antennæ have the clubs hooked at the extremities. Pl. LXXXII., 1.

Times of Appearance.—June and August.

Habitat.—Woods throughout Europe (excepting the Polar Regions), Northern and Western Asia, the Altai, and the Amur. In Britain it is common (the Large Skipper).

Larva.—Dull green, dorsal stripe darker, dotted with black. Head large and brown. Feeds in April on *Triticum repens, Holcus lanatus,* and other grasses.

6.—H. Comma, Linn. Syst. Nat. x. 484 ; Faun. Suec. 285; Esp. 23, 1 ; Hüb. 479-81 ; O. i. 2, 224 ; Godt. i. 12 ; Frr. 646, 1 *a-d.*

Expands from 1·20 to 1·40 in. About the size of the last, but has the hind margins more inclined to concavity in outline, especially in the male. The male has the basal and discal portions of the fore wing fulvous ; the hind margin and apex are dark brown ; there are some square fulvous spots near the apex ; the black line is very distinct, and is straighter than in *H. sylvanus.* Hind wings dark brown, with a submarginal row of fulvous spots, and a fulvous spot near the base. The female is larger than the male, and is somewhat variable as regards the ground colour, which ranges from dull fulvous to dark greenish brown ; the fulvous spots are more discrete than in the male, and the black streak is absent from the fore wings. Under side : Fore wings as above, but greenish towards the apex. Hind wings greenish or yellowish grey, with a submarginal row of seven or eight pure white spots, rather square in shape, and often outlined with blackish ; near the base are two or three similar spots. Clubs of antennæ terminating in a hook. Pl. LXXXII., 2.

Times of Appearance.—July and August.

Habitat.—Dry places throughout the whole region of the European fauna, with exception of North Africa and the Canaries. In England it is somewhat local, frequenting chalk hills (the Pearl Skipper).

Larva.—Olive-green. Head large and black. There are two white spots on each side of the tenth and eleventh segments. Feeds on *Coronilla, Ornithopus, Lotus,* and other *Leguminosæ,* in June and July.

2 z

Catæna, Stgr. Stett. Ent. Z. 1861, p. 357.—Has the hind wings on the under side green, and the white spots larger and distinctly edged with black.

Habitat.—Lapland.

7.—H. Nostrodamus, Fab. Ent. Syst. 328 (1793); Ramb. Cat.
   p. 89; Dup. i. 41, 4-6.
   Pumilio, Hffmgg. Ill. Mag. iii. p. 202 (1804); Hüb, Text.
   p. 72; O. i. 2, 216; Z. Is. 1847, p. 291; Frr. 513,
   2, 3; Stgr. Hor. 1870, p. 88.
   Pygmæus, Cyr. Ent. Nap. i. 51, 5 (1787 sed *Pygmæus*, Fab.
   1775 al. erat pap.); Hüb. 458-60; Esp. 99, 3.
   Lefebvrei, Ramb. Cat. p. 90; Bell, Am. Soc. Tr. p. 679,
   1860.

Expands from 1·10 to 1·30 in. All the wings dark brown, especially towards the base; fringes lighter. The female is somewhat lighter than the male, and has a few white dots on the fore wing. Under side paler than above; there are some indistinct white spots near the apex of the fore wing, and sometimes some submarginal spots on the hind wing. Head and prothorax lighter than the rest of the body. Pl. LXXXII., 3.

Time of Appearance.—August.

Habitat.—Dry places in Spain, Italy, Greece, and the South of Europe generally; also in North Africa and Asia Minor. It does not occur north of the Alps.

Larva.—Unknown.

N.B.—*H. Ætna* of Boisduval, Frr., and H. S., is the name given to a species at one time supposed to inhabit Sicily, but nothing is known of it as European beyond the figures, which in all probability represent an American species (*vide* H. S. 226-28).

Genus 5.—**CYCLOPIDES,** Hüb. Verz. bek. Schmett. p. 111
           (1816); Westw. Gen. Diurn. Lep. p. 520 (1852).
   Steropes, Boisd. Gen. Ind. Meth. 34 (1840).

Characters.—Larva pubescent, elongate, striped longitudinally; the head large. Inhabiting rolled-up leaves.

Pupa.—Spun up in silk threads, conical at the abdominal extremity ; the eyes are prominent, and the head pointed.

Imago.—Head not larger than the thorax. Clubs of antennæ short, ovoid, and without terminal hooks. Palpi separate, hairy, the last joint slender and pointed. Wings ample in proportion to the body, which is much more slender than in the last genus. The males are without the subdiscoidal line found in the last genus ; the under side of the hind wings marked with white spots. Marginal fringe of fore wings spotted above. There is only one European species, and another somewhat similar coexists with it in the Amur district.

The males of this and the next genus are without the oblique black line on the fore wings.

1.—**C. Morpheus**, Pall. Reis. i. p. 471 (1771); Wernb. Btr. i. p. 359.

Steropes, S. V. p. 160 (1776); Esp. 41, 1, 80, 4; Hüb. 473-4; O. i. 2, 217.

Aracinthus, Fab. Gen. p. 271 (1777); Godt. i. 12.

Speculum, Rott. Naturf. vi. 31 (1775); Schrk. Fn. B. ii. 1, p. 161 (1801).

Expands from 1·30 to 1·45 in. All the wings are uniform dark brown. The fore wings have a light yellow or yellowish white spot on the costa, and below it another less distinct ; there are three indistinct light spots near the apex. The hind wings are uniformly dark brown ; the fringes are whitish, that of the fore wings indistinctly spotted with brown. Under side : Fore wings dark brown, with a marginal row of small yellow crescentic spots, and one or two other yellow spots near the apex. Hind wings yellow, with a submarginal row of seven large oval white spots bordered with black ; these spots form a continuous band ; between these and the base are five more similar spots, but detached ; there are three central and two basal spots. The sexes are similar in coloration and pattern, except that the female sometimes has an extra yellow spot on the costa of the fore wings above. Pl. LXXXII., 4.

TIMES OF APPEARANCE.—The end of June and the beginning of July.

HABITAT.—Woods in Central Europe (but not in Britain, Denmark, or Belgium), Livonia, North Italy, Central and South Russia, Armenia, Siberia, and the Amur. Though absent from Great Britain and Belgium, it is found in many parts of France, even in the Bois de Boulogne. The French call it " le mirroir," from the beautiful pattern of the under side of the hind wings.

LARVA.—Light green, with a darker dorsal line ; lateral lines yellowish white. Feeds on grasses in May and June.

Genus 6. — **CARTEROCEPHALUS**, Lederer, z. b. v. ii. p. 26 (1853).

CYCLOPIDES, STEROPES aut HETEROPTERUS auctorum.

The species constituting this genus are separated from the last on account of the great width of the head and the prominence of the eyes, and the posterior tibiæ having only two spines. In other respects they resemble *Cyclopides*, except that the wings are chequered with dark brown and fulvous above instead of being uniform brown. There are two European species, one of which occurs in England.

1.—C. Palæmon, Pall. Reis. i. p. 471 (1771); Wern. Btr. i. p. 359.

PANISCUS, Fab. Syst. Ent. 351 (1775); Esp. 28, 2 ; O. i. 2, 219; Godt. i. 12, 1, 2 ; Frr. 513, 1.

BRONTES, Hüb. 475-6.

Expands from 1 to 1·20 in. Fore wings blackish brown, with a marginal row of fulvous dots, and with several large irregular fulvous spots scattered over the area of the wing, and producing a chequered pattern ; the basal spot is larger than the rest. Hind wings with a marginal row of fulvous dots similar to those on the fore wings ; between this and the base are three or four more fulvous spots, large and rounded. Under side : Fore wings yellow, with greyish black spots. Hind wings yellow, with the spots of the upper side paler, sometimes almost white, and surrounded by dark rings. The sexes are similar, except that the female is somewhat lighter than the male. Fringes yellowish and unspotted. The antennæ are bright fulvous beneath.

TIMES OF APPEARANCE.—May and June.

HABITAT.—Woods in Central Europe (except North Germany and Denmark), Finland, Central and South Russia, North Italy, Dalmatia, Armenia, Siberia, and the Amur. It is a very local species, and in Britain it is confined to a few localities in the Midland Counties (the Chequered Skipper). Pl. LXXXII., 5.

LARVA.—Brown, with yellow lateral stripes; the head is dark brown, the first segment being orange, the legs reddish brown. It feeds on plantain (*Plantago major*) in April. Pl. LXXX., see note to *H Thaumas*.

2.—**C. Sylvius**, Knoch. Btr. i. 5, 1, 2 (1781); Esp. 80, 5, 6; Hüb. 477-8, 641-4; Godt. ii. 27, 1, 2; O. i. 2, 221 (*Silvius*), Frr. 226, 2, 3, 691, 1.

Expands from 1 to 1·10 in. Yellowish fulvous; hind margins of all the wings with narrow borders of dark brown. Fore wings with a marginal row of small black spots; between these and the base four larger black spots, all more or less rounded. In the female there is a black patch on the inner margin near the base. Hind wings nearly similar to those of *Palæmon*, but the fulvous spots are proportionately larger, especially those along the hind margin. Under side : Fore wings yellowish fulvous, with black spots as above. Hind wings greenish grey, with fulvous spots as above. Pl. LXXXII., 6.

TIMES OF APPEARANCE.—May and June.

HABITAT. — Woods in South Norway and Sweden, North Germany, Finland, Central Russia, the Ural, Armenia, Siberia, and the Amur.

LARVA.—Yellowish or brownish white, with reddish dorsal and lateral lines; head light brown. Feeds on grasses.

OTHER SPECIES OF *HESPERIDÆ* INCLUDED IN STAUDINGER'S
CATALOGUE AS BELONGING TO THE EUROPEAN FAUNA.

Genus *SYRICHTHUS.*

S. Gigas, Brem. Lep. O. S. p. 96, T. viii. 3.—Very close to
Tessellum in colour and markings, but is larger; probably a variety
of that species. Habitat, Eastern Amur.

S. Tessellum var. *Nomas*, Led. z. b. V. 1855, p. 193, T. 1, 7.—
Has the hind wings uniformly white beneath. Habitat, Syria.

S. Poggei, Led. Wein. Mts. 1858, p. 141.—The size of *S. Proto*,
but more resembles *Phlomidis* in its markings, &c., as far as I can
judge from Lederer's description. I have not seen a specimen.
Habitat, Syria (Damascus).

S. Maculatus, Brem. & Grey, p. 11; Brem. Lep. O. S. p. 31.
—Expands 1·50 in. Wings above blackish grey; the fore wings
with two interrupted white bands and a white central lunule; the
hind wings with a double row of white dots. Under side: Fore
wings the same as above, but with the apex brown. Hind wings
banded with white and brown. Habitat, Amur.

S. Inachus, Mén. Schrk. p. 46, T. iv. 2.—Expands ·90 in.
Wings purplish brown, glossy. Fore wings with an angu-
larly bent stripe and two small spots of white. Hind wings
with two obscure white lines. Under side similarly marked, but
the ground colour is green mixed with red. Habitat, Amur.

Genus *NISONIADES.*

N. Montanus, Brem. Lep. O. S. p. 31, T. ii. 4.—Expands from
1·02 to 1·10 in. Fore wings olive-grey, with a darker olive basal
spot, two oblique bands, and marginal lunules of the same colour;
on the anterior margin are three white dots. Hind wings blackish,
with a whitish median lunule or stripe and marginal rows of yellow
spots. Under side: Fore wings brownish black, with three rows
of yellow spots. Hind wings the same as the upper side. Habitat,
Amur.

N. Popovianus, Nord. Bull. Mosc. 1851, iv. p. 443, T. 12, 3, 4.
—Size of *Tages*. Wings entire, the fringes not spotted. Fore
wings with a white dot, and the costa and a marginal row of white

dots on a nearly black border. Hind wings with distinct white dots, forming submarginal and marginal rows, and two more near the base. Fringes of fore wings grey, that of the hind wings white. Habitat, South-Eastern Siberia (Kiachta).

N. Guttatus, Brem. & Grey, Schmett. Faun. Nord. Chine, p. 10; Mén. Schrk. 46.—Expands 1·50 in. Wings dark brown above, the bases and the body tinged with green. Fore wings with two transparent spots in the centre, and with an apical band composed of six transparent spots. Hind wings with four central transparent spots arranged in a row. Under side the same as above. Habitat, Amur (Oussouri River) in August.

N. Thetys, Mén. En. ii. p. 126, T. x. 8.—Expands 1·50 in. The wings are very dark brown above. Fore wings ornamented on their disc with six white transparent spots, three being placed near the costa and three beneath them, the lowest nearly touching the hind margin. Hind wings with a slight trace of a lightish central band; hind margin slightly dentate, and with fringe spotted with white. Habitat, Amur.

Genus *HESPERIA*.

H. Sylvatica, Brem. Lep. O. S. p. 34, T. iii. 10.—Expands ·80 in. Wings above ochreous, broadly dark brown along the hind margin; neuration dark. Fore wings with a central dark spot. Under side ochreous, with the nervures dark brown. Wings dark towards the base; fore wings with an obscure spot on the inner margin. Habitat, Amur.

H. Ochracea, Brem. Lep. Ost. Sib. T. i. f. 11.—Expands ·80 in. Wings ochreous above, the neuration black. The fore wings have the hind margin broadly fuscous, with a fuscous transverse stripe reaching from the costa to the inner margin. Hind wings broadly fuscous along the hind margin. Under side the same as above, but paler. Habitat, Amur.

H. Hyrax, Led. Wien. Mts. 1861, p. 149, T. 1, 6.—Rather larger than *Thaumas*, light fulvous above; the subdiscoidal line in the male very short. Under side: Fore wings with a broad black line near the inner margin. Hind wings uniform greenish grey. Habitat, Syria.

H. Thrax, Linn. Syst. Nat. xii. 794; Led. z. b. V. 1855, p. 190, T. 1, 9, 10.—Wings with the hind margins entire, dark brown. Fore wings with three transparent spots, the outer one smaller than the rest. Habitat, Syria and Asia Minor.

H. Zelleri, Led. z. b. V. 1855, p. 194.—Expands from 1·30 to 1·40 in. Wings rather light brown, with a central row of semi-transparent white spots reaching from the costa to the inner margin, the two lowest being much larger than the rest, which are mere dots; in the discoidal cell are two small white spots placed one above the other. Hind wings without markings. Under side as above. Habitat, Syria.

H. Alcides, H. S. 41, 42, vi. p. 38.—Expands 1·50 in. Dark brown. Fore wings with four yellow spots placed in a row reaching from the centre nearly to the inner margin; near the apex is a row of lighter and smaller spots, and there are two more placed one above the other at the extremity of the discoidal cell. Hind wings with two small yellow spots near the centre. Under side: Fore wings bluish grey, basal portion brownish, the yellow spots arranged as above. Hind wings bluish grey, the neuration blackish. Habitat, Asia Minor.

Genus *CYCLOPIDES*.

C. Ornatus, Brem. Lep. O. S. p. 33, T. ii. 5.—Expands from ·80 to ·90 in. Wings greyish black above, the marginal fringes yellowish grey. Under side: Fore wings with the costa black, the apex and hind margin ochreous. Hind wings ochreous, with a straight silvery stripe reaching from the base to the hind margin; the inner margin often silvery. Habitat, Amur.

Genus *CARTEROCEPHALUS*.

C. Argyrostigma, Ev. Bull. Mosc. 1851, ii. p. 624; Nord. Bull. Mosc. 1851, iv. p. 442; T. 12, 1, 2.—Slightly smaller than *Palæmon*. Wings brownish black. Fore wings with three large light yellow spots touching the costa, reaching to the centre of the wing and decreasing in size towards the base; below these and alternating with them so as to produce a chequered pattern are two similar spots touching the inner margin. Hind wings with three small yellowish ochreous spots near the hind margin and a continuous band of smaller spots towards the base. Under side: Fore wings as above, but the ochreous spots are larger. Hind wings reddish brown, with two large central silver spots and some smaller ones near the base and the hind margin. Marginal fringes orange; head and collar yellowish orange; thorax dark brown; abdomen black, with an orange anal tuft. Habitat, South-Eastern Siberia (Kiachta).

NORTH AMERICAN SPECIES OF *HESPERIDÆ* ALLIED TO THOSE OF THE PALÆARCTIC REGION.

The Nearctic Region is numerically richer in *Hesperidæ* than the Palæarctic, and several genera are represented of which no species occur in the European region.

The genus *Spilothyrus* is not represented in North America.

Syrichthus Oneko, Scudder, Proc. Ess. Inst. iii. p. 176, 1862.

S. Hegon, Scudd. *loc. cit.*—Both these inhabit the States.

S. Wydanot, Edw. Proc. Ent. Soc. Phil. ii. p. 21, 1863. Habitat, Washington.

S. Ricara, Edw. *loc. cit.* iv. p. 203. Habitat, Colorado.

S. Ridingsii, Reak, *loc. cit.* p. 151 (1866). Habitat, Colorado.

The last three species are placed between *S. Phlomidis* and *Orbifer* in Mr. Kirby's Catalogue.

S. Oileus, Westwood and Humphreys' Brit. Butt. pl. 38 (1841). *Hesperia Syrichthus*, Lat. Ency. Méth. ix. (1823).—This species was erroneously accounted British, and is described as such by Westwood and Humphreys; it inhabits the Southern States of North America.

S. Ruralis, *S. Cœspitalis*, and *S. Scriptura*, Boisduval, Ann. Soc. Ent. Fr. 1852, p. 311-313.—All inhabit California.

The following species of *Nisoniades* occur in North America:—

N. Briso, Boisd. Lec. Lep. Amer. Sept. t. 66 (1832), &c. Habitat, the United States.

N. Martialis, Scudder, Proc. Bost. Nat. Hist. Soc. xiii. 1870. Habitat, the Northern States.

N. Terentius, Scudd. *loc. cit.*, and *N. Ovidius*, Scudd. *loc. cit.* Habitat, Florida.

N. Funeralis, Scudd. *loc. cit.* Habitat, Texas.

N. Juvenalis, Fabr. Ent. Syst. iii. 1, p. 339, no. 291 (1793). Habitat, the States.

N. Ennius, Scudd., *N. Persius*, Scudd., and *N. Lucillus*, Scudd., Proc. Bost. Nat. Hist. Soc. xiii. 1870. Habitat, Massachusetts.

Several other species are also described by Scudder: *N. Propertius* and *N. Tibullus* from California, allied to *N. Tristis*, Boisd., from the same locality; *N. Plantus*, from Florida.

N. Catullus, Fabr. Ent. Syst. iii. p. 348 (1793).—Is widely distributed through the States.

About forty-five species of North American species of *Pamphila* appear in Mr. Kirby's Catalogue; one of these, *P. Vitellius*, Smith, Lep. Ins. Georg. i. t. 17 (1797), inhabiting Georgia, was admitted as British by Stephens (Ill. Brit. Ent. Haust. i. p. 102, 1828), said to have been taken at Barnstaple. It is also figured in Wood's 'Index Entomologicus' under the name of *P. Bucephalus*, St., with the localities Devonshire and Surrey. It is possible it may have been accidentally imported at that time.

Thymelicus Numitor, Fabr. Ent. Syst. iii. 1, 324, *T. Marginatus*, Harris, Ins. Mass. 1862, *T. Delaware*, Edw. Proc. Ent. Soc. 1863, are the only North American species of this genus recorded in Mr. Kirby's Catalogue. The genus *Thymelicus*, together with *Pamphila*, would correspond to the genus *Hesperia* of this work.

Two species of *Carterocephalus* occur in North America:—

C. Mandan, Edwards, Proc. Ent. Soc. Phil. ii. p. 20 (1863). Habitat, Lake Winnipeg.

C. Mesapano, Scudd. Proc. Bost. Nat. Hist. Soc. xi. p. 383, n. 94 (1868). Habitat, Massachusetts.

They are both allied to *C. Palæmon*, Pall.

ADDENDA AND NOTES TO THE EUROPEAN SPECIES.

Lycæna Minima, Fuess. p. 128, add as a synonym :—
 Alsus, F. Mant. 73 ; Hüb. 278-9 ; O. i. 2, 22 ; Godt.
 ii. 26, 5, 6.

After the description, on p. 107, of *Lycæna Zephyrus*, Frr.,
insert the following :—

Lycæna Lycidas, Trapp; G. H. Jäggi, Mitheilungen die Schweizer
 Entomologischen Gesellschaft, Feb. 1881.

Expands from 1·20 to 1·25 in. Male blue ; hind margins
brownish black. Hind wings with some marginal black dots ;
inner margin greyish white. Female brown. Fore wings with a
narrow discoidal black spot. Hind wings with some orange-yellow
lunules near the anal angle. Fringes of all the wings white in
both sexes. Under side grey, somewhat brownish in the female.
Fore wings with a discoidal and a central row of black spots ;
hind margin with a double row of black spots, between which is a
row of orange lunules. Hind wings with three basal black spots,
an irregular central row, external to which is a whitish band, and
a submarginal orange band placed between two rows of black spots.
Pl. LXXXII., 7 (supplementary).
 Time of Appearance.—End of June.
 Habitat.—The Simplon and other elevated places in Switzerland.

Page 176. **Vanessa Antiopa**, Linn.

Note.—It is sometimes said that southern specimens of this
species have the hind marginal border yellow, whilst those from
the more northern parts of Europe, and especially British examples,
have the border white. I found the two specimens in the collection
made by the late Sir Sidney Saunders, in Albania, both with white
borders. It may be mentioned that the same collection contained
specimens of *Polyommatus Phlæas* and *Pararge Egerides* exactly
resembling those found in Britain, no special Turkish forms of
these species being present.

Page 178.   **Vanessa Callirhoe v. Vulcanica.**

Note.—" I have found this species in Madeira in 1879.   It is common in gardens in February.   It flies like *Atalanta*, and is often seen settling on a path or road with expanded wings, basking in the sun, in which position it is easy to approach.   The larva lives on nettle (*Urtica dioica*), and resembles that of *Vanessa Atalanta;* it lives, like *Atalanta*, in a rolled-up leaf, in which it changes to a pupa.   I have now an empty pupa-case before me; it resembles that of *Atalanta*, but is perhaps a little lighter in colour.   I have found the larva and pupa in April."—T. D. A. Cockerell, ' Science Gossip,' July 2nd, 1883.

Page 202, after the description of *Argynnis Polaris*, Boisd., insert:—

Argynnis Improba, Butler, Markham's ' Polar Reconnaissance,' p. 351 (1881).

Expands 1·60 in.   Very dull fulvous, suffused with fuscous towards the base of all the wings.   Fore wings with a marginal and submarginal row of black spots and some larger ones in the centre of the wing placed irregularly.   Hind wings fulvous along the hind margin, basal and central portion fuscous; there is a submarginal row of black spots.   Under side somewhat as above, but lighter; there are some black transverse lines in the discoidal cell.   Hind wings light fulvous on their outer or hind marginal half, with some reddish spots; the basal half reddish brown, with a bluish white spot near the centre of the costa; beneath this is a central yellowish band not reaching as far as the inner margin.   Drawn from the specimens in the British Museum. Pl. LXXVII., 8 (supplementary).

HABITAT.—Novaya Zemlya.   This species seems to be very distinct, but resembles *Thore* or *Frigga* perhaps more than any other *Argynnis.*

ADDENDA AND NOTES TO THE DESCRIPTIONS OF NON-EUROPEAN
SPECIES OF THE PALÆARCTIC FAUNA.

Genus *PAPILIO.*

P. Podalirius var. *Latteri*, Austat. Petit. nouv. Mem. ii. p. 293.
—Expands from 2·80 to 3·30 in. Differs from the variety *Feisthamelii* in being larger, and having the ground colour of the wings
whiter ; the tails of the hind wings are also longer in proportion
than those of *Podalirius.* The abdomen is white, as in var.
Zancleus. Habitat, Algeria.

Genus *HYPERMNESTRA.*

Note.—H. Helios var. *Maxima*, Staud.—A variety differing from
the type in the greater expanse of the wings and in the intensity of
the markings. *H. Helios* is, according to Erschoff, very common
in the Deserts of Turkestan, especially in the neighbourhood of
the Sea of Aral. He says the ground colour of the wing varies
from light to dark, and that the red and black markings vary very
much in size and beauty.

Genus *PARNASSIUS.*

Note.—P. Corybas, F. de W. 242.—The original species described
by Fischer de Waldheim as coming from Kamtchatka is not the same
as that figured by Erschoff, Lep. Turkestan, p. 2, pl. i. figs. 1, 2 ;
the latter species has been named *P. Discobalus* by Staudinger, and
the former described in Staudinger's Catalogue as *P. Delius ?* var.
Corybas is now considered a small variety of *P. Discobalus*, and
named var. *Minor.* *P. Discobalus* much resembles *P. Corybas*,
described at p. 23 of the present work, but is larger, lighter in
colour, and has the red spots round and clear.

Genus *PIERIS.*

P. Mesentina, Cram. Pap. Exot. 370 ; Godt. Ency. Méth.
p. 130 ; Boisd. Sp. Gen. p. 501.—Expands from 2 to 2·20 in.
Wings white. Fore wings rather pointed ; in the male with a
black discoidal lunule ; half-way between this and the apex a broad
black wavy line reaching to about the centre of the hind margin ;

from this run broad black lines to the margin of the wing, following the course of the nervules. Hind wings with a row of black crescents along the hind margin. Female with all these markings more intense, the apex and hind margin being black, with two apical white spots, the discoidal spot oblong, and the inner half of the costa black. Under side marked as above, but the apex of the fore wings is yellowish veined with black; the hind wings yellowish, with the neuration broadly black. Habitat, Syria.

P. Leucodice, Ev. Bull. Mosc. 1843, iii. p. 541; H. S. 619-20. —Expands from 1·50 to 1·80 in. The shape of *P. Rapæ*. Wings white. Fore wings with a submarginal black band reaching from the costa as far as the third median nervule; from this black lines extend to the margin; there is an oblong black discoidal spot. Under side tinged with yellowish. Fore wings marked as above. Hind wings with the neuration marked with blackish, and with a central blackish line or narrow band. Habitat, Persia.

Genus *COLIAS.*

C. Eogene, Ersch. Lep. Turkest. p. 6.—About the size of *Chrysotheme.* Orange, with a decidedly red tinge; the black border as in *Edusa*, but straighter; discoidal spot narrow. The hind wings have the border straight, the discoidal spot indistinct from its resemblance in colour to the rest of the wing. Under side: Fore wings greenish, orange towards the inner margin. Hind wings yellowish, the discoidal spot pearly white. Habitat, Turkestan.

C. Olga, Romanoff, Memoires Lep. 1883.—The size of *Aurorina.* The male is bright orange, the fore wings have the border as in *Edusa*, that of the hind wings being the same as in *Myrmidone;* the wings have strong violet reflections. Under side as in *Edusa*, but brighter in colour. The female has only been found in the white form; it closely resembles *C. Aurora* var. *Chloë*, but it is altogether lighter, and the discoidal spot of the hind wings is conspicuously reddish orange. Habitat, the Caucasus.

C. Viluiensis, Ménétr. Bull. de l'Acad. xvii. p. 213; Men. Schrenk's Reisen, p. 18, pl. i. fig. 7.—Expands 2 in. Wings pale orange, hind margins with rather a narrow border finely powdered with yellow ; discal spot of the fore wings very faint, being merely a small black ring, that of the hind wings being represented by a faint orange ring. Under side appears to resemble that of *C. Hecla*, but is paler, the hind wings being yellowish green; the pearly discal spot is double. Described from Ménétries's figure. I have never seen a specimen, and therefore cannot say to which group of the genus this species rightly belongs ; the figure puts one in mind of *Hecla* more than of any other species. Habitat, near the Vilui River, Central Siberia.

C. Aurora, var. fœm. *Chloë*, Evers. Bull. Mosc. 1847, ii. p. 73, t. 4; H. S. 457-8.—Bears the same relation to *Aurora* as *Helice* to *Edusa*. It resembles the orange female of *Aurora*, with the exception of the ground colour, which is white tinged with greenish yellow, and dusky towards the bases. The discoidal spot of the hind wings is orange. Habitat, Amur.

C. Alpheraki, Stgr.— Slightly larger than *Phicomone ;* the colour of the wings is brighter, and the border is darker. The female is clearer and greener in colour in the discal portions of the wings. Under side pale greenish grey, almost without any markings ; there is a central pearly spot. Habitat, Turkestan.

C. Wiskolti, Stgr.—Size of *Aurorina*. The wings are greenish. Fore wings with a very broad border extending over a third of the width of the wing; the discoidal spot is long and narrow. The hind wings have a broad border similar to that of the fore wings, but veined with greenish yellow. Under side greenish. Fore wings with a black spot and some obscure spots near the anal angle. Hind wings greenish ; discoidal spot small and pearly. Fringes of all the wings greenish yellow. Habitat, Samarkand.

C. Staudingeri (in coll. Godman).—Very like *Hecla* above, but larger. Fore wings with the discoidal spot long and narrow. The hind margins of all the wings have narrow borders, as in *Hecla ;* a narrow rosy red fringe extends round the borders of all the wings. Under side : Fore wings greenish, except along the inner margin, which is orange mixed with grey. Hind wings greenish ; discoidal spot small and pink. Habitat, Kuldja.

The above three species of *Colias* are described from specimens in Mr. Godman's collection.

Genus *THECLA.*

T. Lunulata, Ersch. Lep. Turkest. p. 7, pl. i. 5.—Expands 1·10 in. Wings above unicolorous dark brown ; the hind wings with rather long tails ; fringes white. Under side ashy grey ; a white central streak reaches along the entire length of the wings ; the fore wings have a submarginal row of four black spots, and the hind wings a row of black spots, above which are some white lunules ; hind margins white. Habitat, Sarafschan in May and June.

T. Mirabilis, Ersch. Lep. Turkest. p. 7, pl. i. 4.—Very close to *Lunulata,* but the colour of the under side is darker, and the white streak does not reach for more than half the length of fore wings. The hind wings have a submarginal row of black spots enclosed in white rings, and not surmounted by white lunules. Habitat, Sarafschan at the end of May.

Genus *THESTOR.*

T. Fedtschenkoi, Ersch. Lep. Turkest. p. 8.—Fringes of wings white. Anterior wings fulvous, with obscure fuscous spots ; the hind margins fuscous. Hind wings entirely dark brown (the male with two fulvous spots at the anal angle). Under side of all the wings light green ; disc of the fore wings fulvous, spotted with dark brown. Habitat, Maracanda in April and May.

Genus *POLYOMMATUS.*

P. Solskyi, Ersch. Lep. Turkest. p. 8. — The male has all the wings brilliant copper. Fore wings with the hind margin and apex broadly black. Under side as in *Ochimus,* but paler ; the central spots not enclosed in light rings ; the transverse series of submarginal black spots on the hind wings not contained in a red band, but those of the inner series alone distinctly bordered with red externally. The female differs considerably from *P. Ochimus.* (The female is not, however, further described). Habitat, Maracanda, August.

P. Sultan, Stgr. (in coll. Godman).—About the size of *Phlœas,* but dark brown with violet reflections in the male. There are no black spots. The hind wings have long tails. Habitat, Samarkand.

Genus LYCÆNA.

The following new species of *Lycæna* are described from the specimens, mostly received from Dr. Staudinger, in Mr. Godman's collection, which he has kindly permitted me to examine.

It will be seen, on referring to the list of Rhopalocera at the end of the volume, that there are many species of this genus that are not noticed beyond the mention of their names. I have, however, endeavoured to give short descriptions of all the recently received species in Mr. Godman's collection; they are for the most part from Turkestan and other parts of Central Asia, and are enumerated in Staudinger's MS. Catalogue for the current year :—

L. Alcedo, Stgr. MS. Cat. 1883.—Expands from 1 to 1·20 in. Male dull violet-blue, broadly bordered with dark brown. Fore wings with a narrow black discoidal spot. Hind wings with a hind marginal row of black spots and traces of an orange band. Female brown. Fore wings with the discoidal spot very small. Hind wings with a very slight trace of marginal spots. Under side pale grey ; marginal bands pale yellow ; the spots forming the central row on the fore wings very large ; some silvery spots on the hind marginal row of the hind wings. Habitat, Persia.

L. Iris, Stgr. MS. Cat. 1883.—Expands from 1 to 1·20 in. Close to *Hycana.* The male has the wings dark blue. Fore wings with a black discoidal spot. All the wings rather broadly bordered with black. Under side brownish grey. Fore wings with central and discoidal black spots, and with a marginal double row of smaller spots. Hind wings almost as in the female of *Cytis*, but the base is less blue. The female is larger than the male. Fore wings brown above, with a black discoidal spot. Hind wings brown, with a marginal row of black spots in blue rings. Under side : Spots as in the male, but larger. The hind wings have three black spots near the anal angle enclosed in metallic silvery blue rings. Habitat, Samarkand.

L. Cytis, Stgr. MS. Cat. 1883.—About the size of *Lysimon.* Male dark blue (the colour of *Arion*) ; fringes plain. Fore wings

3 B

with a submarginal row of four black spots and a discoidal spot. Hind wings without spots. Under side grey. All the wings with a row of black spots. Fore wings with a black discoidal spot. Hind wings blue at the base; a narrow discoidal and two basal spots. The female resembles the male, but the ground colour is brown instead of blue. A variety occurs in which the wings are unspotted above, and the hind wings beneath have a marginal border of black spots. Habitat, Samarkand.

The two following species are close to the European *L. Rhymnus*, Ev. :—

L. Pretiosa, Stgr. MS. Cat. 1883.—Differs from *Rhymnus* in having the fore wings more rounded. The fringes are brown, spotted with white in place of being brown. Under side : Fore wings like *Rhymnus*, but the white streaks are bordered with black. Hind wings with two wavy white streaks extending across the wing and bordered with black; there are three marginal rows of spots, those of the inner row being white, the second black, and the outer orange ; next to the hind margin is a narrow white line. Habitat, Morgelan, Russian Turkestan.

L. Anthracias, Stgr. MS. Cat. 1883.—Size of *Rhymnus*. Fore wings rather pointed. The wings are uniform brown above, the fringes being broad and chequered with brown and white. Under side : Fore wings with three large square black subcostal spots ; another similar one below the external one of these ; hind margin with a row of whitish spots bordered with black. Hind wings greyish black, with a black basal spot and central and hind marginal rows, all in white rings. Habitat, Kuldja.

L. Eversmanni, Stgr. (in coll. Godman).—Expands from 1·25 to 1·30 in. The male is dark rich blue, with an obscurely defined brown border. Fore wings with an obscure linear discoidal spot. Hind wings with four round black marginal spots, the second from the anal angle being the largest. Under side brownish grey. Fore wings with submarginal and discoidal spots very black and distinct, and enclosed in white rings ; there is a double row of brown marginal spots. Hind wings with four basal spots arranged in a square ; the submarginal row like that of the fore wings ; there is a marginal orange band between two rows of black crescents, and near the anal angle are several silvery blue spots. Habitat, Samarkand.

L. Gigas, Stgr. (in coll. Godman), *Lowei*, var. — Expands 1·30 in. Male brilliant blue. Fore wings with no discoidal spot; hind margins of all the wings narrowly blackish brown. Hind wings with a few hind marginal black spots. Under side light grey, marked as in *Pylaon*, but the spots are larger, and there are no orange marginal bands; but near the anal angle of the hind wings are two very bright reddish orange spots with shining silvery blue centres. This magnificent insect is described from a single specimen in Mr. Godman's collection. Habitat, Taurus.

L. Scylla, Stgr. MS. Cat. 1883.—Size of *Alcon*. The male has the wings dark purplish blue, deeply bordered with dark brown. Fore wings sometimes with an obscure black discoidal spot above; fringes bluish white, not spotted. Under side bluish grey. Fore wings with a linear discoidal spot, and an even marginal band of round black spots increasing in size from the costa to the inner margin. Hind wings bluish at the base; there is only one basal spot near the costa; central row rather wavy towards the anal angle, but the spots are round and distinct. Female brown above; otherwise the same as the male. Habitat, Amur.

L. Eros var. *Amor*, Stgr. MS. Cat. 1883.—Size of the type. The male is light blue; all the wings with a narrow black border. Hind wings with a hind marginal row of black spots; fringes white. Under side brownish grey. Fore wings with a discoidal and with a submarginal row of black spots; basal spots absent. Hind wings with three basal black spots and a submarginal row; discoidal spot white; a marginal orange band between two rows of black spots. Female brown, bluish at the base. Fore wings with a discoidal black spot and an indistinct marginal brownish band on the fore wings. Hind wings with a marginal orange band spotted with black. The under side much resembles that of *Bavius*. Habitat, Samarkand.

L. Poseidon, Ld. (*vide* p. 144).—Dr. Staudinger apparently now considers this to constitute a good species. The specimens in Mr. Godman's collection are very bright blue above; fringes white; the hind margins slightly dusky. The female resembles that of

Menaclas. There is a white basal streak beneath in both sexes. *L. Xerxes*, Stgr., appears to be a variety of *L. Icarus;* the colour of the wings above in the male is intense blue; there is no basal streak beneath. Habitat, Persia.

L. Phyllis, Stgr. MS. Cat. 1883.—Size of *Damocles.* All the wings greenish rather than blue, the neuration dusky. The under side resembles that of *Damocles.* Habitat, North Persia.

L. Phryxis, Stgr. MS. Cat. 1883.—The size of *Icarus*, which the male greatly resembles above, but beneath the ground colour is very pale brownish grey, and the black spots are much smaller. The fore wings of the male have basal spots beneath; the hind wings are very bright blue at the base. The female has the fore wings uniform dark brown, with an obscure discoidal spot. Hind wings with four marginal black spots surmounted by orange dots. Under side without basal spots on the fore wings; the other spots are large and distinct on all the wings. Habitat, Samarkand.

L. Pryeri, Murray, Ent. Mo. Mag. x. p. 126, 1873.—Expands from 1·40 to 2 in. Dark brown; basal portions of all the wings violet-blue dashed with white, especially in the female. Fore wings with a narrow black discoidal spot. Hind wings with a similar discoidal spot. Under side bluish white; all the wings with a very narrow linear discoidal spot, and with marginal and submarginal rows of black spots, the submarginal spots being longer than the external ones. Habitat, the Amur and Japan.

Obs.—This is the largest species of the genus, and has hitherto been thought to be confined to Japan. Mr. Godman, however, has a specimen received from Dr. Staudinger from the Amur; it is slightly smaller than the Japanese specimens.

Genus *LIMENITIS.*

L. Lepechini, Ersch. Lep. Turkest. p. 14.—Expands from 1·60 to 1·80 in. The wings are dark brown, central band white; there are two rows of submarginal yellow spots. Under side greenish yellow, the central band white. Habitat, Maracanda, June.

Genus *VANESSA.*

V. Callirhoe, Hüb. Sam. Ex. Schmett. (1806). *Atalanta Indica*, Herbst, Nat. Schmett. vii. t. 180 (1794).—The form described on

p. 177 is the variety *Vulcanica* of Godt. The typical form of the
species inhabits Northern India, China, and the Amur; it is larger
than *Atalanta*, and has the band of the fore wings more wavy and
broader, whilst the colour of the band is yellowish orange rather
than red, more like that of *Cardui;* the apical white spots are
smaller than in *Atalanta*, but not so small as in *Vulcanica*. I have
received through Dr. Staudinger specimens from the Amur.

Genus *JUNONIA.*

J. Here, Lang, Entom. Sept. 1884, p. 206.—Expands from 1·80
to 2 in. The male has the fore wings black, costa and hind margin
white; there is a central white band reaching to about midway
between the costa and the inner margin; below this and reaching
the inner margin is a broad blue band; straight on its inner edge,
and containing an obscurely ocellated spot. In the discoidal cell
are some blue spots nearly touching the costa. Hind wings blue,
black at the base, with two black spots, the lower one obscurely
ocellated; hind margin blue, with black lines. Under side pale
stone-colour. Fore wings with black and blue markings. Hind
wings with a few brown lines; near the anal angle is a small blue
dot. The female has the wings slate-colour, and the black spots
are conspicuously ocellated, and with red rings above. In other
respects it resembles the male. Habitat, Turkey in Asia (Bagdad).

I have described this species in the 'Entomologist' for
September, 1884. It differs from *J. Orithya*, L., with which
it has been hitherto confounded, in several important particulars.
It has occurred in the locality mentioned above, and recently
has been taken at Aden by Major Yerbury. It is much
smaller than *Orithya*, L.; it has no red discoidal spots on the fore
wings; the black spots are not plainly ocellated in the male; the
colour of the under side is much paler and entirely without
the reddish tinge found in that species.

Genus *EPINEPHELE.*

E. Janiroides, H. S. 533-4.—Expands from 1·75 to 2 in. The
male has the fore wings dark brown, with a submarginal band of
light fulvous containing an oval black spot at its upper part,

in which is a white dot. Hind wings light fulvous, brown at the base and along the hind margins. Female similarly coloured, but the fore wings are more or less fulvous throughout, excepting along the hind margin ; the black spot has two white dots. Hind wings the same as in the male. Under side : Fore wings fulvous ; hind margins and costa light brown; the black spot as above. Hind wings brown, the basal portion darker than the rest; there is a submarginal row of black dots in yellowish rings placed on a lightish band. Habitat, North Africa. Described from specimens received from Dr. Staudinger ; they are apparently the same as *Janiroides*, H. S.

E. Pulchella, Stgr. MS. Cat. 1883. — I have received a specimen of this species from Dr. Staudinger. The male is about the size of *E. Amardæa*. The fore wings are fulvous, with a black apical spot ; the hind margins dark brown. Hind wings uniform dark brown, not dentate, but rather prominent at the anal angle ; fringes white. Under side almost exactly as in *Naubidensis*, but the fore wings are paler fulvous. Habitat, Turkestan.

BIBLIOGRAPHICAL LIST.

N.B.—*Some of the Works here mentioned I have not actually seen; these are mostly copied from Staudinger's table, and are followed by (Stgr.). A few Works whose titles are given in full in the text are not again referred to in this List.*

Ann. Soc. Esp.   Anales de la Sociedad Espanola de Historia Natural.   Madrid.
Ann. S. Belg.   Annales de la Société Entomologique Belge Bruxelles, 1857, &c.
Ann. S. Fr.   Annales de la Société Entomologique de France.   Paris, 1856, &c.
Ann. S. Lin. Paris.   Annales de la Société Linnéene de Paris, 1822, &c.
B. Gen.   Boisduval, Genera et Index methodicus.   Parisiis, 1840.
B. Ic.   Boisduval, Icones Historiques de Lépidoptères nouveaux ou peu connus. Paris, 1832-34.
B. Rbr. et Grasl. Coll. Ic.   Boisduval, Rambur et Graslin, Collection Icono-graphique et Historique des Chenilles d'Europe.   Paris, 1832.
B. Spec. Gen.   Boisduval, Species Général des Lépidoptères, Tome i.   Paris, 1836.
Bell.   Bellier de la Chavignerie.   Le Nouveau Guide de l'Amateur d'Insectes.
Bergstr. Nom.   Bergsträsser, Nomenclatur und Beschreibung der Insecten in der Grafschaft Hanau.   Hanau, 1779.
Berge Schm.   Berge's Schmetterlingsbuch, &c.   Stuttgart, 1863.   (See Kirby, Eur. Butt. & Mo.)
Berl. E. Z.   Berliner Entomologische Zeitschrift.   Berlin, 1857, &c.
Bienart.   Reise nach Persien, &c.   Leipzig, 1870.   (Stgr.).
Bkh. (i.-v.).   Borkhausen Naturgeschichte der Europaischen Schmetterlinge. Frankfurt, 1788-1794.
Bon. Descr.   Bonelli Descrizione di sei nouve specie d'Insetti, &c.   (Mém. d'R. Accad. d. Sc. di Torino, Tome xxx., 1824.
Brem. & Grey.   Beitrage zur Schmetterlings Fauna des Nœrdlichen Chinás von Otto Bremer und William Grey.   St. Petersburg, 1853.
Brem. Lep. O. S.   Bremer, Lepidopteren Ost-Sibiriens, insbesondere des Amur-landes (Mém. d' l'Acad. Imp. des Sc. de S. Petersburg, 1864).
Bull. M.   Bulletin de la Société Impériale des Naturalistes de Moscou, 1829, &c.
Bull. S. E. It.   Bulletino della Società Entomologica Italiana.   Florence, 1869, &c.
Butl. Cat.   A. G. Butler, F.L.S., &c., Catalogue of the Satyridæ.   Lond. 1868.
Cr. P. Ex.   Cramer, Papiliones Exotiques des trois parties du Monde.   Utrecht, 1775-1782.   Supplement by Stoll, 1787-1790.
Chenu.   Encyclopedie d'Histoire Naturelle.   Vol. 1., containing Rhopalocera and Sphingina.
Chr. *or* Christp.   Christoph, Hor. Soc. Ent. Ross. &c.
Costa, Faun. Nap.   Costa, Fauna del regno di Napoli, &c.   Napoli, 1832-6.   (Stgr.).
Curtis, B. E.   British Entomology.   London, 1835-1810.
Curt. Descr. App. Nar.   Description of the Insects in the Appendix to the Narra-tive of a Second Voyage in search of a North-West Passage.   London, 1835.
Cyr. Ent. Neap.   Cyrilli Entomologiæ Neopolitanæ.   Neapoli, 1787.   (Stgr.).
Dalm. Anal. Ent.   Dalman, Analecta Entomologica.   Holmiæ, 1823.
Dalm. Förs.   Dalman, Försök till sistematisk Uppställning af Sveriges Fjärilar, 1816.   (Stgr.).

Dbld. Gen. D. L.   Doubleday (Westwood and Hewitson), the Genera of Diurnal Lepidoptera. London, 1846-1852.
Don. Nat. Hist. Donovan, Natural History of British Insects. London, 1792.
Donz. Donzel, Hugues.
Drury. Drury, Illustrations on Natural History. London, 1770.
Duncan. Duncan, British Butterflies. Edinburgh, 1835.
Dup. Duponchel, Histoire Naturelle des Lépidoptères, &c., 1826-1838. Supplement, 1832-1848. (Continuation of Godart's work).
Dup. Cat. Duponchel, Catalogue Méthodique des Lépidoptères d'Europe. Paris, 1844.
Dup. Ic. Duponchel, Iconographie des Chenilles. Paris, 1832.
Entom. The Entomologist, edited by Newman and J. T. Carrington, F.L.S., 1839-84.
Ent. Month. Mag. The Entomologist's Monthly Magazine.
Depuiset & Deyrolle. Catalogue Méthodique des Lépidoptères d'Europe. Paris.
Ersch. Lep. Turkest. Erschoff, Lepidoptera of Turkestan. A work written in Russian, with the exception of the Latin diagnoses. Many species new at the time of publication are described and illustrated with finely coloured plates. No date appears on the title-page, but it was published after 1871.
Ernst & Engr. Ernst & Engramelle, Papillons d'Europe. Paris, 1779-92.
Esp. Esper, Die Schmetterlinge in Abbildungen nach der Natur. Erlangen, 1777-1794.
Ev. Lep. Ross. Eversmann (Fischer de Waldheim), Entomographia Imperii Rossici, Tom v. Moscow, 1857.
Ev. Faun. Lep. V. U. Eversmann, Fauna Lepidopterologica Volgo Uralensis. Casani, 1884.
Fab. E. S. Fabricius, Entomologica Systematica Emendata et Aucta. Paris, 1793.
Fab. Gen. Fabricius, Genera Insectorum. Chilionii, 1777.
Fab. Mant. Fabricius, Mantissa Insectorum, Tom ii. Hafniæ, 1787.
F. S. E. Fabricius, Systema Entomologiæ. Flensburgi et Lipsiæ, 1775.
F. d. W. Ent. or Fischer. Fischer de Waldheim, Entomographie de la Russie (Entomographica Imperii Russici). Moscow, 1820-22.
Feisth. Feisthamel, Le Baron de.
Friv. Imr. Frivaldszky (Emerich von) m. Akademiai L. Tag üjabb Közlései as altala eszközlote, &c. Budan ii. 1835. (Stgr.)
Frr. (n. Btr.). Freyer, Neure Beiträge zur Schmetterlings kunde Augsburg, i.-vii. 1831-58.
Frr. B. or Frr. Freyer, Beiträge zur Geschichte Europäischer Schmetterlinge. Augsburg, i.-iii., 1827-31.
Fuesl. Verz. Fuessly, Verzeichniss der ihm bekaunten schweizerischen Insecten. Zurich & Winterthus, 1775.
Géné. De quibusdam insectis Sardiniæ novis aut minus cognitis (Memorie della R. Acad. d. Sc. di Torino, Tom 39, 1836).
Gerh. Gerhard, Versuch einer Monographie der Europäischen Schmetterlings-arten ; Thecla, Polyommatus, Lycæna, Nemeobius. Hamburg, 1853.
Ghil. Elenco. Ghiliani, Elenco delle specie di Lepidotteri riconosciuti esistenti negli Stati Sardi. Torino, 1852 (Mém. R. Ac. Tor.)
Girard. Girard, Maurice ; Traite Elémentaire d'Entomologie, Tome iii. Fascicule Premier. Paris, Baillière et Fils, 1882. (This recent work contains some very interesting matter on the habits and structure of Lepidoptera).
Gn. et Vill. or V. G. De Villiers and Guenée, Tableaux Synoptiques des Lepidoptères d'Europe, Tome i., Diurni, 1835.
Godt. i.-v. Godt. Histoire Naturelle des Lepidoptères de France, Tome i.-v. Paris, 1821-24 (reprint, 1837).

Godt. Ency. Méth. Encyclopédie Methodique, Tome ix., Article Papillon. Paris, 1819.

Hüb. Hübner, Sammlung Europäischer Schmetterlinge. Augsburg, 1793-1827. (This, with Herrich-Schäffer's Supplement (*vide infra*, H. S.), is considered by all to be the standard work on the Lepidoptera of Europe. I have made very free use of it in the present work, the majority of the figures of larvæ and pupæ having been copied from it.)

H.-G. Hübner, Dassalbe Werk, Fortsetzung von Geyer, 1827-1841. (Stgr.)

Hüb. Btr. Hübner, Beiträge zur Geschicte der Schmetterlinge. Augsburg, 1786-89.

Hüb. Verz. Hübner, Verzeichniss bekanter Schmetterlinge. Augsburg, 1818-25,

Hüb. Vog. Hübner, Sammlung Auserlesener Vogel und Schmetterlinge, &c. Augsburg, 1793.

Hüb. Zutr. Hübner, Zuträge zur Sammlung exotischer Schmetterlinge. Augsburg, 1818-25.

Hbst. Herbst und Jablonsky, Natursystem, &c., aller bekannten Schmetterlinge, 1785-1806.

Hdnr. Cat. Heydenreich, Lepidopterorum Europæorum Catalogus Methodicus. Leipzig, 1851.

Hffmgg. Hor. Hoffmansegg, E. von, Horæ Societatis Entomologicæ Rossicæ. St. Petersburg, 1870. &c.

Hor. (Trudy). Horæ Societas Ent. Ross (Russian edition).

H. S. Herrich-Schäffer, Systematische Bearbeitung der Schmetterlinge von Europa, i.-vi. Regensburg, 1843-1856. Text and plates, 4to. (A supplement and revision of Hübner's great work, with which it is the standard work on the Lepidoptera of Europe. The figures are perfect in accuracy of drawing and colour. I am much indebted to it in the present work for the figures of some half a dozen species.)

Hufn. Berl. M. Hufnagel, Tabellen von den Tage Abend und Nachivögeln der hiesigen Gegend. Berlin (in the Berlinisches Magazin, &c.), 1766-69. (Stgr.)

Haw. Lep. Brit. Haworth, Lepidoptera Britannica. London, 1803-1829.

Ill. Mag. Illiger, Magazin für-Insecten kunde. Braunschweig, i.-vi., 1801-1807.

Ill. S. V. Illiger, Systematisches Verzeichniss von den Schmetterlingen der Weiner Gegend. Braunschweig, 1801.

Is. Isis, Encyclopädische Zeitschrift, herausgegeben von Oken. Leipzig, 1817-1848.

Kef. Keferstein (Ueber den und unmittelbaren Nutzen der Insecten. Erfurt, 1827).

Kirby, W. F., Europ. Butt., &c. European Butterflies and Moths. Messrs. Cassell & Co., 1879-83. Published in monthly parts. (A cheap and useful work on Lepidoptera, based on Berge's Schmetterlinges-buch.)

Kirby Man. Kirby, W. F., A Manual of European Butterflies. London, 1862. (British entomologists owe a deep debt of gratitude to its author for this little book, which was the first step taken in this country to elucidate the subject of which it treats.)

Kirby Cat. Kirby, W. F., A Synonymic Catalogue of Diurnal Lepidoptera. London, 1871-1877. (No comment on this work is necessary; it is the universally recognised Catalogue of the Butterflies of the world.)

Klug, Symb. Phys. Klug, Symbolæ Physicæ seu Icones, &c. Berlin, 1832.

Kn. Btr. Knoch, Beiträge zur Insecten geschichte, i.-iii. Leipzig, 1781-8.

Koll. Ins. Pers. Kollar, Ueber den Character der Insectenfauna von Süd-Persica. Vienna, 1848.

L. or Linn. Linnæus or Linné, with initials of particular work as follows :—
F. S. or Faun. Suec. Fauna Sueccia, i. 1746; ii. 176.

3 c

S. N. or Syst. Nat. X.   Systema Naturæ, editio decima.   Holmiæ, 1758.
S. N. XII.   Systema Naturæ, editio duodecima.   Holmiæ, 1767.
Mus. Ul.   Museum Ludovicæ Ulricæ Reginæ.   Holmiæ, 1764.
Lang Verz.   Lang, Verzeichniss seiner Schmetterlinge, &c.   Augsburg, 1789.
   (Stgr.)
Latr. Gen.   Latreille, Genera Crustaceorum et Insectorum secundem ordinem
   naturalem in familias disposita.   Paris, 1802-1809.
Latr. Fam. Nat.   Familles Naturelles du Règne Animal.   Paris, 1825.   (Stgr.)
Ler. or Led.   Lederer, Jules.
Lef.   Lefebvre, Alexandre.
Lepechin.   Tagebuch der Reise durch verschiedene Provinzen des russischen
   Reiches in den Jahren, 1868-1770.   Altenburg, 1774.   (Stgr.)
Lew. Ins.   Lewin, The Insects of Great Britain, &c.   Lepidoptera.   London, 1795.
Proc. Linn. Soc. Lond.   Proceedings of the Linnean Society of London.
Luc. Expl. Alg.   Lucas, Exploration Scientifique de l'Algérie, Tome iii.   Lépi-
   doptères.   Paris, 1848.
Meig.   Meigen, Systematische Bearbeitung der Europäischen Schmetterlinge,
   i.-iii.   Aachen and Leipzig, 1829-32.   (Stgr.)
Mém. Mosc.   Mémoires de la Société Impériale des Naturalistes de Moscou,
   i.-iii., 1806-1812.
Mém. Ac. Imp. St. Pet.   Mémoires de l'Académie Impériale des Sciences de
   St. Petersburg.
Mén. Cat. Rais.   Ménétriès, Catalogue Raisonnée des objets de Zoologie, recueillis
   dans un voyage au Caucase et jusqu'aux frontières actuelles de la Perse.
   St. Petersburg, 1832.
Mén. En.   Ménétriès, Enumeration corporum animalium Musei Imp., &c.
   Petropoli, pars i., Lep. Diurna.   1855.
Mén. feu Leh.   Ménétriès, Descriptions des Insects recueillis par feu M.
   Lehmann.   St. Petersbourg, 1848 (Mém. Acad. Imp. St. Petersbourg, T. vi.)
Mén. Lep. Sib.   Ménétriès, Lépidoptères de la Sibérie orientale, et en particulier
   des rives de l'Amour.   (Mélanges biol. Ac., St. Pét., 1859, T. i.)
Mén. Schrk.   Ménétriès, Dr. L. v. Schrenck's Reisen und Forschungen im Amur-
   Lande, Band ii., Lepidopteren.   St. Petersburg, 1859.
M. D., or Meyer-Dür, Tgf.   Meyer-Dür, Verzeichniss der Schmetterlinge der
   Schweitz I.   Tagfalter.   (Stgr.)
Mill. Ic.   Millière, Iconographie et description de Chenilles et Lépidoptères
   inédits.   (The figures in this work are of the finest excellence, and will
   probably never be surpassed.)
Müll. Fn. Fr.   (Müller, Otto Fried), Fauna Insectorum Friedrichsdalina Hafniæ
   et Lipsiæ, 1764.   (Stgr.)
Nick.   Nickerl.
Newman, Edw.   An illustrated Natural History of British Butterflies, by
   Edward Newman, F.L.S., F.Z.S.   London, 186— (no date.)
Nordm.   Nordmann, Alexander von.
Nouv. Mém. Mosc.   Nouveaux Mémoires de la Société Impériale des Naturalistes
   de Moscou, Tome 1.   Moscow, 1832.
Oberth.   Oberthür.
O. i., 1, 2.   Ochsenheimer, Die Schmetterlinge von Europa.   Leipzig, (1) 1807,
   (2) 1808.
Pall. Ic.   Pallas, Icones Insectorum præsertim Russiæ Sibiriæque peculiarum,
   quæ collegit et descriptionibus illustravit.   Erlangen, 1781-98.
Pall. Reis.   Pallas, Reisen durch verschiedene Provinzen des russischen Reiches
   in den Jahren 1768-1774, St. Petersburg, Tom. i., 1771.
Petit. Nouv.   Petites Nouvelles Entomologiques.   Paris, 1869, &c.

Pier. Pierret, Alexandre.

Poda Ins. Poda, Insecta Musei Græcensis, quæ in ordines, genera et species, juxta systema naturæ Caroli Linnæi digressit. (Stgr.)

Praun Abb. Praun, Abbildungen und Beschreibungen europäischer Schmetterlinge in Systematischen Reihen folge. Nürnberg, 1858, &c.

Proc. Phil. Proceedings of the Entomological Society of Philadelphia, 1861, &c.

Proc. Bost. Proceedings of the Society of Natural History of Boston, 1847, &c.

Prun. Lep. Ped. Prunner (Leonardo di), Lepidoptera Pedemontana. Augusta Taurinoram, 1798. (Stgr.)

Quens, Act. Hol. Quensel, Acta Holmiæ, 1791.

Rbr. Cat. S. And. Rambur, Catalogue Systématique des Lépidoptères de l'Andalousie. Paris, Pt. i., 1858.

Rbr. Faun. And. Rambur, Faune Entomologique de l'Andalousie. Paris, 1838–39.

Romanoff Mém. Lep. Mémoires sur les Lépidoptères rédigues, par H. N. Romanoff. St. Petersburg, 1883.

Rott. Rottenburg, Anmer Kungen zur den Hufnagel'schen Tabellen der Schmetterlinge (1775–1776).

Sc. Ent. Carn. Scopoli, Entomologia Carniolica, exhibens Insecta Carniolicæ indigena, &c. Vindobonæ, 1763.

Schiff. S. V., or S. V. Schiffermiller, Systematisches Verzeichniss, &c. (See Wien. Verz.)

Schn. N. Mag. Schneider Neustes Magazin für die Liebhaber der Entomologie. Stralsund, 1791–94. (Stgr.)

Schn. Syst. B. Schneider, Systematisches Beschreibung der Europäischen Schmetterlinge. Halle, 1787.

Schrk. Fn. B. Schranck, Fauna Boica. Nürnberg, 1801–2.

Scriba Journ. Scriba, Journal für die Liebhaber der Entomologie. Frankfurt, 1790. (Stgr.)

Scudder Rev. Revision of the hitherto known species of the genus Chionobas in North America. Philadelphia, 1865.

Selys Longch. Cat. Selys Longchamps, Catalogue des Lépidoptères de la Belgique. Liège, 1837. (Stgr.)

Sepp. Beschouwing der Wonderen Gods in de minstgeachte schepzelen of Nederlandsche Insecten, &c., 1762, &c.

Stett. E. Z. Entomologische Zeitung Herausgegeben von dem entomologischen Verein zu Stettin.

Stgr. or Staud, Cat. Staudinger, O., Ph.D., Catalogue ou enumeration methodique des Lépidoptères qui habitant le Territoire de la Faune Européenne, I. Macro-lepidoptera. Dresde, 1871. (The arrangement, nomenclature and synonymy of the above Catalogue are for the most part followed in the present work. Together with Dr. Wocke's Catalogue of the Micro-lepidoptera, which forms Part ii. of the same volume, it is without doubt the most elaborate and accurate guide to the Lepidoptera of the European Fauna that has yet come into the hands of Entomologists.)

Steph. (Ill.) Haust. Stephens, Illustrations of British Entomology, &c. Haustellata, i.–iv. London, 1827–1835.

Samouelle, George. The Entomologist's Useful Compendium, or an Introduction to the Knowledge of British Insects. London, 1819.

S. M. Stainton, A Manual of British Butterflies and Moths, Vol. i., 1857.

Sulz. Abg. G. Sulzer, Abgekürzte Geschichte der Insecten nach dem Linnäischen System. Winterthur, 1776. (Stgr.)

Tgstr. Cat. Tengström, Catalogus Lepidopterorum Faunæ Fenniæ præcursorius. Helsingforsiæ, 1869. (Stgr.)

Spr.   Speyer (Dr. Adolf.)

Thnb. *or* Thunb. Diss. Ent.   Thurnberg, Dissertatio Entomologica sistens Insecta Succica, &c., i.–vii. Upsaliæ, 1784–1794.

Tr.   Treitschke, Die Schmetterlinge von Europa, v.–x. Leipzig, 1825–35. A continuation of the work of Ochsenheimer.

Trans. E. S. Lond.   Transactions of the Entomological Society of London.

V. G.   See Gn. et Vill.

Wallengren.   Skandinaviens Dag-fjärilar.   Lepidoptera Scandinaviæ Rhopalocera deposita et descripta.   Auctore H. D. Wallengren.   Malmogiæ, 1853.

W. V. *or* Wien Verz.   Systematisches Verzeichniss der Schmetterlinge der Weiner Gegend, herausgegeben von einigen Lehrern.   Wien, 1776.   The celebrated Vienna Catalogue.   (See also Schiff. S. V.)

Wernb. Btr.   Werneberg, Beiträge zur Schmetterlingeskunde.   Erfurt, i. ii., 1864.   (Stgr.)

Westwood Br. Butt.   Westwood and Humphrey's British Butterflies.   London, 1840–1861.

Wien Mts.   Wiener Entomologische Monatschrift.   Wien, i.–viii., 1857–64.

Wilson, O.   Wilson, Owen, The Larvæ of the British Lepidoptera and their Food-plants.   London, L. Reeve & Co., 1879 (coloured plates.)

Z. B. V.   Verhandlungen des Zoologisch botanischen Vereins in Wien.   Wien.

Zett. Ins. Lap.   Zetterstedt, Insecta Lapponica.   Lipsiæ, 1840.

Z.   Zeller (in Is, Stett., E. Zeit. &c.)

A SYSTEMATIC LIST OF THE BUTTERFLIES OF THE EUROPEAN FAUNA,

GIVING THE NAMES OF SPECIES AND VARIETAL FORMS.

(The latter are printed in small type below the specific name, and include aberrations, varieties, and seasonal variations.)

The species or varieties to which a † is added are those of which I have not seen either specimens or descriptions: they are taken from Dr. Staudinger's MS. list for the present year. Some species named in the above, although I have seen specimens, are followed by (*S.*); in these cases I am not certain as to the original describer.

SUCCINCTI, *Boisd.*

I. PAPILIONIDÆ.
Papilio, *L.*
1 PODALIRIUS, *L.*
  Feisthamelii, *Dup.*
  Zancleus, *Z.*
  Virgatus, *Butl.*
  Latteri, *Aust.*
2 ALEXANOR, *Esp.*
3 MACHAON, *L.*
  Sphyrus, *Hüb.*
4 HOSPITON, *Gn.*
5 XUTHUS, *L.*
6 XUTHULUS, *Brem.*
7 MAACKII, *Men.*
8 RADDEI, *Brem.*

Luedorfia, *Crgr.*
9 PUTZILOI, *Ersch.*

Thais, *F.*
10 CERISYI, *B.*
  Caucasica, *Ld.*
  Deyrollei, *Oberth.*
11 POLYXENA, *S. V.*
  Cassandra, *Hüb.*
  Ochracea, *Stgr.*
12 RUMINA, *L.*
  Canteneri, *Stgr.*
  Medesicaste, *Ill.*
  Honoratii, *B.*

Hypermnestra, *Mn.*
13 HELIOS, *Nick.*
  Maxima, *S.*†

Doritis, *O.*
14 APOLLINUS, *Hbst.*

Parnassius, *Lat.*
14 APOLLO, *L.*
  Sibiricus, *Nord.*
16 NOMION, *F. d. W.*
17 BREMERI, *Brem.*
18 APOLLONIUS, *Ev.*
  Flavo-maculata (*S.*)†
19 DELIUS, *Esp.*
  Intermedius, *Men.*
20 CORYBAS, *F. d. W.*
21 DISCOBALUS, *Stgr.*
  Minor, *Stgr.*
22 STAUDINGERI (*S.*) †
23 ACTIUS, *Ev.*
24 DELPHIUS, *Ev.*
  Infernalis, *S.*†
25 EVERSMANNI, *Men.*
26 FELDERI, *Brem.*
27 TENEDIUS, *Ev.*
28 CLARIUS, *Ev.*
  Deutata, *Stgr.*
29 NORDMANNI, *Nord.*
30 MNEMOSYNE, *L.*
31 STUBBENDORFII, *Men.*

II. PIERIDÆ.
Aporia, *Hüb.*
32 CRATÆGI, *L.*
33 HIPPIA, *Brem.*

Pieris.
34 MELETE, *Men.*
35 CHEIRANTHI, *Hüb.*

36 BRASSICÆ, *L.*
37 KRUEPERI, *Stgr.*
  Vernalis, *Stgr.*
38 MESENTINA, *Cr.*
39 RAPÆ, *L.*
  Orientalis, *Oberth.*
40 ERGANE, *H.-G.*
41 NAPI, *L.*
  Napææ, *Esp.*
  Bryoniæ, *O.*
42 FRIGIDA, *Scud.*
43 CALLIDICE, *Esp.*
  Chrysidice, *H. S.*
44 LEUCODICE, *Ev.*
45 DAPLIDICE, *L.*
  Bellidice, *O.*
  Raphani, *Esp.*
46 CHLORIDICE, *Hüb.*

Euchloe, *Hüb.*
Anthocharis, *B.*
47 BELEMIA, *Esp.*
  Glauce, *Hüb.*
48 FALLOUI, *Allard.*
49 BELIA, *Cr.*
  Ausonia, *Hüb.*
  Simplonia, *Frr.*
  Algirica, *Luc.*
50 TAGIS, *Hüb.*
  Bellezina, *B.*
  Insularis, *Stgr.*
51 CHARLONIA, *Donz.*
52 LEVAILLANTII, *Luc.*
53 CARDAMINES, *L.*
  Turritis, *O.*
54 GRUNERI, *H. S.*
55 DAMONE, *Feisth.*

56 Eupheno, *L.*
57 Euphenoides, *Stgr.*
58 Pyrothoe, *Ev.*

Callosune, *Dbld.*
59 Nouna, *Luc.*

Zegris, *Rmbr.*
60 Eupheme, *Esp.*
61 Menestho, *Mén.*
62 Fausti, *Chr.*

Leucophasia, *Steph.*
63 Sinapis, *L.*
    Lathyri, *Hüb.*
    Diniensis, *B.*
    Erysimi, *Bkh.*
64 Amurensis, *Mén.*
65 Duponcheli, *Stgr.*
    Æstiva, *Stgr.*

Idmais, *Boisd.*
Terracolus, *Swainson.*
66 Fausta, *Oliv.*

Callidryas, *B.*
67 Pyrene, *Swainson.*
non *Pyranthe, L.*

Colias, *F.*
68 Palæno, *L.*
    Lapponica, *Stgr.*
    Werdandi, *H. S.*
69 Pelidne, *B.*
70 Nastes, *B.*
71 Werdandi, *Zett.*
72 Phicomone, *Esp.*
73 Alpheraki, *Stgr.*
74 Sagartia, *Ld.*
75 Wiscolti, *Stgr.*
76 Melinos, *Ev.*
77 Hyale, *L.*
    Sareptensis, *Stgr.*
78 Erate, *Esp.*
    Pallida, *Stgr.*
79 Boothii, *Curtis.*
80 Staudingeri (*S.*)
81 Hecla, *Lef.*
    Glacialis, *M'Lach.*
82 Chrysotheme, *Esp.*
83 Thisoa, *Mén.*
84 Viluiensis, *Mén.*
85 Eogene, *Fcld.*
86 Myrmidone, *Esp.*
    Alba, *Stgr.*
87 Olga, *Romanoff.*
    (? Caucasica, *Stgr.*)
88 Aurora, *Esp.*
    Chloë, *Ev.*
89 Aurorina, *H. S.*
    Libanotica, *Ld.*
90 Heldreichii, *Stgr.*

91 Edusa, *F.*
    Helice, *Hüb.*

Gonepteryx, *Lch.*
Rhodocera, *Boisd.*
92 Rhamni, *L.*
    Farinosa, *Z.*
93 Aspasia, *Mén.*
94 Cleopatra, *L.*
    Cleobule, *Hüb.*

III. LYCÆNIDÆ.
Thecla, *F.*
95 Betulæ, *L.*
96 Spini, *L.*
    Lynceus, *Hüb.*
    Melantho, *Klug.*
97 W-Album, *Kn.*
98 Ilicis, *Esp.*
    Caudatula, *L.*
    Cerri, *Hüb.*
    Æsculi, *Hüb.*
99 Pruni, *L.*
100 Mirabilis, *Ersch.*
101 Taxila, *Brem.*
102 Acaciæ, *F.*
    Abdominalis, *Ersch.*
103 Myrtale, *Klug.*
104 Ledereri, *B.*
105 Smaragdina, *Brem.*
106 Arata, *Brem.*
107 Quercus, *L.*
108 Fasciata (*S.*) †
109 Diamantina (*S.*) †
110 Sassanides, *Koll.*
111 Frivaldskyi, *Ld.*
112 Fusca, *Brem.*
113 Lunulata, *Ersch.*
114 Rubi, *L.*

Læosopis, *Rmbr.*
115 Roboris, *Esp.*

Thestor, *Hüb.*
116 Ballus, *F.*
117 Mauretanicus, *Luc.*
118 Nogellii, *H. S.*
119 Callimachus, *Ev.*
120 Fedtschenkoi, *Ersch.*

Polyommatus, *Latr.*
121 Virgaureæ, *L.*
    Meigii, *Vogel.*
    Oranula, *Frr.*
    Esthonica, *Stgr.*
    Zermattensis, *Fall.*
122 Thetis, *Klug.*
    Caudata, *Stgr.*
123 Ochimus, *H. S.*
124 Caspius, *Ld.*
125 Phœnicurus, *Ld.*

126 Solskyi, *Ersch.*
127 Thersamon, *Esp.*
    Omphale, *Klug.*
128 Lampon, *Ld.*
129 Asabinus, *H. S.*
    Tauricus, *Stgr.*
    Satraps, *Stgr.*
130 Dispar, *Haw.*
    Rutilus, *Wernb.*
131 Ottomanus, *Lef.*
132 Hippothoe, *L.*
    Confluens, *Gerh.*
    Eurybia, *O.*
    Stieberi, *Gerh.*
    Candens, *H. S.*
    Amurensis, *Stgr.*†
133 Alciphron, *Rott.*
134 Gordius, *Sulz.*
135 Dorilis, *Hüb.*
    Subalpina, *Spr.*
136 Phlæas, *L.*
    Eleus, *F.*
    Schmidtii, *Gerh.*
137 Amphidamas, *Esp.*
    Obscura, *Stgr.*
138 Athamanthis, *Ev.*
139 Sultan (*S.*)   [*gen.?*
140 Tamerlana (*S.*) † *hujus*

Cigaritis, *Luc.*
141 Acamas, *Klug.*
142 Cilissa, *Ld.*
143 Siphax, *Luc.*
144 Zohra, *Donz.*
145 Massinissa, *Luc.*

Lycæna, *F.*
146 Bætica, *L.*
147 Telicanus, *Lang.*
148 Balcanica, *Frr.*
149 Theophrastus, *F.*
150 Fortunata, *Stgr.*
151 Attila, *Brem.*
152 Gamra, *Ld.*
153 Argiades, *Pall.*
    Coretas, *O.*
    Polysphercon, *Stgr.*
154 Fischeri, *Ev.*
155 Trochylus, *Frr.*
156 Galba, *Chr.*
157 Ægon, *S. V.*
    Bella, *H. S.*
158 Argus, *L.*
    Argyronomon, *Bgst.*
    Ægidion, *Meiss.*
    Hypochiona, *Rbr.*
159 Cleobis, *Brem.*
160 Alcedo, *Chr.*
161 Hyrcana, *Ld.*
162 Iris, *Stgr.*
163 Eversmanni, *Stgr.*

164 Sieversi, *Chr.*
165 Optilete, *Kn.*
166 Loewii, *Z.*
  ? Gigas, *Stgr.*
167 Zephyrus, *Frr.*
  Hesperica, *Rbr.*
168 Subsolana, *Ev.*
169 Lycidas, *Trap.*
170 Pylaon, *Fisch.*
171 Zephyrinus (*S.*)†
172 Eurypilus, *Frr.*
173 Orion, *Pall.*
174 Bavius, *Ev.*
175 Baton, *Berg.*
  Panoptes, *Hüb.*
176 Staudingeri, *Chr.*
177 Prosecusa, *Ersch.*
178 Palæstina, *Stgr.*†
179 Scylla, *Stgr.*
180 Panope, *Ev.*
181 Panagæa, *H. S.*
  Anisophthalma, *Koll.*
182 Cytis, *Chr.*
183 Lysimon, *Hüb.*
184 Tengstroemi, *Chr.*
  Maxima, *Stgr.*†
185 Rhymnus, *Ev.*
186 Psylorita, *Frr.*
187 Anthracias (*S.*)
188 Pretiosa, *Stgr.*
189 Cæretes, *Hüb.*
190 Pheretiades, *Ev.*
191 Pheretulus (*S.*)†
192 Orbitulus, *Prun.*
  Pyrenaica, *B.*
  Aquilo, *B.*
  Dardanus, *Frr.*
  Wosnesenskii, *Mén.*
193 Ægargus, *Chr.*
194 Miris, *Stgr.*
195 Astrarche, *Bgst.*
  Allous, *Hüb.*
  Æstiva, *Stgr.*
  Artaxerxes, *F.*
  Salmacis, *Steph.*
196 Anteros, *Frr.*
197 Idas, *Rbr.*
198 Isaurica, *Ld.*
199 Eros, *O.*
  Eroides, *Frr.*
  Myrrha, *H. S.*
  Amor, *Stgr.*
  Altaica, *Stgr.*
  Candalus, *H. S.*
200 Mirza, *Stgr.*
201 Icarus, *Rott.*
  Icarinus, *Scriba.*
  ? Xerxes, *Stgr.*
  Persica, *Brem.*
  Cærulea, *Stgr.*
202 Eumedon, *Esp.*
203 Amanda, *Schm.*

204 Escheri, *Hüb.*
205 Phryxis, *Stgr.*
206 Bellargus, *Rott.*
  Ceronus, *Esp.*
  Cinnus, *Hüb.*
  Polonus, *Z.*
207 Corydon, *Poda.*
  Syngrapha, *Kef.*
  Apennina, *Z.*
  Hispana, *H. S.*
  Albicans, *H. S.*
  Corydonius, *H. S.*
  Caucasica, *Ld.*
208 Marcida, *Ld.*
209 Hylas, *Esp.*
  Armena, *Stgr.*
  Nivescens, *Kef.*
210 Meleager, *Esp.*
  Steveni, *Tr.*
211 Admetus, *Esp.*
  Ripartii, *Frr.*
212 Mithridates (*S.*)†
213 Dolus, *Hüb.*
214 Menaclas, *Frr.*
215 Hopfferi, *H. S.*
216 Phyllis, *Stgr.*
217 Damon, *Schiff.*
218 Damone, *Ev.*
  Damonides, *Stgr.*
219 Poseidon, *Ld.*
  Cærulea, *Stgr.*
220 { Kindermanni, *Ld.*
  { Damocles, *H. S.*
  Iphigenia, *H. S.*
  Actis, *H. S.*
221 Erschoffii, *Ld.*
222 Donzelii, *H. S.*
  Hyacinthus, *H. S.*
223 Biton, *Brem.*
224 Argiolus, *L.*
  Hypoleuca, *Koll.*
225 Sebrus, *B.*
226 { Minima, *Fuess.*
  { Alsus, *F.*
  Lorquinii, *H. S.*
227 Persephatta (*S.*) †
228 Semiargus, *Rott.*
  Bellis, *Frr.*
  Parnassia, *Stgr.*
  Helena, *Stgr.*
  Antiochena, *Ld.*
229 Glaucias, *Ld.*
230 Scylla, *Stgr.*
231 Cœlestina, *Ev.*
232 Cyllarus, *Rott.*
  Tristis, *Gerh.*
233 Melanops, *B.*
  Marchandii, *B.*
234 Astræa, *Fbr.*
235 Iolas, *O.*
236 Alcon, *Fab.*
237 Euphemus, *Hüb.*

238 Arion, *L.*
  Obscura, *Frey.*
  Caucasica (*S.*)†
  Cyanecula, *Ev.*
239 Arionides (*S.*) †
240 Arcas, *Rott.*
241 Pryeri, *Murray.*

IV. ERYCINIDÆ.
Nemeobius, *Steph.*
242 Lucina, *L.*

SUSPENSI, *Boisd.*

V. LIBYTHEIDÆ.
Libythea, *Steph.*
243 Celtis, *L.*

VI. APATURIDÆ.
Charaxes, *O.*
244 Jasius, *L.*

Apatura, *F.*
245 Iris, *L.*
  Iole, *S. V.*
246 Ilia, *S. V.*
  Clytie, *S. V.*
  Metis, *Frr.*
  Bunea, *H. S.*

VII. NYMPHALIDÆ.
Limenitis, *F.*
247 Schrenckii, *Mén.*
248 Populi, *L.*
  Tremulæ, *Esp.*
249 Camilla, *S. V.*
  Pythonissa, *Mill.*
250 Sibilla, *L.*
251 Sydyi, *Ld.*
252 Latefasciata, *Mén.*
253 Helmanni, *Ld.*
254 Amphyssa, *Mén.*
255 Homeyeri, *Ersch.*†
256 Lepechini, *Ersch.*

Neptis, *F.*
257 Nycteis, *Mén.*
258 Lucilla, *S. V.*
  Ludmilla, *H. S.*
259 Aceris, *Lepsch.*
260 Philyra, *Mén.*
261 Thisbe, *Mén.*
262 Raddei, *Brem.*

Vanessa, *F.*
(Araschnia, *Hüb.*)
263 Levana, *L.*
  Porima, *O.*
  Prorsa, *L.*

264 Burejana, *Brem.*

(*Grapta,* Doubl.)
265 Egea, *Cr.*
266 C-Album, *L.*
     F-Album, *Esp.*

(*Vanessa,* Doubl.)
267 Polychloros, *L.*
     Testudo, *Esp.*
     Pyromelas, *Frr.*
268 Xanthomelas, *S. V.*
269 Vau-Album, *S. V.*
270 Urticæ, *L.*
     Turcica, *Stgr.*
     Polaris, *Stgr.*
     Ichnusa, *Bon.*
271 Io, *L.*
     Ioides, *O.*
     Sardoa, *Stgr.*
272 Antiopa, *L.*
     Hygiæa, *Heldr.*
273 Charonia. *Drury.*

(*Pyrameis,* Hüb.)
274 Atalanta, *L.*
275 Callirrhoe, *F.*
     Vulcanica, *Gdt.*
276 Cardui, *L.*
     Elymi, *Rbr.*
277 Huntera, *F.*

Junonia, *Hüb.*
278 Œnone, *L.*
279 Here, n. sp.

Thaleropis, *Stgr.*
280 Ionia, *Ev.*

Melitæa, *F.*
281 Cynthia, *Hüb.*
282 Iduna, *Dalm.*
283 Maturna, *L.*
     Uralensis, *Ld.*
284 { Aurinia, *Rott.*
     { Artemis, *Hüb.*
     Merope, *Prun.*
     Provincialis, *B.*
     Sibirica, *Stgr.*
     Desfontainii, *Gdt.*
     Orientalis, *H. S.*
285 Boetica, *Rbr.*
286 Cinxia, *L.*
287 Arduinna, *Esp.*
     Rhodopensis, *Frr.*
     Fulminans (*S.*) †
288 Phœbe, *Kn.*
     Ætheria, *Ev.*
     Melanina, *Bon.*
     Caucasica, *Stgr.*

Occitanica, *Stgr.*
     Ætherie, *Hüb.*
     Sibina, *Alph.* (*S.*) †
289 Casta, *Koll.*
     Collina, *Ld.*
290 Trivia, *S. V.*
     Fascelis, *Esp.*
     Nana, *Stgr.*
     Sibirica (*S.*)†
     Scotosia, *Butl.*†
291 Didyma, *O.*
     Alpina, *Stgr.*
     Meridionalis, *Stgr.*
     Neera, *F. de W.*
     Occidentalis, *Stgr.*
     Turanica (*S.*) †
     Persea, *Koll.*
     Græca, *Stgr.*
292 Athene (*S.*) †
293 Fergana (*S.*) †
     Maracandica (*S.*) †
294 Dictynna, *Esp.*
     Erycina, *Ld.*
295 Protomedia, *Mén.*
296 Arcesia, *Brem.*
     Baikalensis, *Brem.*
297 Athalia, *Rott.*
     Corythalia, *Hüb.*
     Navarina, *Selys.*
     Caucasica, *Stgr.*
298 Minerva (*S.*) †
299 Aurelia, *Nick.*
     Britomartis, *Assm.*
     Plotina, *Brem.*
300 Parthenie, *Bkh.*
     Varia, *Meyer.*
     Atlantica, *Stgr.*†
301 Asteria, *Frr.*

Argynnis, *F.*
302 Aphirape, *Hüb.*
     Ossianus, *Hbst.*
     Triclaris, *Hüb.*
303 Selenis, *Ev.*
304 Selene, *S. V.*
     Hela, *Stgr.*
305 Oscarus, *Ev.*
306 Euphrosyne, *L.*
     Fingal, *Hbst.*
307 Pales, *S. V.*
     { Isis, *Hüb.*
     { Napæa, *Hüb.*
     Lapponica, *Stgr.*
     Arsilache, *Esp.*
     Caucasica, *Stgr.*
     Græca, *Stgr.*
308 Chariclea, *Schn.*
     Boisduvalii, *Dup.*
309 Improba, *Butl.*
310 Polaris, *B.*
311 Freija, *Thnb.*

312 Dia, *L.*
313 Amathusia, *Esp.*
314 Frigga, *Thnb.*
315 Thore, *Hüb.*
     Borealis, *Stgr.*
316 Amphilochus, *Mén.*
317 Daphne, *S. V.*
318 Angarensis, *Ersch.*
319 Ino, *Esp.*
     Maxima (*S.*)†
320 Hecate, *Esp.*
     Caucasica, *Stgr.*
321 Eugenia, *Ev.*
322 Lathonia, *L.*
     Valdensis, *Esp.*
323 Eliza, *Godt.*
324 Alexandra, *Mén.*
325 Aglaia, *L.*
326 Niobe, *L.*
     Eris, *Meig.*
     Pelopia, *Bkh.*
     Gigantea, *Stgr.*
327 Adippe, *L.*
     Cleodoxa, *O.*
     Chlorodippe, *H. S.*
     Cleodippe, *Stgr.*
328 Laodice, *Pall.*
329 Anadyomene. *Mén.*
330 Sagana, *Dbdl.*
331 Paphia, *L.*
     Valezina, *Esp.*
     Anargyra, *Stgr.*
332 Pandora, *S. V.*

VIII. DANAIDÆ.
Danais, *F.*
333 Chrysippus, *L.*
     Alcippus, *F.*
334 Dorippus, *Klug.*

IX. SATYRIDÆ.
Melanargia, *Meig.*
335 Galatæa, *L.*
     Leucomelas, *Esp.*
     Galene, *O.*
     Procida, *Hbst.*
     Turca, *Mill.*
336 Mauritanica, *Stgr.*
337 Lachesis, *Hüb.*
     Cataleuca, *Stgr.*
338 Titea, *Klug.*
     Teneates, *Mén.*
339 Halimede, *Mén.*
340 Epimede, *Stgr.*
341 Larissa, *H. G.*
     Astanda, *Nord.*
     Hertha, *H. G.*
342 Hylata, *Mén.*
343 Japygia, *Cyr.*
     Cleanthe, *B.*

Caucasica, *Nord.*
Suwarovius, *Hbst.*
344 PARCE, *(S.)*†
345 SYLLIUS, *Hbst.*
Ixora, *B.*
346 PHERUSA, *B.*
Plesaura, *Bell.*
347 ARGE, *Sulz.*
348 INES, *Hffgg.*

Erebia, *B.*
349 EPIPHRON, *Kn.*
Cassiope, *F.*
Nelamus, *B.*
Pyrenaica, *H. S.*
350 HABERHAUSERI *(S.)*†
351 MELAMPUS, *Fuessl.*
Sudetica, *Frr.*†
352 ERIPHYLE, *Stgr.*
353 KEFERSTEINII, *Ev.*
354 ARETE, *F.*
355 MAURISIUS, *Esp.*
356 TURANICA, *Ersch.*
357 MNÆSTRA, *Hüb.*
Gorgophone, *Bell.*
358 PAWLOWSKYI, *Mén.*
359 PHARTE, *Hüb.*
360 THEANO, *Tausch.*
361 MANTO, *Esp.*
Cæcilia, *Hüb.*
Bubastis, *Meissn.*
362 CETO, *Hüb.*
363 MEDUSA, *S. V.*
Psodea, *Hüb.*
Uralensis, *Stgr.*
Polaris, *Stgr.*
Hippomedusa, *O.*
364 ŒME, *Hüb.*
Spodia, *Stgr.*
Cœca, *(S.)*†
365 STYGNE, *O.*
366 NERINE, *Frr.*
Reichlinii, *H. S.*
Morula, *Spr.*
367 EVIAS, *Lef.*
368 MELAS, *Hbst.*
Lefebvrei, *B.*
Hewitsonii, *Ld.*
369 GLACIALIS, *Esp.*
Alecto, *Hüb.*
Pluto, *Esp.*
370 SCIPIO, *B.*
371 EPISTYGNE, *Hb.*
372 AFRA, *Esp.*
373 PARMENIO, *Boeb.*
374 LAPPONA, *Esp.*
Pollux, *Esp.*
375 OCNUS, *Ev.*
376 RADIANS, *Stgr.*
377 CALMUCCA, *Stgr.*
378 TIANCHANICA, *Stgr.*
379 TYNDARUS, *Esp.*

Cæcodromus, *Gn.*
Hispania, *Butl.*
Ottomana, *H. S.*
380 GORGE, *Esp.*
Triopes, *Spr.*
Gorgone, *B.*
Erynis, *Esp.*
381 GOANTE, *Esp.*
382 PRONOE. *Esp.*
Pitho, *Hüb.*
Pyrenaica, *Stgr.*
383 MELANCHOLICA, *H. S.*
384 NEORIDAS, *B.*
385 ZAPATERI, *Oberth.*
386 SEDAKOVII, *Ev.*
387 ÆTHIOPS, *Esp.*
Leucotænia, *Stgr.*
Melusina, *H. S.*
388 AJANENSIS, *Mén.*
389 LIGEA, *L.*
Adyte, *Hüb.*
Livonica, *Teich.*
390 EURYALE, *Esp.*
Ocellaris, *Stgr.*
Euryaloides, *Tengst.*
391 CYCLOPIUS, *Ev.*
392 TRISTIS, *Brem.*
393 EMBLA, *Thnb.*
394 DISA, *Thnb.*
395 DISCOIDALIS, *Kirb.*
396 ERO, *Brem.*
397 EDDA, *Mén.*
398 HADES, *Stgr.*

Œneis, *Hüb.*
399 JUTTA, *Hüb.*
Balderi, *Hüb.*
400 AELLO, *Hüb.*
401 TARPEIA, *Pall.*
402 SCULDA, *Ev.*
403 URDA, *Ev.*
404 NORNA, *Thnb.*
Hilda, *Quens.*
Fulla, *Ev.*
405 BORE, *Schn.*
Tayete, *Hüb.*
406 CRAMBIS, *Frr.*
407 SEMIDEA, *Say.*
408 NANNA, *Mén.*†
409 MULLA, *Stgr.*†

Satyrus, *F.*
410 HERMIONE, *L.*
Syriaca, *Stgr.*
411 ALCYONE, *S. V.*
412 CIRCE, *F.*
413 BRISEIS, *L.*
414 PRIEURI, *Pier.*
415 HEYDENREICHI, *Ld.*
Kaufmani, *Ersch.*†
Nana, *(S.)*†
416 ANTHE, *O.*

Hanifa, *Nord.*
Enervata, *Alph. (S.)*†
Ochracea, *(S.)*†
417 AUTONOE, *Esp.*
Sibirica, *Stgr.*
418 SEMELE, *L.*
Aristæus, *Bon.*
Mersina, *Stgr.*
419 BISCHOFFII, *H. S.*
420 PELOPEA, *Klug.*
Mniszechii, *H. S.*
Caucasica, *Ld.*
Mamurra, *H. S.*
421 GRÆCA, *Stgr.*
422 TELEPHASSA, *Hüb.*
Anthelea, *Hüb.*
423 AMALTHEA, *Friv.*
424 HIPPOLYTE, *Esp.*
425 BEROE, *Frr.*
Aurantiaca, *Stgr.*
Rhena, *H. S.*
426 GEYERI, *H. S.*
427 NEOMIRIS, *Godt.*
428 JOSEPHI, *(S.)*†
429 ARETHUSA, *S. V.*
Erythia, *Hb.*
Boabdil, *Rbr.*
Dentata, *Stgr.*
430 HANSII, *Aust. (S.)*†
431 STATILINUS, *Hufn.*
Allionia, *F.*
Fatua, *Frr.*
Sichæa, *Ld.*
432 SYLVICOLA, *Aust. (S.)*†
433 FIDIA, *L.*
434 PISIDICE, *Klug.*
435 STULTA, *Stgr.*
436 PARISATIS, *Koll.*
437 DRYAS, *Sc.*
Sibirica, *Stgr.*
438 CORDULA, *F.*
439 ACTÆA, *Esp.*
Podarce, *O.*
Peas, *Hüb.*
Bryce, *Hüb.*
Virbius, *H. S.*
Amasina, *Stgr.*
Parthica, *Ld.*
440 ABDELKADER, *Pier.*

Ypthima, *Dbdl.*
441 ASTEROPE, *Klug.*
442 AMPHITHEA, *Mén.*
443 BALDUS, *F.*
444 MOTSCHULSKYI, *Brem.*

Pararge, *Hüb.*
445 EVERSMANNI, *Ev.*
446 ROXELANA, *Cr.*
447 CLIMENE, *Esp.*
Roxandra, *H. S.*
448 MÆRA, *L.*

3 D

Adrasta, *Hüb.*
Adrastoides, *Brem.*
449 HIERA, *F.*
450 NASHREDDINI, (*S.*)
451 MEGÆRA, *L.*
Lyssa, *B.*
Tigellius, *Bon.*
452 EGERIA, *L.*
Egerides, *Stgr.*
453 XIPHIA, *F.*
Xiphioides, *Stgr.*
454 MAACKII, *Brem.*
455 DEIDAMIA, *Ev.*
456 EPIMENIDES, *Stgr.*
457 EPAMINONDAS, *Stgr.*
458 ACHINE, *Sc.*
459 SCHRENCKII, *Mén.*

Epinephele, *Hüb.*
460 WAGNERI, *H. S.*
Mandane, *Koll.*
461 COMARA, *Ld.*
462 DYSDORA, *Ld.*
463 INTERPOSITA, *Ersch.*
464 NARICA, *Hüb.*
465 KIRGHISICA, *Alph.* (*S.*)
466 HABERHAUSERI, (*S.*)†
467 AMARDÆA, *Ld.*
468 CADUSIA, *Ld.*
Naubidensis, *Ersch.*
469 PULCHELLA, (*S.*)†
470 CAPELLA, *Ersch.*
471 LYCAON, *Rott.*
Lupinus, *Cost.*
Sibrica, *Stgr.*
472 JANIRA, *L.*
Hispulla, *Hüb.*
Telmessia, *Z.*
473 JANIROIDES, *H. S.*
474 NURAG, *Ghil.*
475 IDA, *Esp.*
776 TITHONUS, *L.*
477 PASIPHAE, *Esp.*
Philippina, (*S.*)†
478 HYPERANTHUS, *L.*
Arete, *Müll.*

Cœnonympha, *Hüb.*
479 ŒDIPUS, *F.*
Miris, *F.*
480 HERO, *L.*
Perseis, *Ld.*
481 LEANDER, *Esp.*
Obscura, *Stgr.*
482 IPHIS, *S. V.*

Iphioides, *Stgr.*
483 ARCANIAS, *L.*
Darwiniana, *Stgr.*
484 SATYRION, *Esp.*
485 NOLKENII, *Ersch.*
486 ARCANOIDES, *Pier.*
487 CORINNA, *Hüb.*
488 THYRSIS, *Frr.*
489 DORUS, *Esp.*
490 SAADI, *Koll.*
491 AMARYLLIS, *Cr.*
492 PAMPHILUS, *L.*
493 SYMPHITA, *Ld.*
494 RINDA, *Mén.*
495 TIPHON, *Rott.*
Laidion, *Bkh.*
Philoxenus, *Esp.*
Isis, *Thnb.*
496 SUNBECCA, *Ev.*

Triphysa, *Z.*
497 PHRYNE, *Pall.*
Dohrnii, *Z.*

INVOLUTI, *Boisd.*

IX. HESPERIDÆ.
Spilothyrus, *Dup.*
498 ALCEÆ, *Esp.*
Australis, *Z.*
499 ALTHEÆ, *Hüb.*
Bœticus, *Rbr.*
500 LAVATERÆ, *Esp.*

Syrichthus, *B.*
501 PROTO, *Esp.*
502 TESSELLUM, *Hüb.*
Nomas, *Ld.*
503 GIGAS, *Brem.*
504 NOBILIS, (*S.*)†
505 STAUDINGERI, (*S.*)†
506 ANTONIA, *Spr.* (*S.*)†
507 CRIBRELLUM, *Ev.*
508 CYNARÆ, *Rbr.*
509 SIDÆ, *Esp.*
510 CARTHAMI, *Hüb.*
Moeschleri, *H. S.*
511 ALVEUS, *Hüb.*
Fritillum, *Hüb.*
Cirsii, *Rbr.*
Carlinæ, *Rbr.*
Onopordii, *Rbr.*
512 SPEYERI, (*S.*)†

513 SERRATULÆ, *Rbr.*
Cœcus, *Frr.*
Major, (*S.*)†
514 CACALIÆ, *Rbr.*
515 ANDROMEDÆ, *Wallg.*
516 CENTAUREÆ, *Rbr.*
517 MALVÆ, *L.*
Taras, *Meig.*
Melotis, *Dup.*
518 PHLOMIDIS, *H. S.*
519 POGGEI, *Ld.*
520 ORBIFER, *Hüb.*
521 SAO, *Hüb.*
Therapne, *Rbr.*
Eucrate, *O.*
522 MACULATUS, *Brem.*
523 INACHUS, *Mén.*

Nisoniades, *Hüb.*
524 MONTANUS, *Brem.*
525 TAGES, *L.*
Cervantes, *Grasl.*
Unicolor, *Frr.*
526 POPOVIANUS, *Nord.*
527 GUTTATUS, *Brem.*
528 AQUILINA, *Spr.* (*S.*)†
529 MARLOYI, *B.*
530 THETYS, *Mén.*

Hesperia, *B.*
531 THAUMAS, *Hufn.*
532 LUDOVICIÆ, *Mabille.*
533 LINEOLA, *O.*
534 HAMYRA, *Oberth.* (*S.*)†
535 ACTÆON, *Esp.*
536 STIGMA, (*S.*)†
537 SYLVATICA, *Brem.*
538 OCHRACEA, *Brem.*
539 SYLVANUS, *Esp.*
540 COMMA, *L.*
Catena, *Stgr.*
541 HYRAX, *Ld.*
542 NOSTRODAMUS, *F.*
543 THRAX, *F.*
544 ZELLERI, *Ld.*
545 ALCIDES, *H. S.*

Cyclopides, *Hüb.*
546 MORPHEUS, *Pall.*
547 ORNATUS, *Brem.*

Carterocephalus, *Ld.*
548 PALÆMON, *Pall.*
549 SILVIUS, *Knoch.*
550 ARGYROSTIGMA, *Ev.*

INDEX OF FAMILIES AND GENERA.

The former are in larger type than the latter. All synonymic names are in italics.
Those genera to which a * is prefixed contain no European species.

	PAGE
Anteos, Hüb.	64
Anthocharis, B.	85
APATURA, *F.*	155
APATURIDÆ, *B.*	153
APORIA, *Hub.*	26, 67
Araschnia, Hüb.	166
Argus, B.	97
Arge, Hüb.	229
ARGYNNIS, *F.*	194, 220
Argynnis, Lat.	179
Aurotis, Dalm.	82
Cænonympha, Butl.	303
*CALLOSUNE, *Dbd.*	69
*CALLIDRYAS, *B.*	69
CARTEROCEPHALUS, *Ld.*	356, 360, 362
CHARAXES, *O.*	153
Charcharodus, Hüb.	335
Chionobas, B.	264
Chortobius, Gn.	303
Chrysophanus, Hüb.	86
*CIGARITIS, *Luc.*	138
CŒNONYMPHA, *Hub.*	303, 330
COLIAS, *F.*	47, 69, 72, 366
Cupido, Schk.	75, 97
CYCLOPIDES, *Hub.*	354, 360
DANAI CANDIDI, L.	27
DANAIDÆ, *Doubl*	225
DANAIS, *F.*	225
Danaus, Lat.	225
DORITIS, *O.*	14
Doritis, F.	15
Doxocampa, Hüb.	154
DUMICOLES, Dup.	270
EPINEPHELE, *Hub.*	295, 329, 373
EREBIA, *Dalm.*	239, 316, 332
ERICICOLES, Dup.	270
ERYCINIDÆ, *Swainson*	149
Erynnis, Schrk.	335, 347
EUCHLOE, *Hub.*	35, 68
Eugonia, Hüb.	165
GONEPTERYX, *Leach*	64, 71
Grapta, Ddb.	166
HERBICOLES, Dup.	295
HESPERIA, *B.*	349
Hesperia, Fab.	338, 359
HESPERIDÆ, *Leach*	334
HESPERIOIDÆ, Wall.	334
Hipparchia, Lat.	270
,, Fab.	295, 303
*HYPERMNESTRA, *Men.*	22, 365
*IDMAIS, *B.*	69
Inachis, Hüb.	165
Ismene, Nick.	22
*JUNONIA, *F.*	373
LÆOSOPIS, *Rbr.*	82
LEMONIIDÆ, subf. *Libytheinæ*, Kirb.	151
Lasiommata, Westw.	288
LEMONIIDÆ, subf. *Nemeobinæ*, Kirb.	149
LEUCOPHASIA, *Steph.*	45, 69
LIBYTHEA, *F.*	152
LIBYTHEIDÆ, *Westw.*	151
LIMENITIS, *F.*	160, 215, 223, 372
*LUEDORFIA, *Cruger*	21
LYCÆNA, *Fab.*	97, 139, 369
LYCÆNIDÆ, *Steph.*	74
Maniola, Schk.	239, 295
MELANARGIA, *Meig.*	229, 315
MELITÆA, F.	179, 219
NEMEOBIUS, *Steph.*	150
NEPTIS, *F.*	63, 217
NISONIADES, *Hüb.*	347, 358, 361
NYMPHALIDÆ, *Sw.*	159
Nymphalis, Lat.	160
ŒNEIS, *Hüb.*	264, 320, 333
Pamphila, F.	349
PAPILIO, *L.*	5, 20, 25
PAPILIONIDÆ, *Steph.*	5
PARARGE, *Hüb.*	288, 326

	PAGE
PARNASSIUS, *Lat.*	15, 22, 25, 365
PIERIDÆ, *Steph.*	26, 73
Phryne, H. S.	313
PIERIS, *Schk.*	27, 67, 72, 365
Polygonia, Hüb.	165
Polyommatus, Lat.	97
POLYOMMATUS, *Lat.*	86, 137, 368
Pontia, Fab.	26, 27, 35
Pronophila, Mén.	328
Pyrameis, Hüb.	167
Pyrgus, Hüb.	338
RAMICOLES, Dup.	288, 295
Rhodocera, B.	64
RUPICOLES, Dup.	270
SATYRIDÆ, *Swains.*	228
Satyrinæ, Bates,	228
SATYRUS, *F.*	270, 322, 333
Satyrus, Lat.	295, 303

	PAGE
SPILOTHYRUS, *Dup.*	335
Steropes, Boisd.	354–356
SYRICHTHUS, *B.*	338, 358, 361
THAIS, *F.*	9
Thais, Lat.	14
*THALEROPIS, Stgr.	219
Thannaos, B.	347
THECLA, *F.*,	75, 135, 368
THESTOR, *Hüb.*	83, 137, 368
Thymelicus, Hüb.	349
Tomares, Rbr.	83
TRIPHYSA, *Z.*	313
VANESSA, F.	165, 218, 223, 372
VICICOLES, Dup.	288
*Ypthima, Dbdl.	325
ZEGRIS, *Rbr.*	43, 69, 72
Zephyrus, Dal.	76

INDEX OF SPECIES, VARIETIES, AND SYNONYMS.

NOTE.—This Index does not contain the names of species or varieties marked † in the Systematic List, p. 382; nor of any not alluded to in the body of the work; neither does it include any of the American species mentioned at the end of each family section. Names to which an asterisk is prefixed are those of species and varieties which do not occur in Europe. The names of species are printed in small capitals; those of varieties in roman type; synonyms in italics.

	PAGE
*ABDELKADER, *Pier.*	325
ACACIÆ, *F.* (Thec.)	79
*ACAMAS, *Kl.* (Cigar.)	138
ACERIS, *Lep.* (Nept.)	165
ACHINE, *Sc.* (Par.)	295
Acis, Schiff.	129
Acis, Fab.	127
ACTÆA, *Esp.* (Sat.)	287
ACTÆON, *Rott.* (Hesp.)	352
*ACTIS, *H. S.*	144
*ACTIUS, *Ev.* (Parn.)	28
ADIPPE, *L.* (Arg.)	212
ADMETUS, *Esp.* (Lyc.)	124
Adonis, Hüb.	120
Adrastra, Hüb.	291
Adyte, Hüb.	262
*ÆGARGUS, *Chrstp.*	146
Ægeria, Esp.	294
Ægerides, *Stgr.*	294
Ægidion, Meiss	106
Ægon, Schiff. (Lyc.)	103
AELLO, *Hüb.* (Œn.)	266
Æsculi, Hüb.	79
*ASPASIA, *Men.*	71
Ætheria, Ev.	187
Ætherie, Hüb.	187
ÆTHIOPS, *Esp.* (Ereb.)	260
Æthiops minor, Esp.	256
AFRA, *Esp.* (Ereb.)	254
Agave, Esp.	280
Agestis, Hüb.	114
Agestor, Godt.	119
AGLAJA, *L.* (Argyn.)	209
Alba, *Stgr.*	60
Albicans, *H. S.*	122
ALCEÆ, *Esp.* (Spil.)	336
*ALCEDO, *Stgr.* (Lyc.)	369
*AJANENSIS,*Men.* (Ereb.)	318
*ALCIDES, *H. S.*	360
ALCIPHRON, *Rott.* (Polyom.)	93
Alcippus, *F.*	227
ALCON, *F.* (Lyc.)	132

	PAGE
ALCYONE, *Schiff.* (Sat.)	273
Alecto, Hüb.	252
ALEXANOR, *Esp.* (Pap.)	6
ALEXANDRA, *Men.* (Ag.)	208
Alexis, Esp.	261
Alexis, Hüb.	117
Alexius, Frr.	117
Allionia, *F.*	283
Allous, Hüb.	115
Alpina, Stgr.	189
*ALPHERAKI, *Stgr.*(Col.)	367
Also, Boisd.	321
Alsus, F.	363
ALTHEÆ, *Hüb.*	337
Alveolus, Hüb.	344
ALVEUS, *Hüb.* (Syr.)	341
AMALTHEA, *Friv.* (Sat.)	279
AMANDA, *Schn.* (Lyc.)	119
*AMARDÆA, *Ld.* (Ep.)	329
AMARYLLIS, *Cr.* (Cœn.)	309
Amarillis, Bkh.	300
Amasina, *Stgr.*	325
Ammonia, H. S.	219
Amor, *Stgr.*	371
Amphidimas, Esp.	96
*AMPHILOCHUS, *Men.* (Arg.)	221
Amphion, Esp.	109
*AMPHITHEA, *Men.* (Ypth.)	326
Amphitrite, Hüb.	237
*AMPHYSSA, *Men.* (Lim.)	216
Amurensis, *Men.*	69
Amyntas, Butl. (Cœn.)	305
Amyntas, Fab. (*Lyc.*)	101
Anargyra, *Stgr.*	215
*ANADYOMENE, *Feld.* (Arg.)	221
ANDROMEDÆ, *Wally.* (Syr.)	343
*ANGARENSIS, *Ersch.* (Arg.)	221
*Anisophthalma, *Koll.*	142

	PAGE
*ANTHRACIAS,*Stgr.*(Lyc.)	370
ANTEROS, *Frr.* (Lyc.)	115
ANTHE, *O.* (Satyr.)	276
Anthelea, H. S.	279
Anthelea, Hüb.	323
Anthyale, Hüb.	50
Antiochena, *Ld.*	145
ANTIOPA, *L.* (Van.)	176
Appenina, *Rott.*	122
APHIRAPE, *Hüb.* (Arg.)	196
APOLLINUS,*Hbst.*(Dor.)	14
APOLLO, *L.* (Parn.)	16
*APOLLONIUS,*Ev.*(Parn.)	23
Appenina, *Ld.*	122
Aquilo, *Boisd.*	113
Arachne, Esp.	283
Araehne, Fab.	258
Aracinthus, Fab.	355
*ARATA, *Brem.* (Thec.)	135
Arcania, L. (Cœn.)	306
Arcanius, L.	306
*ARCANOIDES,*Pier.*(Cœn.)	330
ARCAS, *Rott.* (Lyc.)	134
Arcas, Esp.	132
*ARCESIA, *Brem.* (Mel.)	220
Arctica, Zett.	200
ARDUINNA, *Esp.* (Mel.)	186
ARETE, *F.* (Ereb.)	243
Arete, Mull.	302
ARETHUSA, *Esp.* (Sat.)	281
ARGE, *Sulz.* (Melanarg.)	237
ARGIADES, *Pall.* (Lyc.)	101
ARGIOLUS, *L.* (Lyc.)	127
Argiolus, Esp.	129
ARGUS, *L.* (Lyc.)	105
*ARGYROSTIGMA, *Ev.* (Cart.)	360
Argyronomon, *Berg.*	105
ARION, *L.* (Lyc.)	133
Argyrotoxus, Bgstr.	103
Aristæus, Bon.	278
Aristolochiæ, Schn.	11
Armena, *Stgr.*	144

Arragonensis, Gerh. . 122
Arsilache, Esp. . . . 199
Artaxerxes, Fab. . . 115
Artemis, Hüb. . . . 183
ASABINUS, H. S. (Pol.). 137
*Astanda, Nordm. . . 315
Astasia, Hüb. . . . 158
ASTERIA, Frr. (Mel.) . 193
Asterie, H. S. . . . 193
*ASTEROPE, Klg. (Ypth.) 325
ASTRARCHE,Byst.(Lyc.) 114
*ASTRÆA, Frr. (Lyc.) . 145
ATALANTA, L. (Van.) . 177
Atalanta Indica,Hbst. 177
ATHALIA, Rott. (Mel.) . 191
Athalia minor, Esp. . 192
*ATHAMANTHIS, Ev. }
        (Pol.) } 138
Atratus, Esp. . . . 252
Atropos, Hüb. . . . 234
*ATTILLA, Brem.(Lyc.). 140
Atys, Hüb. 112
*Aurantiaca, Stgr. . . 324
AURELIA, Nick. (Mel.). 192
AURINIA, Rott. (Mel.) . 183
*AURORA, Esp. (Col.) . 71
*AURORINA, H. S. (Col.) 70
Aurorina var.Heldreichi 63
Ausonia, Hüb. . . . 37
AUTONOE, Esp. (Sat.) . 276
Australis, Z. . . . 336
BALCANICA, Frr. (Lyc.) 100
Balder, H. S. . . . 265
Balderi, Hüb. . . . 266
*BALDUS, F. (Ypth.) . 326
BALLUS, F. (Thest.) . 84
Balthus, F. 108
BATON, Berg. (Lyc.) . 109
Battus, Hüb. . . . 108
BAVIUS, Ev. (Lyc.). . 108
BELEMIA, Esp.(Euchl.) 36
BELIA, Cr. (Euchl.). . 37
Bella, H. S. 105
BELLARGUS,Rott.(Lyc.) 120
Bellidice, O. 34
Bellezina, Boisd. . . 38
Bellis, Frr. 130
Bellus, Gerh.. . . . 81
BEROE, Frr. (Sat.) . . 323
Beroë, Fab. (Apat.) . 157
BETULÆ, L. (Thec.) . 75
*BISCHOFFII, H. S. (Sat.) 322
Blandina, Fab. . . . 260
*BITON, Brem. (Lyc.) . 145
Biton, Esp. 126
Boabdil, Ramb. . . . 282
BOETICA, Ramb. (Mel.) 184
BOETICA, L. (Lyc.) . . 99
Bæticus, Rbr. . . . 337
Boisduvalli, Dup. . . 200
Boisduvalli, H.S.(Lyc.) 116
Boötes, Boisd. . . . 270

*BOOTHII, Curtis (Col.) 70
Boothii, B. 55
BORE, Schn. (Œneis) . 269
Borealis, Stgr. . . . 205
BRASSICÆ, L. (Pier.) . 28
*BREMERI,Brem.(Parn.) 23
BRISEIS, L. (Sat.) . . 275
Britomartis, Assm. . . 192
Bonellii, Hüb. . . . 250
Brontes, Hüb. . . . 356
Bryce, Hüb. 287
Bryoniæ, O. 32
Bunea, H. S. 159
*BUREJANA,Brem.(Van.) 218
CACALIÆ, Rbr. (Syr.) . 343
*CADUSIA, Ld. (Epin.) . 330
Coecilia, Hüb. . . . 245
Cœcus, Frr. 343
Cœcodromus, Gn. . . 256
Calais, Scudd. . . . 270
C-ALBUM, L. (Van.) . 170
Callenphenia, Butl. . 41
CALLIDICE, Esp. (Pier.) 32
CALLIMACHUS, Ev. }
        (Thest.) } 85
CALLIROHOE,F. (Van.) } 177
                      } 372
*CALMUCCA, St. (Ereb.) 320
Camilla, Esp. . . . 164
CAMILLA, Schiff. (Lim.) 161
Camilla, L. 162
*Candalus, H. S. . . . 143
Candens, H. S. . . . 93
Canteneri, Staud. . . 13
*CAPELLA, Chr. (Epin.). 330
CARDAMINES, L., Euchl. 39
CARDUI, L. (Van.) . . 178
Carlinæ, Rbr. . . . 342
Carmon, H. S. . . . 144
CARTHAMI, Hüb. (Syr.) 341
*CASPIUS, Ld. (Pol.) . . 137
Cassandra, Hüb. . . 12
Cassioides, Esp. . . 255
Cassiope, F. 242
*CASTA, Koll. (Mel.) . . 220
Castor, Esp. 255
Catalcuca, Stgr. . . 232
Catena, Stgr. . . . 354
*Caucasica, Ld. (Lyc.) . 143
Caucasica, Ld. (Sat.) . 323
Caucasica,Ld.(Thais). 10
Caucasica, Stgr. }
  (Didyma var.)} 189
Caucasica, Stgr. }
  (Hecate var.)} 207
Caucasica, Stgr. }
  (Pales var.) } 199
Caucasica, Stgr. }
  (Phœbe var.)} 187
*Caudatula, L. . . . 135
Celæno, Hüb. . . . 268
Celimene, Cr. . . . 267

CELTIS, Esp. (Libyth.) 152
CENTAUREÆ, Rb. (Syr.) 344
CERISYI, B. (Thais.) . 9
Ceronus, Esp. . . . 121
Cerri, Hüb. 79
Cervantes, Grasl. . . 349
Cethosia, Hüb. . . . 213
CETO, Hüb. (Erebia) . 246
CHARICLEA, Schn.(Arg.) 200
*CHARLONIA, Donz. }
        (Euchloe)} 68
Charlotta, Haw. . . 210
*CHARONIA, Drur. (Van.) 218
*CHEIRANTHI, Hüb. (Pier.) 34
Chiron, Rott. . . . 118
CHLORIDICE, Hüb.(Pier.) 67
Chlorodippe, H. S. . . 212
*Chloë, Evers. . . . 367
Chryseis, Bkh. . . . 92
*Chrysidice, H. S. . . 33
CHRYSIPPUS, L. (Dan.) 226
CHRYSOTHEME,Esp.(Col.) 56
Chrysothememe, Stev. 63
*CILISSA, Ld. (Cigar.) . 138
Cinnus, Hüb. . . . 121
CINXIA, L. (Melitæa) . 185
Cinxia-major, Esp. . 185
CIRCE, Fab. (Sat.) . . 273
Circe, Schiff. . . . 94
Cirsii, Rb. 342
*CLARIUS, Ev. (Parn.) . 24
Clarius, H. S. . . . 19
Cleanthe, Boisd. . . 234
Cleo, Hüb. 255
*Cleobule, Hüb. . . . 71
Cleobis, Esp. . . . 127
*CLEOBIS, Brem. (Lyc.) 140
- Cleodippe, Stgr. . . 213
Cleodoxa, O. . . . 212
CLEOPATRA, L.(Gonept.) 66
CLIMENE, Esp. (Parar.) 290
Clite, Hüb. 305
Clotho, Hüb. . . . 234
Clymene, O. 290
Clytie, Schiff. . . . 158
Cœcodromus, Gn. . . 256
CŒLESTINA, Ev. (Lyc). 130
Cœcilia, Esp. . . . 245
Collina, Ld. . . . .
Cœcus, Frr. 343
*COMARA, Ld. (Epin.) . 329
COMMA, L. (Hesp.) . . 353
Confluens, Gerh. . . 92
CORDULA, F. (Sat.) . . 286
*CORDULINA, Stgr. (Sat.) 325
Coretas, O. 102
CORINNA, Hüb. (Cœn.) 308
CORYDON, Poda. (Lyc.) 121
*CORYBAS, F. de W.(Parn.) 23
*Corydonius, H. S. . . 143
Corythalia, Hüb. . . 192
Corythalia, Hüb. . . 190

Corythallia, Esp. . . 186
*Crambis, Frr. (Œn.) . 321
Crameri, Butl. . . . 37
Cratægi, L. (Apor.) . 26
Cribrellum, Ev. (Syr.) } 339
*Cyane, Ld. (Lyc.) . . 140
*Cyanecula, Ev. . . . 145
*Cyclopius, Ev. (Ev.) . 319
Cyllarus, Rott. (Lyc.) 130
Cynara, F. 215
Cynthia, Hüb. (Mel.) . 180
Cynthia, Esp. . . . 182
Cyparissus, Hüb. . . 107
Cynaræ, Ramb. (Syr.) 340
Cyrene, Bon. . . . 208
*Cytis, Stgr. (Lyc.) . . 369
Dalmata, Godt. . . . 254
*Damocles, H. S. . . 144
Damœtas, Hüb. . . . 130
Damon, Schiff. (Lyc.) . 126
Damone, Ev. . . . 127
Damone, B. (Euchl.) . 40
Daphne, Schiff. (Arg.) 205
Daphnis, Bgstr. . . 123
Daplidice, L. (Pier.) . 33
Daplidice var. Russiæ, Esp. } 34
Dardanus, Frr. . . . 114
Darwiniana, Stgr. . . 307
Davus, F. 311
Dejanira, L. 295
*Deidamia, Ev. (Parg.) 327
Deione, H. G. (Mel.) . 191
Delia, Bkh. 185
Delius, Esp. (Parn.) . 17
*Delphius, Ev. (Parn.) 24
Dentata, Stgr. . . . 282
Desfontainesi, B. . . 184
Desfontainesii, Ev. . 184
Desfontainii, Godt. . . 184
*Deyrollei, Oberth. . . 10
Dia, L. (Arg.) . . . 203
Dia lapponica, Esp. . 202
Dia major, Esp. . . 203
Diana, Hüb. 203
Dictynna, Esp. (Mel.) 190
Dictymna, Hüb. . .
*Didymoides, Ev. . . 220
Didyma, O. (Mel.) . . 188
Diniensis, B. 46
*Diodorus, Brem. (Lyc.) 142
Diomedes, Bkh. . . . 133
Dioxippe, Hüb. . . . 268
Disa, Thnb. (Ereb.) . 263
*Discoidalis, Kirb. (Ereb.) } 319
Dispar, Brem. . . .
*Discobalus, Stgr. (Parn.) } 365
Dispar, Haw. (Pol.) . 90
Dohrnii, Zell. . . . 314

Dolus, Hüb. (Lyc.) . 125
Donzelii, B. (Lyc.) . 127
Dorilis, Bkh. . . . 309
Dorilis, Hufn. (Pol.) . 94
Dorion, Hüb. . . . 309
*Dorippus, Klg. (Dan.) 227
Dorus, Esp. (Cœn.) . 309
Dorylas, Hüb. . . . 122
*Douei, Pier. 68
Dromus, Fab. . . . 255
Dromus, H. S. . . . 256
Dryas, Sc. (Sat.) . . 285
Duponchelii, Stgr. (Leuc.) } 46
*Dysdora, Ld. (Epin.) . 329
*Edda, Mén. (Ereb.) . 320
Edusa, F. (Col.) . . 61
Egea, Cr. (Van.) . . 169
Egerides, Stgr. . . . 294
Egeria (r), L. (Pararg.) 293
Electra, Lewin. . . . 61
Eleus, F. 96
*Elvira, Ev. (Lyc.) . . 141
Eliza, Godt. (Arg.) . 208
*Ella, Brem. (Arg.) . . 221
Embla, Thnb. (Ereb.) . 263
Embla, Boisd. . . . 263
Epidolus, B. 125
*Epargyros, Ev. . . . 188
*Epimede, Stgr. (Melan.) 315
*Epimenides, Mén. (Pararg.) } 328
*Epaminondas, Stgr. (Pararg.) } 328
Epiphania, B. . . . 85
Epiphron, Kn. (Ereb.) 241
Epistigne, Hüb. (Ereb.) 253
Erate, Esp. (Col.) . . 54
Erebus, Kn. 134
Ergane, Hüb. (Pier.) . 30
Erina, Fab. 245
*Eogene, Ersch. (Col.) 366
Eriphyle, Frr. (Ereb.) 243
Eris, Meig. 211
*Ero, Brem. (Ereb.) . 319
Eroides, Friv. . . . 116
Eros, O. (Lyc.) . . . 116
Erothoë, Ev. 43
*Erschoffii, Ld. (Lyc.) 141
*Erycina, Ld. 220
Erynis, Esp. 257
Erysimi, Bkh. . . . 46
Erythia, Hüb. . . . 282
Escheri, Hüb. (Lyc.) . 119
Esculi, Hüb. 79
Ethus, Fab. 263
Eudora, Esp. . . . 207
Eucrate, O. 347
Eucrate, Tr. 346
Eugenia, Ev. (Arg.) . 221
Eumedon, Esp. (Lyc.) 118
*Eunomia, Esp. . . . 196

*Eunomia, Mén. . . . 318
Eupheme, Esp. (Zegr.) 43
Euphemus, Godt. . . 132
Euphemus, Hüb. . . 133
*Eupheno, L. (Euchl.) 68
Eupheno, Esp. . . . 41
Euphenoides, Stgr. (Euchl.) } 41
Euphrosyne, L. (Arg.) 198
Euridice, Hüb. . . . 92
Europomene, Esp. . . 49
Euryaloides, Tengs. . 263
Euryale, Esp. (Ereb.) 262
Eurybia, O. 92
Eurydice, Rott. . . . 92
*Eurypilus, Frr. (Lyc.) 142
*Eversmanni, Ev. (Parg.) 326
Everos, Dup. 116
*Eversmanni, Mén. (Parn.) 24
*Eversmanni, Stgr. (Lyc.) 370
Evias, Lef. (Ereb.) . 250
Evippus, Hüb. . . . 82
*Falloui, All. (Euchl.) 68
Farinosa, Z. 66
Fascelis, Esp. . . . 188
Fatua, Fr. (Sat.) . . 284
Fauna, Hüb. 283
*Fausta, Ol. (Idm.) . 69
*Fausti, Christp. (Zegris) 69
*Fedtschenkoi, Ersch. (Thest.) } 368
Feisthamelii, Dup. . . 6
Fidia, L. (Sat.) . . . 284
Fingal, Hbst. . . . 198
Fischeri, Ev. (Lyc.) . 102
*Fortunata, Stgr. (Lyc.) 139
*Franklinii, Curt. . . 113
Frelja, Thnb. (Argyn.) 202
Frigga, Thnb. (Arg.) 204
*Frigida, Scud. (Pier.) 67
Fritillum, Hüb. . . . 342
*Frivaldszkyi, Ld. (Thec.) } 136
Fulla, Ev. 268
*Fusca, Brem. (Thec.) 136
Galatæa, D. (Melan.) 230
Galene, O. 231
*Galba, Ld. (Lyc.) . . 140
*Gamra, Ld. (Lyc.) . . 140
Gefion, Quens. . . . 263
Geticus, Esp. . . . 303
Gemina, Ld. 337
*Geyeri, H. S. (Sat.) . 324
*Gigantea, Stgr. . . . 221
*Gigas, Brem. (Syr.) . 358
Glacialis, M'Lach. (Col.) 56
Glacialis, Esp. (Ereb.) 251
Glauce, Boisd. . . . 36
*Glaucias, Ld. (Lyc.) . 146
Goante, Esp. (Ereb.) . 257
Gordius, Sulz. (Pol.) . 94
Gorge, Esp. (Ereb.) . 256

*Gigas, Stgr. 371
Gorgone, B. 257
Græca, Stgr. (Arg.) . 199
Græca, Stgr. (Mel.) . 189
GRÆCA, Stgr. (Sat.) . 278
Griela, Fab. . . . 263
GRUNERI, H. S. (Euchl.) 40
*GUTTATUS, Brem. (Nis.) 357
*IIADES, Stgr. (Ereb.) . 317
Hafis, Koll. 85
*HALIMEDE, Mén.
(Melan). } 315
Hanifa, Nord. . . . 276
HECATE, Esp. (Arg.) . 206
HECLA, Lef. (Col.) . . 55
Hela, Stgr. 197
HELDREICHI, Stgr. (Col.) 63
Helena, H. S. . . . 61
Helena, Stgr. . . . 130
Helice, Hüb. . . . 62
*Helictha, Ld. . . . 55
*HELIOS, Nick. (Hyperm.) 22
Helius, H. S. . . . 137
Helle, Hüb. 96
*HELMANNI, Ld. (Lim.) 216
HERMIONE, L. (Sat.) . 272
Hermione major, Esp. 272
Hermione minor, Esp. 273
*HERE, Lang (Junon.) . 373
HERO, Hüb. (Cœn.) . 304
Herse, Bkh. 255
Herse, W. V. . . . 300
Hertha, H. G. . . . 233
Hesperica, Rbr. (Lyc.) 107
*HEYDENREICHI, Ld.(Sat.)322
*Hewitsoni, Ld. . . . 317
HIERA, F. (Pararge) . 292
Hiere, F. 93
Hilda, Quens. . . . 268
Hille, F. 96
*HIPPIA, Brem. (Apor.) 67
Hippodice, Hüb. . . 286
HIPPOLYTE, Esp. (Sat.) 280
Hippomedusa, O.
(vide errata)} 247
Hipponoë, Esp. . . 93
HIPPOTHOE, Ld. (Pol.) 92
Hippothoë, Lew. . . 90
Hispana, H. S. . . . 122
Hispania, Butl. . . . 256
Hispulla, Hüb. . . . 298
Honoratii, B. . . . 14
*HOPFFERI, H. S. (Lyc.) 144
HOSPITON, Gené. (Pap.) 8
*HUNTERA, Fab. (Van.) 218
HYALE, L. (Col.), Esp. 53
*Hyacinthus, H. S. . . 145
Hygiæa, Hldrch. . . 176
HYLAS, Esp. (Lyc.) . 122
*HYLATA, Mén. (Melan.) 316
Hylus, F. 109
HYPERANTHUS, Linn. . 302

Hypsipyle, F. . . . 11
Hypochiona, Rbr. . . 106
*Hypoleuca, Koll. . . 145
Hypoleucos, Ld. . . 345
Hypoxanthe, Kirb. . 95
*HYRAX, Ld. (Hesp.) . 359
*HYRCANA, Ld. (Lyc.) . 141
Icarinus, Scriba. . . 117
Icarius, Esp. . . . 119
ICARUS, Rott. (Lyc.) . 117
Ichnusa, Bon. . . . 174
IDA, Esp. (Epineuh.) . 300
IDAS, Rbr. (Lyc.) . . 118
IDAS, Lew. 114
IDUNA, Dalm. (Mel.) . 181
Ignitus, H. S. . . . 88
ILIA, Schiff. (Apat.) . 157
ILICIS, Esp. (Thecla) . 78
IMPROBA, Butl. (Arg.) 364
*INACHUS, Mén. (Syric.) 357
INES, Hffgg. (Melan.) . 238
INO, Rott. (Arg.) . . 206
Insularis, Stgr. . . . 38
*Intermedius, Mén. . . 23
*INTERPOSITA, Ersch.
(Epin.) } 330
Io, L. (Van.) . . . 175
Iodes, O. 175
IOLAS, O. (Lyc.) . . 132
Iolaus, Bon. (Sat.) . 280
Iolaus, Hüb. (Lyc.) . 132
Iole, Schiff. . . . 157
*IONIA, Ev. (Thal.) . . 219
*Iphias, Ev. 331
Iphigenia, Esp. . . . 187
*Iphigenia, H. S. . . 144
Iphigenus, Hbst. . . 303
Iphioides, Stgr. . . . 306
IPHIS, Schiff. (Cœn.) . 305
Irene, Hüb. 248
IRIS, L. (Apat.) . . 156
Iris rubescens, Esp. . 158
*IRIS, Stgr. (Lyc.) . . 369
*ISAURICA, Stgr. (Lyc.) 143
Isis, Hüb. (Arg.) . . 199
Isis, Thnb. (Cœn.) . . 313
Ixora, B. 237
JANIRA, L. (Epi.) . . 298
*JANIROIDES, H. S. 299, 373
Janirula, Esp. . . . 297
Janthe, H. S. . . . 242
JAPYGIA, Cyr. (Melan.) 234
JASIUS, L. (Char.) . . 154
Jason, L. 154
Jurtina, L. 298
JUTTA, Hüb. (Œneis) . 265
*KAUFMANI, Ersch.
(Sat.) } 325
*KEFERSTEINII, Ev.
(Ereb.) } 316
Kefersteinii, Gerh. (Pol.) 89
*KIRGHISICA, Stgr.(Epin.)330

KREUPERI, Stgr.(Pier.) 29
LACHESIS, Hüb.(Melan.) 232
Laidion, Bkh. . . . 312
L-Album, Esp. . . . 172
Lampetie, Hüb. . . 93
LAODICE, Pall. (Arg.) . 213
LAPPONA, Esp. (Ereb.) 255
Lapponica, Stgr.(Arg.) 199
Lapponica, Stgr. (Col.) 50
*LAMPON, Ld. (Pol.) . 137
LARISSA, H. G. (Melan.) 233
*Latogenia, H. S. . . 220
*Latefasciata, Mén. . . 216
LATHONIA, L. (Arg.) . 207
Lathyri, Hüb. . . . 46
Lathyri, Dup. . . . 46
LAVATERÆ, Esp. (Spil.) 337
*Latteri, Aust. . . . 365
LEANDER, Esp. (Cœn.) 305
*LEDERERI, Boisd.
(Thec.)} 135
Lefebvrei, B. (Ereb.) . 251
Lefebvrei, H. S. . . 248
Lefebvrei, Godt. (Lyc.) 125
Lefebvrei, Rbr. (Hesp.) 354
Legeri, Frr. 87
*LEPECHINI, Ersch.(Lim.)372
Leucippe, Schn. . . 191
*LEUCODICE, Ev. (Pier.) 366
Leucomelus, Esp. . . 231
Leucotænia, Stgr. . . 261
*LEVAILLANTII, Luc.
(Euchl.) } 68
LEVANA, L. (Van.) . . 167
*Libanotica, Ld. . . 71
Ligea, Esp. 246
LIGEA, L. (Ereb.) . . 261
Linceus, Hüb. . . . 78
LINEA, F. 350
LINEOLA, O. (Hesp.) . 351
Livonica, Teich. . . 262
*LOEWII, Z. (Lyc.) . . 141
Lorquinii, H. S. . . 129
Lucilla, Esp. (Lim.) . 161
LUCILLA, F. (Nep.) . 164
LUCINA, L. (Nem.) . 150
*Ludmilla, H. S. . . 217
LUDOVICIÆ, Mab.
(Hesper.) } 351
Lupinus, Costa. . . 297
LYCAON, Rott. (Epin.) 297
LYCIDAS, Trap. (Lyc.). 363
Lyc, Bgst. 183
Lyllus, Esp. 310
Lynceus, Esp. . . . 76
Lyssa, B. 293
LYSIMON, Hüb. (Lyc.) . 111
*MAACKII, Brem.(Pararg.)328
*MAACKII, Mén. (Pap.) . 21
MACHAON, L. (Pap.) . 7
*Macrophthalmus, Ev. 324
*MACULATUS, Brem.(Syr.)358

Mæra, L. (Pararge) . 290
Maia, Cr. 215
Malvæ, L. (Syr.) . . 344
Malvæ, Hüb. . . . 336
Malvæ, Esp. 341
Malvarum, Hffg. . . 336
*Mamurra, H. S. . . 323
*Mandane. Koll. (Epin.) 329
Manto, Esp. (Ereb.) . 245
Manto, Fab. 255
*Maracandica, Ersch. }
(Ereb.) } 319
Marchandi, Gerh. . . 131
Marchandii, B. . . . 132
Marchandii, Hüb. . . 132
*Marcida, Ld. (Lyc.) . 144
Marloyi, B. (Nis.) . 349
Marmoræ, Hüb. . . 280
Marrubii, H. S. . . 337
Marrubii, Kirby . . 337
Martiani, H. S. . . . 284
*Massinissa, Luc. (Cig.) 139
Maturna, Esp. . . . 183
Maturna, L. (Mel.) . 182
Maturna, Hüb. . . 181-191
*Maurisius, Esp.(Ereb.) 316
*Mauritanica, Stgr. }
(Melan.) } 315
*Maxima, Stgr. . . . 365
*Mauritanicus, Luc. }
(Thest.) } 137
Maurus, Esp. . . . 250
Medea, Hüb. . . . 260
Medea, Bkh. 246
Medesicaste, Ill. . . 14
Medon, Esp. 114
Medusa, F. 246
Megæra, L. (Pararge) 292
Melampus, Fuess. }
(Ereb.) } 42
*Melancholica, H. S. }
(Ereb.) } 318
Melanops, B. (Lyc.) . 131
*Melantho, Kl. . . . 135
Melas, Hbst. (Ereb.) . 250
Meleager, Esp. (Lyc.) 123
Meleager, Hüb. . . . 113
*Melete, Mén. (Pier.) . 67
*Melinos, Ev. (Col.) . 69
Melotis, Dup. . . . 345
Menaclas, Frr. . . . 125
Menaclas, Poda. . . 310
Menesthio, Mén. (Zegr.) 44
Meone, Esp. 293
Meridionalis,Ld.(Zegris) 44
Meridionalis,Stgr.(Mel.)189
Merope, Prun. . . . 183
*Mersina, Stgr. . . . 322
*Mesentina, Cr. (Pieris) 365
Metis, Frr. 158
Miegii, Vogel. . . . 87
Minima, Fuess. (Lyc.) 128

*Minor, Stgr. 365
*Mirabilis, Ersch. }
(Thec.) } 368
*Mirza, Stgr. (Lyc.) . 147
Mnemosyne,L.(Parnass.) 21
Miris, Fab.? 304
Mnestra, Hüb. (Ereb.) 244
*Mniszechii, H. S. . . 323
Moeschleri, H. S. . . 341
*Montanus,Brem. (Nis.) 353
Morula, Spr. 249
Morpheus, Pall. (Cycl.) 355
*Motschulskyi, }
Brem. (Sat.) } 326
*Mulla, Stgr. (Œneis) 321
Myrmidone,Esp.(Col.) 59
Myrmidone, Steph. . 63
*Myrrha, H. S. . . . 143
*Myrtale, Kl. (Thec.) . 136
Nana, Stgr. 188
*Nanna, Stgr. (Œneis) 320
Napæa, Hüb. 199
Napææ, Esp. 32
Napi, L. (Pier.) . . . 31
Narcæa, Frr. 30
Narica, Hüb.(Epineph.) 296
*Naricina, Stgr. . . . 330
*Nashreddini, Stgr. }
(Parge.) } 327
Nastes, B. (Col.) . . 51
Navarina, Selys. . . 192
Neera, F. de W. . . 189
Nelamus, B. 242
Neleus, Frr. 255
Neomiris, Godt. (Sat.) 280
Neoridas, B. (Ereb.) . 259
Nephele, Hfn. (Cœn.) . 310
Neriene, F. de W. . . 54
Nerine, Frr. (Ereb.) . 249
Nevardensis, Stgr. . . 256
Niobe, L. (Arg.) . . 210
Nivescens, Kef. . . . 128
Nogellii, H. S. (Thestor) 84
*Nolckeni,Ersch.(Cœn.) 331
*Nomas, Ld. 358
*Nomion, F. de W. }
(Parnass.) } 22
Noraæ, Bon. . . . 308
Nordmanni, Ndm. }
(Parnass.) } 19
Norna, Hüb. . . . 266
Norna, Thnb. (Œneis) 268
Norna, Quens. . . . 269
Nostrodamus,F.(Hesp.)354
*Nouna, Luc. (Callos.) . 69
Nurag, Ghil. (Epineph.) 299
*Nycteis, Mén. (Nept.) 217
Occidentalis, Stgr. . . 190
Occitanica, Stgr. . . 187
Occitanica, Esp. . . 236
Ocellaris, Stgr. . . . 262
Ochimus, H. S. (Pol.) . 89

*Ochracea,Brcm.(Hesp.)359
Ochracea, Stgr. (Thais.) 13
*Ocnus, Ev. (Ereb.) . 318
Œdipus, F. (Cœn.) . 303
Œme, Hüb. (Ereb.) . 245
*Œno, B. 321
*Œnone, L. (Jun.) . . 218
*Olga, Roman. (Col.) . 366
*Omphale, Kl. . . . 138
Onopordii, Rbr. . . . 342
Optilete, Kn. (Lyc.) . 106
Oranula, Frr. . . . 87
Orbifer, Hüb. (Syr.) . 346
Orbitulus,Prun.(Lyc.)113
Orientalis, H. S. . . 184
*Orientalis, Mén. . . 220
Orion, Pall. (Lyc.) . 108
*Ornatus, Brem. }
(Cyclop.) } 360
*Oscarus, Ev. (Arg.) . 220
Ossianus, Hbst. . . 196
Ottomana, H. S. . . 526
Ottomanus, Lef. (Pol.) 87
Palæmon, Pall. (Cart.) 356
Palæno, L. (Col.) . . 49
Palæno, Esp. 53
*Palæstina, Stgr.(Lyc.) 148
Pales, Schiff. (Arg.) . 198
Pallida, Stgr. . . . 55
Pamphila, Hüb. . . 310
Pamphilus, L. (Cœn.) 310
*Panagea, H. S. (Lyc.) 142
Pandora, Schiff. (Arg.) 215
Paniscus, F. 356
Panope, Ev. (Lyc.) . 110
Panoptes, Hüb. . . . 110
Paphia, L. (Arg.) . . 214
*Parisatis, Koll. (Sat.) 324
Parmenio,Böh. (Ereb.) 317
Parnassia, Stgr. . . 130
Parthenie,Bkh.(Mel.) 193
Parthenie, Hbst. . . 192
*Parthica, Ld. . . . 325
Pasiphae, Esp. (Epin.) 301
Paulina, Nord. . . . 222
*Paulowskyi, Mén. }
(Ereb.) } 316
Peas, Hüb. 287
Pelidne, B. (Col.) . . 50
*Pelopea, Kl. (Sat.) . 322
Pelopia, Bkh. . . . 211
Persea, Koll. . . . 190
Perseis, Dup. . . . 330
Persephone,Hüb.(Erc.)252
Persephone,Hüb.(Sat.) 276
*Persica, Bien. . . . 143
Phædra, Esp. (Ep.) . 300
Phædra, L. (Sat.). . 285
Phaëton, Frr. . . . 89
Pharte, Hüb. (Ereb.) 244
Phegea, Bkh. . . . 254
Pheretes, Hüb. (Lyc.) 112

*Pheretiades, Ev.(Lyc.) 142
Pherusa, B. (Melan.). 237
Phicomone, Esp. (Col.) 52
Philea, Hüb. . . . 307
Philomela, Esp. . . 262
Philomene, Hüb. . . 49
Philoxenus, Esp. . . 312
*Philyra, Men. (Nep.) 217
Philæas, L. (Pol.) . . 95
Phlomidis, H.S. (Syr.) 346
*Phryxis, Stgr. (Lye.) 372
*Phyllis, Stgr. (Lyc.) 372
Phoeas, Rott. . . . 94
Phœbe, Kn. (Mel.) . 186
Phœbus, Prun. . . . 17
*Phœnicurus, Ld.(Pol.) 137
Phoreys, Frr. . . . 246
Phryne, Pall. (Triph.) 314
Phryneus, F. . . . 314
Pilosellæ, Esp. . . . 185
Pilosellæ, Fab. . . . 300
Pirata, Esp. 275
Pirene, Hüb. . . . 248
*Pisidice, Klg. (Sat.) . 324
Pitho, Hüb. 259
Plautilla, Hüb. . . . 165
Plesaura, Bell . . . 237
Pluto, Esp. 252
Podarce, O. 287
Podalirius, L. (Pap.) . 5
*Poggei, L. (Syr.) . . 357
Polaris, B. (Arg.) . . 201
Polaris, Stgr. (Er.) . . 247
Polaris, Stgr. (Van.) . 175
Pollux, Esp. 255
*Polona, Bien. . . . 143
Polychloros, L. (Van.) 171
Polymeda, Hüb. . . 302
Polysphercon, Bgst. . 102
Polyxena, Schiff. (Thais) } 11
Pontica, Frr. . . . 279
*Popovianus,Nord. (Nis.)357
Populi L. (Lim.) . . 160
Porima, O. 168
*Poseidon, L. . . 144, 371
*Prieuri, Pier. (Sat.) . 322
Procida, Hbst. . . . 231
*Progne, Cr. (Van.) . 218
Pronoe, Esp. (Ercb.) 258
Prorsa, L.. 185
*Pretiosa, Stgr. (Lyd.) 370
Proto, Esp. (Syr.) . . 338
*Protomedia,Men.(Mel.) 220
*Prosecusa, Erseh. (Lyc.) } 147
Proserpina, Schiff. . 273
Provincialis, B. . . . 187
Pruni, L. (Thec.) . . 84
*Pryeri, Murr. (Lyc.) . 372
Psodea, Fr. 240
Psodea, Hüb. . . . 245

Psyche, Hüb. . . . 236
Psylorita, Frr. (Lyc. 112
*Pulchella, Stgr. (Epin.) } 374
Pumilio, Hffg. . . . 354
*Putziloi,Erseh. (Lued.) 21
Pygmæus, Cyr. . . . 354
Pylaon, F. de W. (Lyc.) 107
Pylarge, Hüb. . . . 303
Pyrenaiea, B. (Lyc.) . 113
Pyrenaica, H. S. (Ereb.) 242
Pyrenaiea,Stgr. (Ereb.) 259
*Pyranthe, L. (Callidr.) 69
*Pyrene, Swain. (Callidr.)
Vide Corrigenda.
Pyromelas, Frr. . . 172
Pyrothoe, Ev. (Euehloë) 42
Pyrrha, F. 245
Pyrrhomelana, Hüb. . 171
Pythonissa, Mill. . . 163
Quercus, L. (Thecla) . 81
*Raddei, Brem. (Nept.) 218
*Raddei, Brem. (Papil.) 21
*Radians, Stgr. (Ereb.) 320
Rapæ, L. (Pier.) . . 30
Reichlini, H. S. . . 249
Rhamni, L. (Gonept.) 65
Rhamnusia, Frr. . . 297
*Raphani, Esp.
Vide Corrigenda.
Rhea, Hüb. 154
*Rhena, H. S. . . . 324
*Rhodopensis, Frr. . . 219
Rhymnus. Ev. (Lyc.) 111
*Rinda, Men. (Cœn.) . 331
Ripartii, Frr. . . . 125
Rippertii, B. 125
Rivularis, Scop. . . 164
Rodoris, Esp. (Læosopis.) } 82
Rothliebii, Stgr. . . 312
*Roxandra, H. S. . . 327
Roxelana, Cr. (Parg.) 289
Rubi, L. (Thecla) . . 81
Rumina, L. (Thais) . 13
*Ruslana, Motsch. . . 221
Rutilus, Wernb. . . 91
Rustan, Koll. . . . 349
*Saadi, Koll. (Cœn.) . 331
*Sagana, Dbdl. (Arg.) . 222
*Sagartia, Ld. (Col.) . 70
Salmaeis, Steph. . . 115
Sao, Hüb. (Syr.) . . 347
Saportæ, Dup. . . . 131
Sappho, Kirby . . . 165
Sappho, Pall. . . . 164
Sardoa, Stgr. . . . 175
Sareptana, Stgr. . . 54
Sareptensis, Stgr. . . 54
*Sassanides, Koll. (Thec.) } 136
Satyrion, Esp. (Cœn.) 304

Schmidtii, Gerh. . . 96
*Schrenckii,Men.(Lim.) 215
*Schrenckii, Men. (Parag.) } 328
Scipio, B. (Ercb.) . . 252
Scæa, Hüb. 257
*Sculda, Ev. (Œneis) . 320
Sebrus, B. (Lyc.) . . 128
*Sedakovii, Ev. (Ereb.) 318
Selene, Sehiff. (Arg.) 197
Selenis, Ev. (Arg.) . 196
Semele, L. (Sat.) . . 277
Semiargus,Rott. (Lyc.) 129
*Semidea, Say. (Œneis) 321
Serieea, Frr. 349
*Scylla,Stgr. (Lye.) 148,371
Serratulæ, Rbr. (Syr.) 342
Sertorius, Hff. . . . 347
Sibylla, L. (Lim.) . . 162
*Sibiricus, Ndm. (Parn.) 22
*Sibirica, Stgr. (Aurinia v.) } 219
*Sibirica, Stgr. (Didyma v.) } 220
*Sibirica, Stgr. (Antonoë v.) } 322
*Sibirica,Stgr. (Dryas v.) 325
*Sichæa, Ld. 324
Sidæ, Esp. (Syr.) . . 340
*Sieversii, Chrstp. (Lyc.) } 146
Silvius, Frr. 357
Simplonia, Frr. . . . 87
Sinapis, L. (Leuc.) . . 45
*Siphax, Luc. (Cigar.) 189
*Smaragdina, Brem. (Thec.) } 185
*Solskyi,Erseh.(Pol.) . 368
Speculum, Rott. . . 355
Sphyrus, Hüb. . . . 8
Spini, Schiff. (Thec.) . 76
Statilnus, Hfn. (Sat.) 263
*Staudingeri, S. (Col.) 367
*Staudingeri,Chrstp. (Lyc.) } 146
Steropes, Schiff. . . 355
Stevenii, Tr. 124
*Stubbendorfii, Men. (Parn.) } 25
*Stulta, Stgr., Sat. . 325
Stygne, O. (Ereb.). . 248
Stygne, Hüb. . . . 253
Subalpina, Spr. . . . 95
*Subsolana, Ev. (Lyc.) 141
*Sultan, Stgr. (Pol.) . 368
*Sunbecca, Ev. (Cœn.) . 332
Suwarovius, Hbst. . . 255
*Sydyi, Ld. (Lim.) . . 216
Syllius, Hbst.(Melan.) 236
Sylvanus, Esp. (Hesp.) 353
Sylvatica,Brem.(Hesp.)359
Sylvius, Kn. (Col.) . 357

*Symphita, Ld. (Cœn.). . 331
Syngrapha, Kef. . . . 122
*Syriaca, Stgr. 322
Tages, L. (Nision) . . . 348
Tagis, Hüb. (Euch·oc) 38
Taras, Meig. 345
Tarpeia, Pall. (Œneis) 267
Taygete, Hüb. . . . 270
*Taxila, Brem. (Thec.) 135
*Telephassa, Hüb. (Sat.) 323
Telephii, Esp. . . . 108
Telicanus, Lang (Lyc.) 100
*Telmessia, Z. . . . 299
*Teneates, Mén. . . . 315
*Tenedius, Ev. (Parn.) 24
*Tengstroemi, Ersch. }
        (Lyc.) } 146
Tessellum, Hüb. (Syr.) 339
Tessellum, O. . . . 341
Tesselloides, H. S. . . 346
Testudo, Esp. . . . 171
Thalia, Esp. . . . 197
Thaumas, Hfn. (Hesp.) 350
*Theano, Tausch. (Er.) 317
*Theophrastus, F. (Lyc.) 140
Therapne, Rbr. . . . 347
Thersamon, Esp. (Pol.) 89
Thersites, B. . . . 117
Thetis, Esp. (Lyc.) . 117
Thetis, Hüb. (Melan.) 238
Thetis, Klg. (Pol.) . 88
*Thetys, Mén. (Nis.) . 359
Thia, Hüb. 14
*Thisbe, Mén. (Nept.) . 217
Thisoa, Mén. (Col.) . 57
Thore, Hüb. (Arg.) . 204
*Thrax, Ld. (Hesp.) . 360
Thyrsis, Frr. (Cœn.) . 308

*Tianchanica, Stgr. }
        (Ereb.) } 320
Tigelius, Bon. . . . 293
Tiphon, Rott., Cœn. . 310
Tiphys, Esp. 121
Tircis, Cr. 314
Tiresias, Rott. . . . 101
Titania, Hüb. . . . 203
*Titea, Kl. (Melan.) . 315
Tithonius, Hbst. . . 300
Tithonus, L. (Epin.) . 300
Tomyris, Hbst. . . . 197
Tremulæ, Esp. . . . 161
Triangulum, Fab. . . 169
*Triclaris, Hüb. . . . 220
Triopes, Spr. 257
*Tristis, Brem. (Ereb.) 319
*Tristis, Gerh. (Lyc.) . 145
Truvia, Schiff. (Mel.) . 187
Trochylus, Frr. (Lyc.) 103
Tschudica, H. S. . . 43
Tullia, Hüb. . . . 311
*Turanica, Ersch. (Ereb.) 317
Turca, Mill. 231
Turcica, B. (Melan.) . 231
Turcica, Stgr. (Van.) . 174
Turritis, O. 39
Tyndarus, Esp. (Ereb.) 255
Unedonis, Hüb. . . . 154
Unicolor, Frr. . . . 348
Uralensis, Ev. (Mel.) . 186
Uralensis, Stgr. (Ereb.) 247
Uralensis, Stgr. (Mel.) 183
*Urda, Ev. (Œneis) . 321
Urticæ, L. (Van.) . . 173
V-Album, Esp. . . . 169
Vau-album, Esp. (Van) 172

Valdensis, Esp. . . . 208
Valezina, Esp. . . . 214
Varia, M.-D. 193
Velleda, Rott. . . . 273
Vernalis, Stgr. . . . 30
*Viluiensis, Mén. (Col.) 366
Virbius, H. S. . . . 288
*Virgatus, Butl. . . . 20
Virgaureæ, L. (Pol.) . 86
Virginiensis, Drury . 218
Virgula, Hüb. . . . 351
Vulcanica, Godt. 177, 364
        Vide p. 372
W-Album, Kn. (Thec.) 76
*Wagneri, H. S. (Epin.) 329
Wanga, Brem. . . . 319
Werdandi, H. S. . . 50
Werdandi, Zett. (Col.) 51
Werdandi (Nastes var.) 51
*Wiskolti, Stgr. (Col.) 367
*Wosneskensii, Mén. . 143
Xanthe, Hüb. . . . 98
Xanthe, F. 94
Xanthe, Lang, Verz. . 96
Xanthomelas, Esp. }
        (Van.) } 172
*Xenia, Fr. 316
*Xiphia, F. (Pararge) . 327
*Xiphiodes, Stgr. . . 327
*Xuthus, L. (Pap.) . . 20
*Xuthulus, Brem. (Pap.) 21
Zancleus, Z. 6
Zapateri, Oberth. (Ereb.) 260
*Zelleri, Ld. (Hesp.) . 360
Zephyrus, Friv. (Lyc.) 107
Zermattensis, Fallou. . 87
*Zohra, Donz. (Cig.) . 139

CORRIGENDA.

Page 1, line 3, *for* not spiny *read* generally not spiny

,, 12, ,, 20, for *Cussandra* read *Cassandra*

,, 18, ,, 3 from bottom, *for* larva *read* larvæ

,, 28, ,, 4, *for* nervures *read* nervules

,, 29, ,, 12, *erase the words* nasturtiums and

,, 45, ,, 4, *for* Stev. *read* Steph.

,, 47, ,, 12 from bottom, *for* smaller *read* larger

,, 58, ,, 18, *for* R. D. *read* F. D.

,, 68, ,, 10, ,, Vouzel *read* Donzel

,, 68. *Insert* Pieris Daplidice var. Raphani, Esp. 84, 3, i. 2, p. 163
Has the spots beneath light yellow. Habitat, Persia.

,, 69, line 6 from bottom, for *Pyranthe*, L., read *Pyrene*, Swainson.

,, 71, ,, 3, *for* 1200 *read* 12,000

,, 74, ,, 14, *after* extensive *insert* families

,, 94, ,, 20, *for* 1·08 *read* 1·80

,, 116, ,, 7 from bottom, *for* Friv. Imm. *read* Friv. Imr.

,, 132, ,, 4, *for* Enc. *read* Eur.

,, 139, ,, 16, ,, medium *read* median

,, 142, ,, 5 from bottom, *for* Ms. *read* Is.

,, 143, ,, 3, for *Wosneseuskii* read *Wosnesenskii*.

,, 144, ,, 22, ,, *Iphegenia* read *Iphigenia*

,, 149, ,, 6, ,, *Scolitandides* read *Scolitantides*

,, 154, ,, 9, *erase* comma *after* ultimæ

,, 157, ,, 1, *for* Scandanavia *read* Scandinavia

,, 186, ,, 3, ,, L. V. *read* Ev.

,, 228, ,, 16, ,, face ,, free

,, 241, ,, 19, ,, Kutz. ,, Kn.

,, 247, bottom line, *for* **Hypomedusa** *read* **Hippomedusa**

,, 272, ,, 1, *erase* comma *after* L.

,, 273, ,, 22, for *Statilimus* read *Statilinus*

,, 291, ,, 20, *for* LXXI., 4. *read* LXXII., 1.

,, 292, ,, 2, *erase* Pl. LXXI., 5.

,, 292, ,, 13, *for* LXXII. *read* LXXI.

,, 293, ,, 5, ,, LXXII. ,, LXXI.

,, 293, ,, 29, ,, ditto ,, ditto

,, 294, ,, 5, ,, LXXIII. ,, LXXII.

,, 294, ,, 27, ,, ditto ,, ditto

,, 295, ,, 18, ,, ditto ,, ditto

,, 296, bottom line, *for* LXXIV. *read* LXXIII.

,, 299, line 4 from bottom, *for* LXXIII. *read* LXXII.

,, 299, ,, 20, *for* section *read* book

,, 334, ,, 21, ,, heteroceous *read* heterocerous

,, 336, ,, 3 from bottom, for *Australis*, L., read *Australis*, Z.

,, 337, ,, 3, *for* vv. *read* V.

The references to the plates have in some instances been accidentally omitted from the text; it will be found, however, that the index to the plates is so arranged as to render almost unnecessary any further allusion to them.

PRINTED BY WEST, NEWMAN AND CO., HATTON GARDEN, LONDON, E.C.

www.ingramcontent.com/pod-product-compliance
Lightning Source LLC
Chambersburg PA
CBHW021352210326
41599CB00011B/838